A Field Guide to the
Wild Flowers
of Britain and Europe

A Field Guide to the
Wild Flowers
of Britain and Europe

Thomas Schauer

Illustrated by
Claus Caspari

Translated and adapted by
Richard Pankhurst

Collins
St James's Place, London

William Collins Sons & Co Ltd

London · Glasgow · Sydney · Auckland

Toronto · Johannesburg

Pflanzenführer was first published in Germany in 1978
by BLV Verlagsgesellschaft mbH, München.

© BLV Verlagsgesellschaft mbH, München 1978

© in the English translation William Collins Sons & Co Ltd 1982

ISBN Paperback edition: 0 00 219257 8
ISBN Hardback edition: 0 00 219256 X

Filmset by Jolly & Barber Ltd, Rugby

Printed in Germany by BLV Verlagsgesellschaft mbH, München

Contents

Translator's note

The area covered by this book is northern Europe, from the British Isles and France to Germany and Scandinavia, and the Alps. The authors have also included some common or interesting plants which occur further south, in the Mediterranean region. The authors designed the book so that it will be equally useful in Britain as in mainland Europe, but nevertheless additional changes have been made during translation for the convenience of British readers, as follows.

The scientific (Latin) names used for species and families have been standardised to agree with those in the *Flora Europaea*, which provides an international reference. As far as possible, the English or common names have been taken from the *English Names of Wild Flowers*, by Dony, Rob and Perring, as recommended by the Botanical Society of the British Isles. Other common names for non-British plants have been taken from other field guides, or in a few cases invented for the first time where they have apparently not been given English names before.

The distribution notes have been edited to make it clear in every case whether or not each species occurs in Britain. The frequency with which a plant occurs may vary across its range, and if so this is indicated, e.g. the Bluebell is common in Britain but becomes rare further east. Additional notes on the habitats of species have been added where necessary to cover the habitats for Britain as well as for the rest of Europe. This has also resulted in some cross-references between the habitat sections being added to the text.

In addition to the species that are both illustrated and described, a number of other, similar species are described in the text alone. In such cases the details given refer to the principal characters of the plant and the description as a whole refers back to the main species, i.e. details such as flowering time, habitat, frequency and distribution are not given where these are the same as for the species illustrated. On a few pages there has not been room to describe every species that is shown on the facing plate, in which case the description can be found below a rule on the next page of text.

The following abbreviations have been used throughout:

C.	Central, e.g. C. Europe.
Fl.	Flowering time, e.g. Fl.6–8 means in flower from June to August.
Lf, lvs	Leaf, leaves.
p., pp.	Page (number).
Pl.	Plate (number).
*	Before the scientific name, means that this species does *not* grow wild in the British Isles.

Introduction

The increase of interest in our native plants may be seen as a reaction to the growing threats and dangers to them and to their environment. To be able to protect them and guard against their destruction, however, one must first know them. The development and broadening of knowledge about plants and their habitats will be the concern of this book. The species are presented according to broad categories of habitat, within which they follow a systematic ordering by families and genera. In order to permit easier recognition, an identification key for the species has been provided as well as a general account of the families, with detailed drawings of the important and characteristic genera, see pp.22–37.

Central Europe was originally a land of widespread forest. However, the tree cover was certainly not thick, uniform and uninterrupted, as is the case, for example, with modern forestry plantations. Fire and lightning, wind throw and the factors of age and disease would have created temporary open spaces, then as now. There would also remain more or less permanently open areas of moorland or bog, stretches of coastline, rock faces and locally occurring pockets of exceptional dryness, in which during years of extreme drought the sparsely growing layer of trees and shrubs would be shrivelled up. Such open areas are otherwise confined to the alpine zone above the tree level. The open spaces were greatly extended by pasture land, dairy and arable farming, and of course also by settlements. At the present day the total area of natural vegetation in Europe has certainly contracted. A cultivated landscape arose from the uncultivated original, and habitats whose vegetation is not somehow marked or altered by man are now almost nowhere to be found, with the possible exception of some rock-crevice associations in the alpine zone and a few remaining patches of undisturbed moorland. The remaining areas are occupied and utilised by man in some way or another: for example, water is used as a source of energy, to enable transportation or as a dump for effluents; the dry, steppe or semi-arid grasslands which were formerly used for grazing are now declared to be barren or derelict land and given over to military training areas or building land for housing, industry or leisure developments; and the forests often bear a resemblance to living factories for wood. In short, the vegetation profile of each habitat is no longer original or natural, but at best semi-natural.

The most strongly man-modified, restricted and exploited habitat is that of the widely occurring fields, waysides and waste and derelict open spaces, although in these places the once rich flora is severely threatened. The next man-created habitats, though considerably poorer in species, are the agriculturally productive (rich) meadows and grazing lands, i.e. agricultural grassland. The habitats of poor and dry grassland, formerly highly cultivated and developed by mowing (once every one or two years) or by grazing, are now reduced to a few surviving areas and are severely threatened. The woodland of the broad lowland river valleys was the first to be developed for agriculture when farming began, because they had rich alluvial soils. These soils were for the most part washed away from the fields, transported by the river and then redeposited when it overflowed its banks during floods. Thus we reach woodland habitats, which range from almost original, many-layered species-rich forest remnants, only slightly influenced by man, to evenly aged monocultures of spruce; and from shady damp ravine or riverside woodlands to sunny, dry forest or scrub margins.

Woodlands are related by transitional habitats to the environment of dry grassland, or to the next habitat: the wetlands, moorlands and marshes. These highly diverse habitats likewise include remains of natural or semi-natural vegetation, such as a few surviving examples of high or low altitude bogs, mountain streams, and perhaps still a few lakes, but the quarries, fishponds and reservoirs that constitute the greater part of this type of habitat are greatly influenced by man or have man-made origins. The next broad category of habitat, the alpine zone, under which the vegetation of various types which occurs above the treeline will be considered, was also affected by disturbance. The grazing and mowing of alpine meadows, and nowadays an increasing degree of road construction, tourist industry developments and sports installations are detrimental to this habitat. Finally there remains only the environment of the seashore and coast, which is so thoroughly altered that often undamaged areas are only to be found on a few short stretches of coast or on small parts of islands.

Every day in Britain about 50 hectares of land are built over, and of this the building of new roads accounts for a great deal. The area of semi-natural habitat which is lost every year by development, including agriculture, corresponds to about one half of the total area designated as nature reserves. It appears all the more necessary, therefore, to protect species which are becoming rarer by guaranteeing the protection of the localities in which they grow. Only a relatively small number of plants (mostly ones having showy flowers or other obvious attractions) are under legal protection, meaning that it is illegal to pick or collect them. However, there are other rare species which are not, or not yet, legally protected and which ought to be preserved as well as their localities.

Details of plant frequency are dealt with in very general terms and are valid only for north-west Europe. Even so, there can be great differences in the abundance of plants between north and south, or east and west.

The protection of wild plants

Rare plants are protected by law in most European countries, although the extent and effectiveness of the legislation varies greatly. No attempt is made in the text to give details for each country in turn, but where a plant is known to be legally protected in at least part of the area covered by this book, this is stated. In Britain, the **Conservation of Wild Creatures and Wild Plants Act, 1975** makes it illegal to uproot any wild plant unless you have the permission of the owner or occupier of the land on which it grows, or unless you own the land yourself. Certain rare and attractive plants (which are noted among those marked 'Protected!' in the text) are totally protected, and it is an offence to remove any part of them.

The Botanical Society of the British Isles distributes a free leaflet, the *Code of Conduct*, which gives good advice to botanists on how best to treat wild flowers while studying them, and includes a list of the protected rare species. Send an SAE to the General Secretary, BSBI, c/o Department of Botany, British Museum (Natural History), Cromwell Road, London SW7 5BD.

Glossary

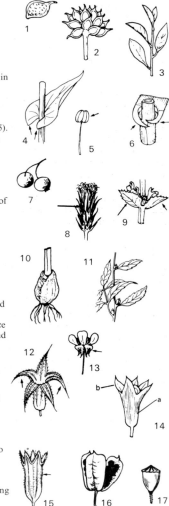

Accessory fruit: a fleshy fruit whose construction is based on some other organ of the flower, e.g. strawberry, which is a fleshy receptacle.

Achene: 1-seeded dry indehiscent fruit (1, 2).

Acicular: needle-shaped.

Actinomorphic: see Symmetric.

Acuminate: long-pointed, with point narrow.

Alternate: with 1 leaf at each stem node and pointing in different directions (3).

Amplexicaul or **semi-amplexicaul:** the leaf base completely or partly surrounds the stem (4).

Angled: showing corners in cross-section.

Anther: the male part of a flower, containing pollen (5).

Apiculate: shortly pointed, the point relatively broad.

Appressed hairs lie flat against a surface.

Arachnoid: cobwebby.

Aristate: with a bristle-like projection.

Auricle: a small, lobe-like appendage at the leaf base; hence auriculate, lobed (6).

Awn: bristle-like projection, as in the flowering parts of grasses (see also Glume, 38).

Berry: fleshy, 1- to many-seeded fruit (7).

Bifid: split deeply in two.

Bract: scale- or leaf-like upper leaf surrounding a flower or inflorescence (8), or from whose axil the flowers arise (9).

Bracteole: small, scale-like, often membranous bract, occurring on flower stalks.

Bulb: underground organ composed of densely packed fleshy scale leaves (10).

Bulbils are bud-like growths in the leaf or inflorescence axils of the parent plant, which fall to the ground and take root there (11).

Caducous: falling off early.

Calcicole: plant usually or always on soils with free calcium carbonate, e.g. limestone or chalk.

Calcifuge: plant not usually found on soils with chalk or limestone; opposite of calcicole.

Calyx: the sepals as a whole; either composed of spreading (12) or reflexed (13) free sepals, usually green in colour, or else combined into a calyx tube (14a), with calyx teeth above (14b). A calyx may also be ribbed (15).

Campanulate: bell-shaped.

Capitate: shaped like a head, head-like.

Capsule: dry fruit, consisting of several carpels, opening by pores or slits (16, 17).

Carpel: female reproductive organ of a flower, consisting of an ovary, style and stigma.

Catkin: spike-like inflorescence with a pendent, flexible main axis and inconspicuous flowers (18).

Ciliate: having hairs on the margin (19).

Circumscissile capsule: capsule that opens by means of a lid which springs off (20).

Claw: the narrow basal part of a petal.

Cone: the 'fruit' of the coniferous trees, consisting of numerous overlapping woody scales (21).

Connate: joined together at the base.

Connivent describes 2 organs that converge.

Cordate: heart-shaped, with the point at the tip (22).

Coriaceous: leathery.

Corm: a swollen underground stem.

Corolla: the petals as a whole, free or combined (23); if combined in a corolla tube (24b), the opening is called the throat (24a).

Corona: corolla-like ring of free or combined appendages inside the perianth (25).

Corymb: a panicle with all the flowers lying in roughly the same plane.

Crenate: with a notched or scalloped margin (22).

Crisped hairs are stiff, flexuous or bent at the tip and long (26).

Culm: flowering stem of grasses.

Cuneate: wedge-shaped (27).

Cupule: an involucre with the bracts combined into a cup-shape, e.g. beech.

Cuspidate: with a short point on a rounded outline.

Cyathium: see Spurge Family, p.32.

Cyme: inflorescence with growing points terminated by flowers, and having terminal flower(s).

Decurrent: leaf blade partly grows down stem (28).

Decussate: the next pair of leaves are positioned at right angles to the last (29).

Dentate: see Toothed.

Dichotomous: branching into two.

Dioecious: when male and female flowers are on different plants.

Disc: disc-shaped or sometimes fleshy and nectar-secreting part of a receptacle (30).

Disc floret: in the Daisy Family, in the centre of the flower head, without a spreading 'petal' (31).

Distichous: arranged in 2 vertical rows on opposite sides of the stem.

Drupe: fruit with a fleshy exterior and inner stone-like wall to the seed (32).

Elliptic: of a leaf or petal shape, see 33.

Emarginate: notched.

Epicalyx: calyx-like structure of several bracts close under the calyx (34).

Excurrent: running out of the tip.

Falcate: sickle-shaped.

Filament: the stalk below the anther(s) in a stamen (35).

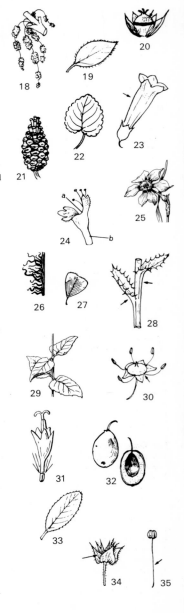

Filiform: thread-like.

Flaccid: soft and weak.

Follicle: fruit consisting of 1 carpel which opens only along 1 side (36).

Furrowed: with longitudinal channels.

Geniculate: bent like a knee.

Glabrous: without hairs of any sort.

Glandular hairs have a small gland at the tip (37).

Glume: tough, membranous leaf-like part of the flowers of grasses, often with an awn (38), occurring as an inner and outer glume (39a). Within the glumes there is the lemma (39b, 40a) and often a fourth glume-like part, the palea (39c, 40b). See also Grass Family, p.24.

Hairs: 1- or many-celled, simple or branched, straight or curved projections from the surface of a plant.

Halophyte: plant of salty soil.

Hastate: like sagittate, but basal lobes horizontal.

Head or **capitulum** flowers are shortly stalked and densely packed into a compact head (41, 42).

Hermaphrodite: flower with both stamens (male) and ovaries (female).

Honey-leaf: leaf-like nectar-secreting gland.

Hyaline: thin, green and translucent.

Hypanthium: see Rose Family, p.34.

Imbricate: overlapping.

Imparipinnate: pinnate with a terminal leaflet; with an odd number of leaflets (43).

Inflorescence: a flower branch, or the part of the stem that carries the flowers.

Internode: see Shoot axis.

Involucre: a collection of bracts.

Keel: see Standard.

Labellum: see Orchid Family, p.22.

Lanceolate: spear-shaped (44).

Leaf axil: angle between leaf and stem (45).

Leaf blade: the broadened part of a leaf, usually flat.

Leaf sheath: the broadened lower part of a leaf which encloses the stem in a tube or pouch (46b).

Leaflet: part of a compound leaf (47b).

Lemma: a bract in the Grass Family. See Glume.

Lenticular: convex on both sides.

Ligule: a small extension at the junction of leaf sheath and blade, e.g. in grasses (46a).

Limb: the broader part of a petal.

Linear: very narrow, as a straight line (48).

Lip: of unsymmetric flowers, the lower part (see also Unsymmetric).

Lobed: divided, but not into separate parts (49). Some leaves have an enlarged terminal lobe (50).

Lyrate: shaped like a lyre.

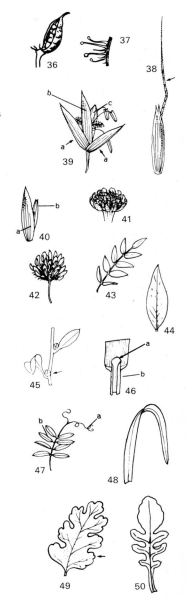

Mericarp: 1-seeded section of a dry fruit. See also Schizocarp.

-merous: when the calyx and corolla each have the same number of parts, e.g. if 5 the flower is said to be 5-merous, etc.

Microspecies: a minor subdivision of a species, other than a subspecies, variety or form.

Monoecious: when both male and female flowers occur on the same plant.

Mucronate: shortly pointed, the point narrow.

Nectar: a sweet liquid.

Nectary: nectar-secreting glands that occur in various parts of the flower and which attract insects (51).

Node: see Shoot axis.

Obcordate: heart-shaped, with the stalk in the notch (52); opposite to cordate.

Obovate: egg-shaped but upside down, with the broadest part above (53).

Ochrea: a sheathing stipule; see the Dock Family, p.32.

Opposite: with 1 leaf on either side of the same stem node (54).

Ovary: (55; 56 in cross-section) consists of several combined carpels and bears 1 or more slender styles with various forms of stigma for the reception of pollen. Different kinds of ovary are distinguished by their position: inferior, when positioned below the point of insertion of the calyx and corolla (57); half-superior, when placed partly within the cup-shaped part of the receptacle (58); superior, when positioned above the point of insertion of the calyx and corolla (59).

Ovate: egg-shaped (60).

Ovule(s): enclosed in the carpel or ovary, after fertilisation develops into the seed.

Palate: a swelling of the lower lip or corolla at the junction of lip and tube (see 108a).

Palea: scale-like membranous bract in the Grass Family. See Glume.

Palmate or **digitate:** branched or lobed like the fingers of a hand (61).

Panduriform: with a constriction near the base, like a violin.

Panicle: branched inflorescence with stalked flowers (62).

Pappus: hairs or bristles on the fruit of many Compositae, taking the place of the calyx (63).

Parallel-veined: of a leaf, see 64.

Parasite: a plant which lives off other living plants to which it is attached.

Paripinnate: pinnate without a terminal leaflet, i.e. the number of leaflets is even (65).

Pedicel: stalk of a flower in an inflorescence.

Peduncle: stalk of a flower or inflorescence.

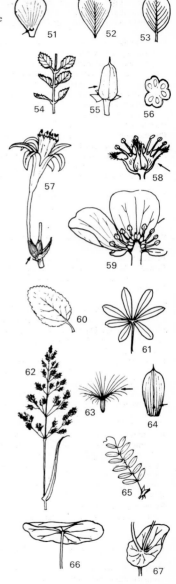

Peltate: of a flat organ, with the stalk placed centrally below, not at the margin (66).

Perfoliate: when the undivided leaf blade completely surrounds the stem (67).

Perianth: the outer, non-sexual part of a flower, consisting of perianth segments (68) which are either separate or combined in a tube. If the perianth consists of 2 rows of segments, the outer is called the calyx (68a) and the inner the corolla (68b).

Petal: an inner perianth segment; petals are usually brightly coloured and surround a flower's reproductive organs.

Petiole: stalk of the leaf blade.

Phylloclade: leaf-like, broadened shoot or branch (69).

Phyllode: a broadened, leaf-like petiole (70).

Pinnate: in 2 opposite rows along a common axis, e.g. of compound leaves (65), or of leaf venation (71).

Pinnatifid: like pinnatisect, but only cut part way to the midrib.

Pinnatisect: pinnately divided almost to the midrib, but not into separate leaflets (72).

Pod: dry fruit consisting of 1 carpel, opening by a seam around both sides (73).

Prickles: hard, prickly outgrowths of the outer surface of leaf and stem.

Procumbent: growing along the ground.

Pruina or **bloom:** a whitish or blue coating on a leaf or stem which can easily be rubbed off.

Punctate: spotted.

Raceme: an elongated inflorescence with stalked flowers, usually without a terminal flower (74).

Ray floret: in the Daisy Family (Compositae), at the outer edge of the flower head with the petal spreading horizontally (75).

Receptacle: the part of the flower base which bears the parts of the flower or, in the Compositae, the individual florets (76a). It may be convex (76b, 77) or sometimes hollowed out in a pitcher shape (78).

Receptacular scale: a small chaffy scale between the florets of many Compositae.

Reniform: kidney-shaped (79).

Reticulate-veined: with veins in a network (80).

Rhizome: rootstock; a creeping underground stem which sprouts again each year (81).

Rosette: an arrangement of leaves, usually at the base of the stem (82).

Rotate: spread out flat like a wheel.

Rugose: wrinkled.

Runner: a stem which creeps over the ground and roots at the nodes to form new plants (83, p. 14).

Sagittate: arrow-head shaped (84, p. 14).

Saprophyte: a plant which lives on decayed organic matter.

Scale-leaf: scale-like leaf at the base of the stem and on underground parts (85, p. 14).

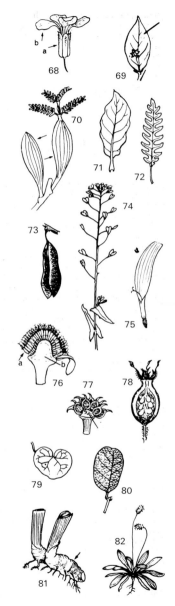

Scarious: thin, dry and not green.

Schizocarp: a dry fruit, which falls into several 1-seeded sections (mericarps) when ripe, e.g. sycamore (86) where each mericarp is also winged (86a); also in the Borage Family (87) and the Carrot Family (88).

Secund: leaning to one side.

Sepal: an outer perianth segment; sepals form a ring immediately below the petals and are collectively known as the calyx. They are usually green or brown, but are sometimes coloured. See also Calyx.

Serrate: saw-edged (e.g. 33).

Serrulate: minutely serrate.

Sessile: without a leaf stalk.

Setaceous: shaped like a bristle.

Shoot axis or **stem** consists of sections (internodes, 89b) which are usually elongated and marked off by the nodes (89a), from which leaves arise.

Siliceous describes a soil rich in silica, usually acid.

Silicula: fruit of the Cruciferae which is not more than 3 times as long as wide (90).

Siliqua: fruit of the Cruciferae which is more than 3 times as long as wide (91, 92).

Silky hairs are thick, appressed and shining.

Sinuate: with a wavy outline.

Sinus: the bay between 2 teeth on the margin of a leaf, petal, etc.

Spadix: a fleshy spike, often with a club-shaped appendage at the tip (93).

Spathe: a sheath enclosing an inflorescence.

Spathulate: spoon-shaped (94).

Sphalerocarpium: a fruit based on a cone whose scales become fleshy, e.g. juniper.

Spike: elongated inflorescence with sessile flowers (95).

Sporophyll: as in conifers, a leaf-like structure bearing the pollen producing organs.

Spur: a hollow projection from the base of a petal or sepal (e.g. 108d).

Squarrose: square in outline.

Stamen: the pollen-bearing male organ of a flower, consisting of a filament and an anther.

Staminode: a modified stamen which does not contain pollen (96).

Standard: an upper petal in, e.g. the Pea Family (Leguminosae) flower (97a); the keel is the lower petal and the wings are the 2 side petals (97b).

Stellate hairs are branched in a star shape (98).

Stigma: tip of the style which receives pollen (99a).

Stipule: scale- or leaf-like appendage(s) at the base of the petiole, usually paired (100).

Stolon: a creeping short-lived stem, usually above ground (101). A runner and a sucker are both forms of stolon.

Striate: with inconspicuous ridges or lines.

Style: the stalk (99b) that connects the ovary and stigma (99a) in the female flower. Styles lie in the centre of the flower, within the ring of stamens.

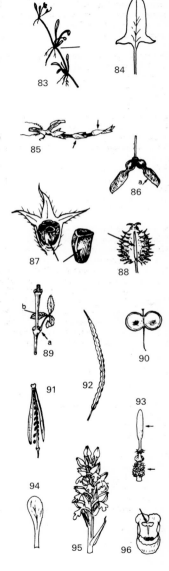

Stylopodium: see Carrot Family, p.27.

Subulate: awl-shaped, finely tapering to a point.

Symmetric or **actinomorphic** petals are equivalent; the flower can be divided into equal parts by cross-sections taken in many different ways (e.g. 68).

Tendril: fragile structure, often spirally twisted, for attachment, developed out of a leaf or leaflet (e.g. 47a).

Terete: circular in cross-section.

Ternate: divided into 3 (102).

Throat: see Corolla.

Toothed or **dentate:** shallowly but sharply lobed (e.g. 72).

Trifoliate: having 3 leaflets (103).

Trigonous: bluntly triangular in cross-section.

Triquetrous: like trigonous, but with sharp angles.

Truncate: cut off square.

Tuber: fleshy, thickened part of root or stem; a food storage organ (104).

Tubercle: a small swelling.

Twice-pinnate: pinnate with the lateral branches again pinnate (105).

Umbel: umbrella-shaped inflorescence, with flower stalks all arising from the same point (106).

Unisexual: a flower either with stamens (male flowers) or with ovaries (female flowers) only.

Unsymmetric or **zygomorphic** flowers can only be divided into 2 equal parts by 1 cross-section; the upper and lower parts are different (107, 108, or 24, 97). Can be 1-lipped with only the lower lip developed (107), or 2-lipped, with the upper lip developed as well (e.g. 24). The 'snapdragon' type of flower occurs in the Figwort Family (Scrophulariaceae) (108), with upper lip (108b), lower lip (108c), palate (108a) and spur (108d).

Urceolate: pitcher-shaped.

Utricle: an achene with a loose, inflated outer coat.

Verrucose: warty.

Vesicle: bladder.

Villous: long woolly hairy, shaggy.

Viscid: sticky

Whorled or **verticillate** describes an arrangement with more than 2 leaves or flowers at each node (109).

Wing: see Standard.

Winged: having outgrowths on the side, e.g. petioles, mericarps.

Woolly hairs are soft, thick and long (110).

Zygomorphic: see Unsymmetric.

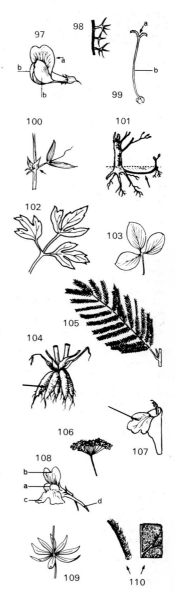

A short description of the habitats

Arable fields, waste and stony places, waysides

Although there is great variation in ruderal habitats (waste places, gravelly and stony areas, waysides, rural waste land), as well as in cultivated land (arable fields, vineyards, gardens) with respect to dampness (dry to moist to waterlogged) and the nature of the soil (gravel, sand, garden topsoil, mud), the plants growing here all require open ground without competition from meadow and woodland species. They also need abundant nutrients, especially nitrogen, which is well tolerated by these species (although also by others as well).

Among these species there are many short-lived annuals, which occur in the first place as colonists or pioneers on these bare soils or on cleared areas free of vegetation. On the humus-rich topsoils of arable fields, vineyards and gardens, these species would be rapidly replaced by meadow and woodland plants, were it not for the optimum conditions for annual field weeds being created by regular cultivation of the ground. On waste ground and rubbish tips, in gravel pits, on railway tracks and rural waste land, unless the land is repeatedly cleared, the annual plants are soon succeeded by perennial shrubs which are also suited to extreme conditions such as very thin soil, very high concentrations of nitrogen, or pressures from vehicles or trampling.

The origins of field weeds are highly various. In the case of arable fields, vineyards and gardens (i.e. weed floras), a distinction is made between weed communities of rootcrops and of cereal crops. In the former, mechanical working of the soil (e.g. hoeing) frequently occurs even in summer, so that only fast-growing species such as knotweed, goosefoot or spurge, which bear ripe fruits soon after germination, are viable. In the latter case, mechanical working of the soil occurs in late autumn or early spring. The species associated with cereal culture, such as Corncockle and most species of poppy, also germinate at a low temperature (about 10 °C), while rootcrop-associated weed species require germination temperatures of at least 15 °C and preferably 20 °C. Therefore in the first place only cereal-crop weeds develop in fields ploughed in a colder season, leaving little space for the later-germinating rootcrop weeds. Nevertheless, the latter have greater chances of survival if their autumn- or spring-germinating competitors are removed in summer by hoe-cultivation. In addition, the rootcrop weeds receive much greater assistance from nitrogen fertilisation. However, because of chemical weed-killers the survival of all wild field species has become problematical, with the result that many plants such as Corncockle, Venus'-looking-glass and Cornflower have almost disappeared.

Within weed floras, i.e. vegetation on rubbish tips, gravel pits and quarries, builder's rubble and roadside verges, numerous plant associations which cannot all be named here can be distinguished. There is the short-lived association of mallow and nettle on nitrogen-rich uncultivated soil in villages, or the goosefoot pioneer communities on nutrient-poor uncultivated soil, which in turn are replaced by other longer-living species, e.g. plants of trampled grassland and other associated communities. Hedge Mustard, Flixweed and Dwarf Mallow represent examples of short-lived communities of waste places. Species which prefer (or tolerate) warmth and dryness, such as Bugloss, Hound's-tongue, Musk Thistle, Mignonette and Evening-primrose, are found in association. In damper habitats or in areas with a cooler, damp climate we find the association of burdocks, Greater Celandine, Hemlock and Dead-nettle. Here may also be mentioned the vegetation on wet, muddy soils at the edges of ponds and ditches near villages and where cattle water. These associations of bank and mud floras consist almost entirely of annual species such as Water-pepper, Bur-marigold and some dock species, all requiring nutrient-rich soils, either well-watered or wet, bare, free of vegetation at germination time, and large amounts of light. Constantly open ground, and hence the conditions for these plant communities, is created by cattle trampling or by longer periods of flooding at intervals of at least once a year (fish ponds, reservoirs).

Dry and poor grassland, stony slopes, walls

In central and western Europe, open dry grassland only exists naturally in places where the soil is too unfavourable to forest and scrub growth. The treeless islands with steppe-like vegetation mostly originated or have been extended through grazing or annual mowing. Most rocky heathlands, steppe lands or other treeless, shrubless grassland communities below the alpine treeline, such as calcareous (poor) grassland on limestone or siliceous grassland on acid, lime-free rock, are secondary habitats in our area, which were once stocked with oak, beech or other dry woodland. An exception are those habitats which are too cold and/or windy for tree growth, e.g. in parts of Scotland and Scandinavia. Therefore their origins are not directly comparable to those of the natural steppes of eastern Europe, or the northern regions where the treeline descends to sea level. In central Europe, dry grassland has its focal point in warm regions with little precipitation. The plants are distinguished by deep root formations, and features which reduce evaporation such as buried stomata (leaf pores), thick hairiness and having low biomass above ground, i.e. through a xeromorphic life-form. In the first instance we have continental dry grassland, whose habitats are distinguished by extensive dryness in the vegetative or resting phase, small annual precipitation and great seasonal and even daily extremes of temperature. We may cite as typical examples the inner alpine rocky steppes (e.g. Aosta valley) with Feather Grass, Tunic Flower and Hairy Milk-vetch. The latter form a transition to the sub-Mediterranean, sub-Atlantic dry grasslands, which have less extreme variations of seasonal temperature and greater precipitation. Plants of these communities prefer warmth, are frequently sensitive to frost, and often grow with species of Downy Oak woodland. In semi-dry grasslands the proportions of precipitation obtaining are similar to those of dry grassland, but their distribution extends over regions with somewhat lower temperatures and fewer variations of temperature. This type of grassland occurs in habitats with deeper soils with balanced water conservation and is even more dependent than dry grasslands on the historical forms of management. In Britain, examples of this occur on chalk downland or limestone hills grazed by sheep and/or rabbits, or on acid soils as in the Breckland of East Anglia. Differences of exposure are also decisive; thus, on south-facing slopes dry grassland may be seen, whereas neighbouring north-facing slopes show semi-dry grassland.

As stated above, the dry and semi-dry grasslands of central Europe may not always be kept free from trees and shrubs unless there is mowing or grazing. Although unbroken grass cover presents adverse conditions for tree germination, species such as Blackthorn, Juniper or Aspen may advance relatively quickly by means of suckers from the margin of a copse and gradually take over uncolonised grassland.

Trees and shrubs have difficulty in gaining a foothold on bare rock, rocky outcrops, on wall tops or in cracks in walls and cliffs where there is little humus formation and minimal water storage capacity. Colonising associations which are rich in lichens and

17

mosses thrive in these places, with many succulent plants such as House-leek and stonecrop species which can store water over a long period in their fleshy leaves, or other species such as Thyme which survive extremes of heat and dryness well.

Rich meadows and pastures

By rich meadows and pastures is meant intensively exploited agricultural land, which does not mean land rich in species. Regular mowing and/or grazing with frequent fertilising allow the development of relatively species-poor, narrowly defined meadow associations whose constituent species tolerate the regular grazing off or mowing down, and are even favoured by it, in so far as other competitors for light are kept away. Depending on the amount of precipitation, the soil humidity or the ground water level, these rich green meadows (contrasting with the predominantly brown, pale and thinly covered dry grasslands) extend from almost dry to damp and wet associations. Many areas on dry, deep soils which once bore a colourful meadow community of False Oat-grass, Meadow Clary, Ox-eye Daisy and Spreading Bellflower have nowadays been transformed into arable land. Frequently, only the somewhat damper meadow communities maintain their hold in the wider valley areas, provided that they are not drained, as has happened to many damp and wet meadows with Marsh-marigold, Marsh Ragwort and some sedges. In a few localities 'water meadows' still survive which have not been fertilised, ploughed or reseeded. These can contain a large number of species, some of which are very colourful, e.g. Fritillary. Such habitats are covered in the next section, since they are not 'rich' in the sense used here. There is still a relatively good degree of conservation over rich alpine meadows with Yellow Oat-grass, Wig Knapweed and Great Masterwort; and in spring with numerous crocus blooms as in the Vosges and southern mountain valleys, or, in the Swiss Jura, Pheasant's-eye Narcissus. Meadows that are intensively grazed almost all the year round by large herds of cattle, as in northern Germany and western Britain, consist principally of species resistant to trampling, such as Perennial Rye-grass, Dog's-tail Grass, White Clover and Common Couch, which can also regenerate quickly.

Waters, moors and marshes

With the exception of woods on moorland and by river margins, this environment is by its very nature largely free of woodland and still has a relatively large number of almost natural plant communities. Water and marsh plants are frequently provided with special adaptations: roots and shoots have a densely branched ventilation system by means of which they may grow on constantly submerged soil. Underwater plants are able to take up gases such as oxygen and carbon dioxide (released from the water) and nutrients through the thin epidermis (outer skin) of the leaves and stems.

Let us begin with the conditions experienced by vegetation of underwater plants found in fast-flowing waters. Owing to the strong current the plant is abundantly supplied with oxygen and nutrients; but only a few (stress-tolerant) species are suited to the considerable mechanical strain. These mostly have floating grass-like or strap-shaped leaves (River Water-crowfoot, Loddon Pondweed), whereas species found in still waters exhibit round or wide leaf shapes. Alongside streams we may find a marginal reed bed of Reed Canary-grass, which is less easily bent by fast currents than are rushes and Common Reed. The vegetation of standing or slowly flowing water is far more abundant. A distinction is made between nutrient-poor (oligotrophic) water with clear visibility to some depth and a small amount of plant production, and nutrient-rich (eutrophic) waters with considerable production of higher plants (which contribute to the accretion of material to the land) and lower organisms (plankton) which cause turbidity. Beginning with the underwater vegetation, we have at a depth of about 2–5 metres a zone of pondweeds and water-milfoils, followed in clear, oligotrophic lakes to about 15 metres down by a layer of stoneworts (Chara), and in an upward direction towards the bank, at a depth of 1–3 metres, by floating-leaf associations with water-lilies, particularly abundant in nutrient-rich lakes; connected to this is a broad belt of Common Reed which merges into a sedge marsh. Here we enter the species-rich zone of land accretion, sometimes used as hay meadow and conserved thereby; for at this point alder wood and/or willow thickets evolve quickly.

Due to the constant collection of organic material which is not decomposed in the oxygen-poor water in wet habitats, a peat layer is formed, which varies in its thickness and rate of growth. Plant associations under which a peat layer has formed are counted as mires (moors, bogs or fens) in contrast to marshes, which are designated wet habitats without peat formation. We distinguish between low or flat mires (fens) with an almost level surface, and high moors or moorland (raised bogs) with a typically rounded, convex surface shape. The former are fed by underground or running water; the latter, uninfluenced by horizontal water movement, are dependent upon precipitation. Therefore raised bogs are only to be found in areas of high rainfall.

The fens, especially those which are calcareous and nutrient-rich, have great numbers of flowering plants, particularly sedges, rushes and spike-rushes and cotton-grasses of various kinds, while the very acid and nutrient-poor moors and bogs show only a few flowering plants such as Cranberry and Common Cotton-grass. The principal contributors to the formation of moorland are the many species of Sphagnum (bog moss), the upper parts of which grow unrestrictedly while the lower, older parts which die off (in the acid, oxygen-poor environment) are not decomposed and gradually contribute to the formation of the peat layer, which may achieve a thickness of up to 10 metres. On the margins of a raised bog, the steeply sloping edges sometimes carry a dense pine or birch forest. In the middle there are frequently one or several small, deep bog pools with brownish water, in which floating islands of Sphagnum, Rannoch-rush, Bog-sedge and other plants with long rhizomes occasionally form, giving what is called a quaking bog. There is frequent alternation on the surface of a raised bog between hummocks and hollows which are seasonally filled with water.

Woods, wood margins, scrub and wet woods

An uncommonly high number of European woodland associations, too many to enumerate here, may be distinguished by various factors: differing soils (dry, damp; calcareous, lime-free or acid; shallow or deep; stony, sandy, loam or clay); altitude (from zero to about 2400 metres above sea level) and climate (low or high rainfall; mild in winter or with great contrasts between summer and winter temperatures). In many instances the woods have been greatly altered by forestry practices or woodland grazing with the result that natural or almost natural woods are seldom to be found, or their natural combination of tree species has to be separated from the grasses and weeds of the woodland floor, the herb layer.

With the **mixed deciduous forests** with deciduous trees such as maple, beech, ash, lime and elm, we distinguish mixed beech woods in habitats that are neither too wet, too dry or too cold; hornbeam-dominant woods in drier habitats; ash and alder woods in damper habitats. Beechwoods in calcareous mountain districts are particularly abundant in species. Woods with a large number of such deciduous trees as listed above display, before leafing out, a luxuriant ground cover of Hepatica, Wood Anemone, Woodruff, Spring Pea, Lungwort and many other spring-flowering plants, the leaves of which soon turn yellow and wither, not only because of poor light but because of the layers near to the ground warming up. In the higher areas (subalpine at 500–1000 metres, alpine at 800–1350 metres) we have Silver Fir and beech forest, to which spruce has been added by human agency, while beech-oak forest on acid soil is often the dominant woodland of low-lying areas of central Europe.

Oak (Pedunculate or Sessile) is a species of tree which prefers warmth, dryness and light, and which forms well-lit forests with a species-rich shrub and herb layer. The Downy Oak is found in very dry areas that are mild in winter and is accompanied by many plants that are widespread in the south and which occur also in dry grassland. In areas of cold winters and continental climate, Pedunculate and Sessile Oaks are associated with Scots Pine. On poor sandy soils, e.g. in southern England and north-western Germany, oak-birch woodland is typical, with Bilberry, Heather, Wavy Hair-grass and bracken indicating acid, nutrient-poor soil.

Next to consider are the **coniferous forests**. The Silver Fir, which is found in the rainy oceanic climate of the north Alps up to 1700 metres in pure stands with an undergrowth of tall shrubs, or occurs as pure fir forest on limestone or siliceous soils in subalpine regions, has already been referred to in connection with beech forest. In the more continental inner Alps there are extensive areas of Scots Pine forest in dry valleys and of larch with Arolla Pine forest in higher places, the latter forming the treeline at about 2400 metres. The modest Scots Pine occurs in the most various habitats as a colonist for woodland, for example as oak-pine forest in warmer areas or as sandy pine forest on acid soil, on raised bogs and on dunes. In many cases these forests, where thin growth allows penetration of light, admit species of dry grassland and heath which otherwise can only flourish on the margins of thicker, shadier forest.

Forestry development nowadays favours the spruce, which is the dominant tree species over large areas of very different habitats, and which occurs naturally at the treeline, i.e. up to 1900 metres at the subalpine level of the northern Alps. It also occurs naturally in valleys with low air temperatures, with a risk of late frost, i.e. areas inhospitable to fir and deciduous trees; on mountain screes; on the edges of raised bogs and on river margins; and scattered in various deciduous- and coniferous-forest associations in which it was never a strong enough element to become dominant.

Finally, **forests by rivers**. Fluctuating ground level and occasional flooding determine the plant associations of river margins. In many localities flooding has been eliminated by river control, the ground water level has dropped and the conditions for river-margin forests are no longer present; agricultural meadows and arable fields reach as far as the river. There still exist segmented remains of natural river-margin associations, which show a sequence of zones proceeding from the river to the higher valley margin, roughly in the following order: willow thicket (White, Grey, Almond or Osier Willow according to area); then alder carr merging into ash forest with alder; at a higher level, further from the (ground) water level and rarely flooded, as mixed hardwood river-margin forests with elm, oak, hornbeam and maple.

Alps

From this very extensive environment with its many aspects and its geographical limitations, a selection has been made of species whose main distribution, with few exceptions, is limited to the Alps, to other high mountains or to lower mountains further north (e.g. the Scottish Highlands) and to the Arctic, and in fact are those found predominantly in the higher situations near and above the treeline. During the ice ages the vegetation of the Alps became mixed with that of the Arctic, with the result that at the present day there are species of arctic-alpine distribution. Besides the forests (previous section) at the subalpine level, we note dwarf shrub associations with Alpenrose, Bearberry or Crowberry, Mountain Pine on drier soils and Green Alder on damper soils. The adjacent alpine level includes the alpine meadow and rock region, which is dominated by very variable but always extreme habitat conditions: for example, conditions for grassland associations occur on calcareous or siliceous rocks; on even-surfaced, stony, sunny slopes; in extremely humus-poor cracks with great extremes of temperature; on screes and corries which are constantly moving, have long periods of snow cover and therefore short growing periods; on ridges and ledges exposed to the wind; or in snow valleys on irregular terrain, with long periods of soaked soil and with an extremely long period of snow cover, say eight to nine months of the year. Almost all alpine flowering plants are noted for their very short vegetative period from germination to flowering and fruiting, and very compact growth.

Seashore and coast

Salt plant associations are relatively poor in species, few being able to withstand the extreme conditions of the sea coast, such as the constantly changing level of water because of tides and violent storms, and high salt content. Apart from Glasswort, the so-called salt plants or halophytes have the best chances of survival in salt-free or only slightly salty environments. Since these species require huge amounts of light and cannot endure shade from other plants, they retreat to these inhospitable habitats in order to survive without competition and usually grow in large populations.

The wide band of shore between high and low water, rich in silt and clay and containing nutrients, is called mudflats, or saltmarsh; its lower parts, from the average high water line to about 30cm below it, are colonised by a thick cover of glassworts. Next follows a zone of Sea Aster, Sea-lavender, Orache and Common Saltmarsh-grass (good for grazing). The latter plant indicates the transition from mudflat to saltmarsh. In the region of Sea-lavender ground cover, which is washed over only by exceptional tides, we find many oxygen-dependent species such as Sea Wormwood and Sea-purslane. Further inland, on higher and somewhat less saline ground, the salt plants are displaced by species of other grassland communities.

On flat seashores, sand is constantly washed up by the surf over the average high water line; subsequently it dries on the surface and is carried inland by storms until it gathers on the lee side of a plant and forms into a sand-dune. Marram grass encourages the formation of dunes and is also capable of growing up through the constantly growing dune. Sea Rocket, Prickly Saltwort, Grass-leaved Orache and Sea-holly colonise this 'young' dune. They are stabilised by the finely branched runners of the root system of Sea Couch. If another new young dune forms in front of the one exposed to storms, then on the lee side more plants find a hold and form humus; the young dune becomes a mature (grey) dune colonised by species of drier sandy grassland such as Sand Sedge, Grey and Crested Hair-grass, later with Sea-buckthorn and Rowan and in newly created wind-made open places with Crowberry.

In addition, inland salt areas have halophyte vegetation which is particularly rich in species in south-eastern Europe, but may be found nearer than this at Lake Neusiedl. This situation is dominated by alkali plants which can also survive in habitats that are drying out, whereas the salt plants of central Europe are more closely linked to damp soils containing common salt.

Family characteristics

Page references for each family are given in the left-hand margin.

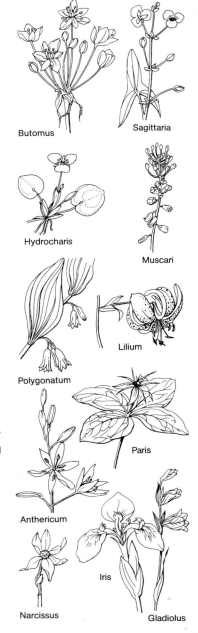

Monocotyledonous plants with conspicuous coloured flowers.

226 **Flowering-rush Family** *Butomaceae*. Marsh and water plants with basal leaves and symmetric flowers in umbels; perianth with 6 segments.

224 **Water-plantain Family** *Alismataceae*. Marsh and water plants with unisexual flowers and sagittate leaves (Sagittaria), or hermaphrodite flowers (Alisma), perianth 3-merous, the outer 3 segments green and the inner 3 coloured.

226 **Frogbit Family** *Hydrocharitaceae*. Water plants; flowers usually unisexual, before flowering enclosed in a leaf sheath (spathe) composed of 1–2 bracts; perianth 6-merous; stamens 3 to many; ovary inferior; leaves rounded (Hydrocharis), lanceolate in whorls of 3 (Elodea) or in a rosette (Stratiotes).

54 **Lily Family** *Liliaceae*. Plants with tubers, bulbs
136 or rhizomes; flowers conspicuous (but small,
206 green in the evergreen Ruscus), usually
246 3-merous with 2 rows of 3 petal-like perianth
302 segments (4-merous in Paris, 2-merous in Maianthemum), symmetric; styles 3 (Tofieldia, Veratrum, Colchicum), but one in the other genera; perianth segments united (e.g. Convallaria, Polygonatum, Muscari) or free (Anthericum, Lilium); ovary superior.

138 **Daffodil Family** *Amaryllidaceae*. Bulbous plants,
206 similar to Liliaceae, but ovary inferior; flowers
308 3-merous (with a corona in Narcissus); leaves all basal.

138 **Iris Family** *Iridaceae*. Plants with tubers or
206 rhizomes, rarely bulbous; leaves sword-shaped
248 (grass-like in Crocus); flowers zygomorphic in Gladiolus, otherwise symmetric; perianth segments 6 in two rows of 3, in Iris the outer 3 are shaped differently from the inner 3; ovary inferior.

140 **Orchid Family** *Orchidaceae*. Plants with leaves
248 spirally arranged or in 2 rows; flowers
310 actinomorphic, in spikes or racemes (Orchis),
412 less often only 1–2 (Cypripedium); perianth segments in whorls with 3 petal-like inner

Butomus

Sagittaria

Hydrocharis

Muscari

Lilium

Polygonatum

Paris

Anthericum

Iris

Narcissus

Gladiolus

segments; one segment is developed as a lip and is often larger and 3-lobed (the labellum); labellum downward-pointing (but directed upwards in Nigritella, Malaxis, Epipogium); flowers often with a spur (Gymnadenia, Orchis); stamens 1–2, grown together into a column with the style; pollen grains united into a stalked pollinium, which is removed as a whole at pollination; ovary inferior; leaves usually green, but whitish and without chlorophyll in the saprophytic orchids such as Neottia, Corallorrhiza and Limodorum.

Monocotyledonous plants with inconspicuous, small flowers which are colourless, greenish, greenish-yellow or whitish; numerous marsh and water plants.

228 **Reed-mace Family** *Typhaceae*. Water and marsh plants with linear leaves and a creeping rhizome; flowers unisexual, male and female flowers in different club-shaped inflorescences with one inflorescence just above the other; perianth in the form of long hairs or scales (Typha).

228 **Bur-reed Family** *Sparganiaceae*. Water and marsh plants; leaves grass-like; flowers unisexual in different heads, which are spherical and 1–3cm across; perianth of 3–6 membranous scales (Sparganium).

222 **Pondweed Family** *Potamogetonaceae*. Floating or submerged water plants, leaves alternate (Potamogeton) or opposite (Groenlandia), the submerged leaves often grass-like or lanceolate; floating leaves elliptic, coriaceous; flowers green, 4-merous, hermaphrodite.

224 **Horned Pondweed Family** *Zannichelliaceae*. Submerged water plants; leaves filiform; flowers unisexual, without perianth, 1–6 together on the same plant in the leaf axils, surrounded by a hyaline spathe (Zannichellia).

226 **Arrow-grass Family** *Juncaginaceae*. Marsh
440 plants; leaves grass-like, in a rosette; stem leafless; flowers hermaphrodite with 6 greenish perianth segments in a raceme; carpels joined together, but separating from the bottom upward on ripening to form a schizocarp of 3–6 parts (Triglochin).

224 **Rannoch-rush Family** *Scheuchzeriaceae* Marsh plants; leaves slightly grooved; stem leafy; flowers hermaphrodite with 6 greenish-yellow perianth segments, in a raceme; carpels 3(–6), joined only at the base (Scheuchzeria).

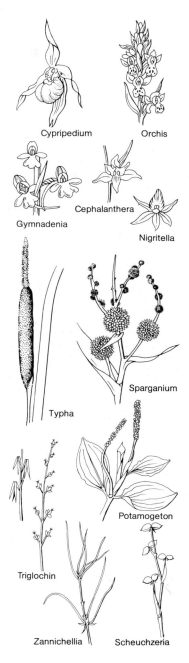

Cypripedium

Orchis

Gymnadenia

Cephalanthera

Nigritella

Typha

Sparganium

Potamogeton

Triglochin

Zannichellia

Scheuchzeria

134 **Sedge Family** *Cyperaceae*. Marsh plants with
232 grass-like, 3-sided leaves and often 3-cornered
298 (trigonous) stems; flowers in 1- or many-
412 flowered spikelets in the axils of glumes;
440 spikelets occur singly or mostly combined in
spikes, panicles or heads in the axils of bracts;
perianth consists of hairs or bristles, or is
lacking; stamens 3, rarely 2; ovary 1, styles 2 or
3. In Carex, flowers unisexual, male and female
in the same or different spikes, ovary enclosed in
a bottle-shaped utricle (see diagram); other
genera with hermaphrodite flowers; in
Trichophorum, Eleocharis, and some
Eriophorum, stem with only a single terminal
spikelet, other genera with several spikelets, in
Scirpus, Schoenoplectus and Cladium (which
have prickly saw-like edges to the leaves) in
panicles, in Schoenus and Rhynchospora
clustered in heads, in Holoschoenus in globose
heads, and in Blysmus in compressed terminal
spikes

54 **Rush Family** *Juncaceae*. Plants grass-like; stems
134 without nodes, usually terete and never sharp-
244 angled; with bristle-like, furrowed rush leaves
300 (Juncus) or with flat leaves (Luzula); flowers
412 hermaphrodite, greenish, brownish or whitish,
440 in panicles, umbels or heads; perianth segments
6, stamens 6, styles 3.

50 **Grass Family** *Gramineae* or *Poaceae*. Plants with
130 hollow stems with solid nodes; leaves alternate,
202 consisting of a tubular leaf sheath and a narrow,
228 spreading leaf blade, with a small appendage
294 (ligule) at the join (the ligule is sometimes
412 replaced by a tuft of hairs); flowers
438 hermaphrodite, occurring singly or combined in
sessile or stalked spikelets; each spikelet consists
of a lower (l.g.) and upper (u.g.) glume and the
flower; each flower consists of the lemma (l.),
often with an awn, and a palea (p.) (sometimes
absent); stamens 3, styles feathery, 2, ovary 1,
superior. According to the arrangement of the
spikelets in the inflorescence, 3 groups can be
distinguished.
　Spiked grasses: spikelets sessile or on very
short, unbranched stalks, in spikes or in spike-
like racemes, which may be terminal and simple,
as in Lolium or Brachypodium, or several
arranged together pinnately or digitately as in
Spartina and Echinochloa.
　Spike-panicle grasses: spikelets in a terminal,
spike-like, dense panicle with very short,
branched stalks (Cynosurus, Alopecurus).
　Panicled grasses: spikelets in racemes or
panicles with long stalks, occasionally spikelets
clustered at the end of longer panicle branches,

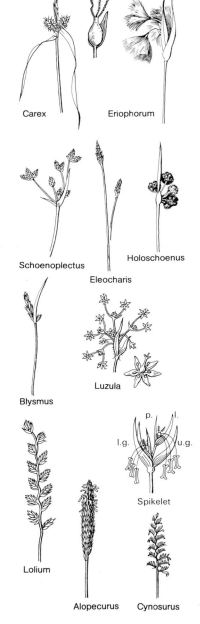

Carex　　　　Eriophorum

Schoenoplectus　　　Holoschoenus

Eleocharis

Luzula

Blysmus

Lolium

p.　l.

l.g.　　u.g.

Spikelet

Alopecurus　Cynosurus

e.g. Dactylis, Agrostis, Anthoxanthum, Apera, Arrhenatherum, Avena, Briza, Bromus, Calamagrostis, Corynephorus, Deschampsia, Festuca, Glyceria, Holcus, Milium, Molinia, Phalaris, Phragmites, Poa, Puccinellia, Stipa, Trisetum.

246 **Arum Family** *Araceae*. Plants with thick rhizome;
302 flowers small, in a thick spadix, often surrounded by a calyx-like leaf sheath or spathe (Arum, Calla) or the spathe stem-like, greenish, not concealing the spadix (Acorus); perianth segments usually wanting.

308 **Yam Family** *Dioscoreaceae*. Climbing plants with heart-shaped leaves, and small, green, axillary flowers and red berries (Tamus).

Dicotyledonous plants with conspicuous symmetric flowers.

146 **Sandalwood Family** *Santalaceae*. Mostly
326 semi-parasites with simple, alternate leaves and small, 4- to 5-merous, white flowers; ovary inferior (Thesium).

56 **Birthwort Family** *Aristolochiaceae*. Plants with
326 alternate, simple leaves, and symmetric (Asarum) or zygomorphic, 2-lipped corolla (Aristolochia); perianth segments united, 3-merous; stamens 6 or 12; ovary inferior.

62 **Pink Family** *Caryophyllaceae*. Plants with
146 simple, opposite or whorled leaves and
208 5-merous corolla (rarely 4-merous as in Sagina)
252 or corolla absent (see page 32); calyx tubular
328 with 5–6 lobes (Agrostemma, Gypsophila,
414 Lychnis, Melandrium, Silene, Viscaria),
442 sometimes with scale-like bracts at the base (Dianthus, Petrorhagia), or calyx lobes free or only joined at the base, 4–5-merous (Arenaria, Cerastium, Minuartia, Moehringia, Sagina, Spergula, Spergularia, Stellaria); stamens 5 or 10; styles 2–5; ovary superior.

68 **Buttercup Family** *Ranunculaceae*. Plants with
150 symmetric flowers (e.g. Adonis, Aquilegia,
208 Ranunculus, Thalictrum) and zygomorphic
254 flowers (Aconitum); leaves alternate (but
328 opposite in Clematis), often palmately divided
416 or lobed; perianth with 4 to many segments, all alike or divided into calyx and corolla; sometimes between the perianth segments and the numerous stamens there are nectaries (or honey-leaves); ovaries many, free, when ripe forming achenes or follicles.

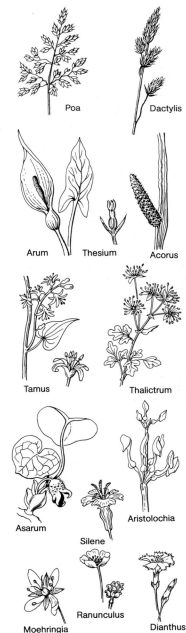

Poa   Dactylis

Arum   Thesium   Acorus

Tamus   Thalictrum

Asarum   Aristolochia

Silene

Moehringia   Ranunculus   Dianthus

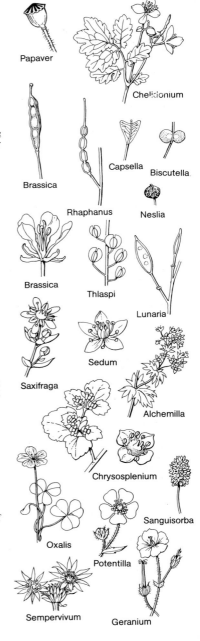

70 **Poppy Family** *Papaveraceae*. Plants with milky
418 juice; leaves alternate, usually pinnate; sepals
442 usually 2, soon falling; petals 4(–6); stamens
usually numerous; stigma disc-shaped with 4–20
rays and fruit a capsule (Papaver) or style
2-lobed and fruit a siliqua (Chelidonium).

72 **Cabbage Family** *Cruciferae* or *Brassicaceae*.
152 Plants with alternate, simple or pinnately
208 divided leaves, and a very consistent flower
258 type; petals 4, sepals 4, stamens usually 6 (4 long
338 and 2 short); ovary superior; fruit a siliqua (over
418 3 times as long as broad), e.g. Brassica, or
442 Raphanus (with its jointed fruit), or else a
silicula (less than 3 times as long as broad), e.g.
Capsella, Thlaspi, Lunaria, Neslia, Biscutella.

154 **Stonecrop Family** *Crassulaceae*. Plants with
420 entire, fleshy leaves, often in basal rosettes;
flowers clustered or in racemes; petals 5
(Sedum) or 8–20 (Sempervivum), stamens as
many as or twice as many as the petals; ovary
superior, carpels usually several, free or united
at the base.

158 **Saxifrage Family** *Saxifragaceae*. Plants, usually
260 with alternate leaves, which are simple or
340 pinnately divided and often in rosettes; perianth
420 4-merous, in one row and greenish-yellow
(Chrysosplenium), or in 2 rows and 5-merous
(Saxifraga); stamens 8 or 10.

82 **Rose Family** *Rosaceae*. Trees, shrubs (see page
158 34), or herbs; flowers usually 5-merous; petals
208 free, usually 5 (but in Dryas 10, Potentilla erecta
260 4, and in Aphanes, Alchemilla, and Sanguisorba
342 absent).
420

86 **Wood Sorrel Family** *Oxalidaceae*. Leaves usually
358 alternate or basal, with 3 leaflets; flowers
5-merous; petals free; stamens 10; styles 5
(Oxalis).

88 **Crane's-bill Family** *Geraniaceae*. Leaves alternate
210 and usually palmately divided (Geranium), or
264 pinnately divided (Erodium); petals 5, free;
360 sepals 5; stamens 10, united at the base; ovary of
3–5 united carpels with elongated, beak-shaped
styles; fruit falling apart into 1-seeded
mericarps.

170 **Flax Family** *Linaceae*. Plants with simple
alternate leaves; petals and sepals free, 5, rarely
4; stamens usually 5, united at the base; styles
3–5; fruit a 5-celled capsule (Linum).

26

92 **Mallow Family** *Malvaceae*. Leaves alternate and
174 palmately divided; sepals and petals always 5,
264 free; flowers have an epicalyx of 6–9 segments
(Althaea, Alcea) or 3 segments (Malva);
stamens many, united into a tube.

Linum

Malva

92 **St. John's-wort Family** *Hypericaceae*. Leaves
174 opposite or whorled, simple, dotted with
370 translucent oil glands; flowers yellow; sepals and
petals free, always 5; stamens many, combined
into 3–5 bundles; ovary superior; styles 3–5
(Hypericum).

Hypericum

Bryonia

Helianthemum

174 **Rock-rose Family** *Cistaceae*. Leaves alternate
and filiform (Fumana) or opposite and ovate
(Helianthemum); sepals and petals free, always
5; stamens many; ovary superior; style one.

94 **Gourd Family** *Cucurbitaceae*. Usually climbing
plants with coiled tendrils and alternate leaves,
which are often palmate or lobed; flowers
unisexual, 5-merous; ovary inferior; fruit a berry
(Bryonia).

264 **Loosestrife Family** *Lythraceae*. Leaves usually
opposite or whorled, entire; flowers 4–6-merous
in long spikes; calyx tubular; stem usually
4-angled (Lythrum).

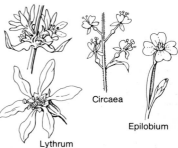

Circaea

Epilobium

Lythrum

94 **Willow-herb Family** *Onagraceae*. Leaves usually
266 opposite; flower stalk widened out into a
370 tubular or cup-shaped calyx-like structure
(hypanthium), which supports either 4 petals
and sepals and 8 stamens (Oenothera,
Epilobium) or 2 each of sepals, petals and
stamens (Circaea).

94 **Carrot Family** *Umbelliferae* or *Apiaceae*. Plants
176 with branched, hollow and angled stems; leaves
212 usually several times pinnate and alternate;
266 petiole sheathing at base; flowers 5-merous,
372 symmetric, sometimes with the outer petals of
424 unequal size (Orlaya); flowers usually in
444 compound umbels (Chaerophyllum), rarely in
heads or in simple umbels (Astrantia,
Eryngium, Hydrocotyle, Sanicula); inflorescence
therefore umbrella-shaped; at the base of the
relatively long main branches of the umbel there
are often scale-like bracts of about 5–10mm
long, and at the base of the secondary umbels
likewise smaller bracteoles; petals 5, often
emarginate; styles 2, often with an enlarged
disc- or cone-shaped base (stylopodium).

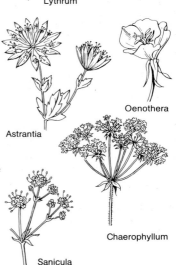

Astrantia

Oenothera

Chaerophyllum

Sanicula

378 **Wintergreen Family** *Pyrolaceae*. Plant with alternate, simple, evergreen leaves (Pyrola), or a saprophyte without chlorophyll and pale-coloured, scale-like leaves (Monotropa); flowers 4–5-merous, usually in racemes; petals free; sepals united; stamens 8 or 10; ovary superior; style 1.

98 **Primrose Family** *Primulaceae*. Plant with basal
272 rosette (Androsace, Primula, Soldanella), or
382 leaves opposite or whorled (Anagallis,
426 Hottonia, Lysimachia, Trientalis), or leaves
444 alternate (Glaux); flowers usually 5-merous (7-merous in Trientalis); corolla bell- or funnel-shaped (Soldanella), or absent (Glaux); ovary superior; style 1.

446 **Sea-lavender Family** *Plumbaginaceae*. Plants with undivided, ovate leaves (Limonium) or grass-like leaves (Armeria); flowers 5-merous in flat-topped panicles or head-like inflorescences; calyx tube scarious and often coloured; ovary superior.

178 **Gentian Family** *Gentianaceae*. Plants with
276 opposite, undivided leaves; flowers 4–5-merous;
386 corolla tubular, lobes twisted in bud; stamens
428 4–5; ovary superior; style 1 (Gentiana, Swertia).

386 **Periwinkle Family** *Apocynaceae*. Plants with opposite, evergreen, simple leaves and 5-merous flowers; corolla lobes united at base into a tube; ovary superior (Vinca).

384 **Milkweed Family** *Asclepiadaceae*. Plants with opposite, simple leaves and 5-merous flowers; stamens 5, united into a column with the superior ovary, with 2 petal-like appendages at the base of the filament, forming a corona (Vincetoxicum).

98 **Bindweed Family** *Convolvulaceae*. Often twining plants, with simple, alternate leaves and 5-merous flowers; corolla lobes united; ovary superior; calyx enclosed in 2 enlarged bracts (Calystegia) or bracts absent (Convolvulus).

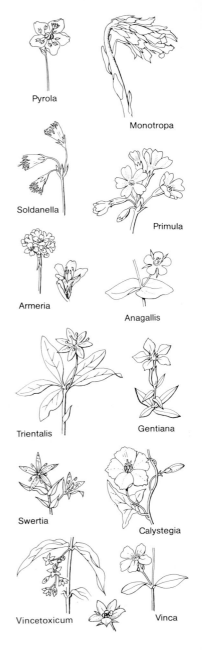

Pyrola

Monotropa

Soldanella

Primula

Armeria

Anagallis

Trientalis

Gentiana

Swertia

Calystegia

Vincetoxicum

Vinca

100 **Borage Family** *Boraginaceae*. Plants usually
182 stiffly hairy; leaves undivided, alternate; flowers
278 5-merous, symmetric, rarely zygomorphic
388 (Echium); in dichotomously branched
430 inflorescences with recurved spikes, which are
446 spirally coiled in bud; corolla often closed by
scales at the throat (e.g. Anchusa,
Cynoglossum, Lappula, Symphytum, Myosotis)
or the throat open (Cerinthe, Echium,
Lithospermum, Nonea, Onosma); fruit
consisting of 4 one-seeded achenes, often with
hooked bristles.

110 **Nightshade Family** *Solanaceae*. Plants with
392 alternate, simple leaves and 5-merous flowers;
corolla lobes united at the base; stamens 5;
corolla rotate (Solanum) or funnel-shaped
(Hyoscyamus, Datura) or bell-shaped (Atropa);
stigma 2-lobed; fruit a 2-celled berry or capsule.

432 **Globularia Family** *Globulariaceae*. Plants with
alternate, undivided leaves and head-shaped
inflorescences; calyx tubular, 5-merous; corolla
with a narrow tube, 2-lipped; stamens 4
(Globularia).

98 **Bedstraw Family** *Rubiaceae*. Plants with
180 opposite or whorled leaves and many-flowered
216 inflorescences; corolla usually small, 3–5 lobed;
276 stamens 3–5; style 1; ovary inferior (Galium).
386

114 **Valerian Family** *Valerianaceae*. Plants with
190 opposite leaves and flowers usually small in
282 dense-flowered cymose panicles; calyx toothed
400 (Valerianella), or enlarging into a pappus when
ripe (Valeriana); corolla tubular, 5-lobed, rarely
spurred (Centranthus); stamens 1–4.

114 **Teasel Family** *Dipsacaceae*. Plants with opposite
190 leaves and dense flower heads with small
216 flowers; flower heads oblong (Dipsacus, with
400 spiny bracts) or hemispherical (Scabiosa,
Knautia, Succisa); flowers 4–5-merous,
somewhat irregular, with epicalyx; ovary
inferior.

116 **Bellflower Family** *Campanulaceae*. Plants with
192 alternate simple leaves and 5-merous flowers;
218 petals linear, free at flowering time or only
282 joined together at the tips (Phyteuma, Jasione)
400 or petals united and corolla bell-shaped
432 (Campanula, Legousia); ovary inferior; fruit a
capsule.

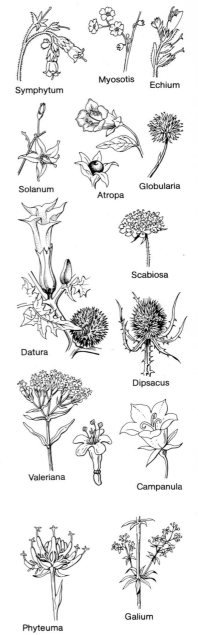

Symphytum

Myosotis

Echium

Solanum

Atropa

Globularia

Scabiosa

Datura

Dipsacus

Valeriana

Campanula

Phyteuma

Galium

116 **Daisy Family** *Compositae* or *Asteraceae*. Plants
194 with alternate leaves, rarely opposite (Arnica);
218 inflorescence very characteristic; flowers small,
284 many together in a head (capitulum) or disc,
402 surrounded by bracts; receptacle often with
434 glume-like scales; calyx of the individual flowers
446 (florets) often replaced by scales or bristles, or
transformed at fruiting time into a ring of hairs
(pappus); pappus hairs can be simple or
pinnately branched, and serve as an organ of
flight for the distribution of seed; plant with
milky juice if flower head has only ray (ligulate)
florets (e.g. Taraxacum, Crepis, Sonchus), or
plant without milky juice if the head has both
disc (tubular) florets and ray florets at the
margin (e.g. Aster, Bellis, Doronicum), or if all
the florets, at the centre and the margin, are
tubular (e.g. Centaurea, Cirsium, Gnaphalium).

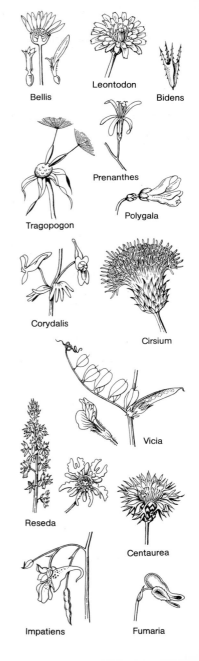

Bellis

Leontodon

Bidens

Prenanthes

Tragopogon

Polygala

Corydalis

Cirsium

Vicia

Reseda

Centaurea

Impatiens

Fumaria

Dicotyledonous plants with
conspicuous, zygomorphic flowers.

Birthwort Family *Aristolochiaceae*. See page 25.

68 **Fumitory Family** *Fumariaceae*. Plants with
152 alternate leaves which are usually divided, and
338 racemose inflorescences; corolla 4-merous, with
a spur; sepals 2, soon falling; stamens 6, 3 of
them fused (Fumaria).

84 **Mignonette Family** *Resedaceae*. Plant with
alternate leaves and small flowers in racemes;
sepals 4–8; petals 4–8, those at the top being the
largest and more deeply divided; ovary superior
(Reseda).

84 **Pea Family** *Fabaceae* or *Papilionaceae*. Plants
160 with alternate leaves; these may be either
210 trifoliate (e.g. Medicago, Trifolium), or
264 imparipinnate (e.g. Astragalus, Hippocrepis,
352 Lotus), or paripinnate, with a tendril or point at
422 the end (Lathyrus, Vicia). For further notes see
444 page 34.

170 **Milkwort Family** *Polygalaceae*. Plants with
alternate, entire and undivided leaves, and pea-
like flowers; sepals 5, overlapping, the 2 inner
larger and petal-like; petals 5, the lowest large
and fringed with appendages; stamens 10; ovary
superior (Polygala).

364 **Balsam Family** *Balsaminaceae*. Leaves simple,
undivided; petals 5, free; sepals 3, one of which
is coloured and has a spur; stamens 5, anthers
fused together; fruit a 5-valved capsule which
springs open on ripening and throws out the
seeds (Impatiens).

92 **Violet Family** *Violaceae*. Plants with entire,
176 alternate leaves; flowers 5-merous,
264 zygomorphic, spurred and usually solitary;
368 pedicels with 2 bracteoles; stamens 5; sepals
424 with appendages; fruit a 3-valved capsule; seeds
with oil-bodies (Viola).

104 **Mint Family** *Labiatae* or *Lamiaceae*. Plants
182 with 4-angled stems and decussate or rarely
214 whorled leaves and branches; flowers apparently
278 in whorls in the axils of the upper leaves, the
390 overall inflorescence spike- or raceme-like;
430 petals 5, united; corolla usually 2-lipped, rarely
apparently 1-lipped (Ajuga, Teucrium); stamens
4 (2 long and 2 short), rarely only 2 stamens
(Salvia); ovary superior, 4-celled.

110 **Figwort Family** *Scrophulariaceae*. Plants whose
186 leaves are usually undivided, alternate or
214 opposite; inflorescences spikes or racemes;
280 flowers 4–5-merous and 2-lipped, rarely almost
394 symmetric (Veronica, Verbascum); in Digitalis
430 and Schrophularia the corolla has a
campanulate tube with a 2-lipped mouth;
stamens usually 2 or 4, rarely 5 (Verbascum);
ovary superior; mouth of the corolla sometimes
closed by an outgrowth of the lower lip (palate),
e.g. Antirrhinum, Cymbalaria, Linaria; corolla
sometimes spurred (e.g. Chaenorrhinum,
Linaria).

188 **Broomrape Family** *Orobanchaceae*. Parasitic
396 plants without chlorophyll and with scale-like,
alternate leaves; flowers in racemes or spikes;
calyx 4–5-divided, or with 2 deeply cut lobes;
corolla tubular, 2-lipped; stamens 4; ovary
superior (Orobanche).

Valerian Family *Valerianaceae*. See page 29.

Dicotyledonous plants with
inconspicuous, green or greenish-white
flowers, or without perianth altogether.

326 **Hemp Family** *Cannabaceae*. Plants dioecious,
erect (Cannabis) or twining (Humulus); leaves
palmately divided; flowers unisexual, 5-merous,
greenish, in panicles or spikes.

56 **Nettle Family** *Urticaceae*. Plants mono- or
dioecious, often with stinging hairs (Urtica);
leaves simple, opposite (Urtica) or alternate
(Parietaria); flowers usually unisexual, in
panicles, spikes or heads; perianth 4–5-merous;
stamens 4, curved inward in bud and at
flowering springing back and flinging out the
pollen.

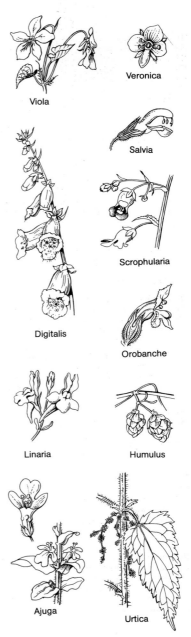

Viola

Veronica

Salvia

Scrophularia

Digitalis

Orobanche

Linaria

Humulus

Ajuga

Urtica

56 **Dock Family** *Polygonaceae*. Plants with angled
206 stems; leaves alternate, with the stipules united
252 into a tubular sheath (ochrea) which encloses
414 the stem; flowers hermaphrodite; perianth
segments 4 (Oxyria), or 5 (Fagopyrum,
Polygonum) or 6 (Rumex); ovary superior: fruit
a 2–3 sided nut.

60 **Goosefoot Family** *Chenopodiaceae*. Many of the
440 family are shore or ruderal plants with thick
and fleshy leaves and stems (Suaeda, Salsola,
Salicornia; flowers alternate; flowers greenish;
perianth segments absent or 1–5, often
persisting and adherent to the fruit; fruit a
1-seeded nut (Atriplex).

62 **Amaranth Family** *Amaranthaceae*. Ruderal
plants with entire, alternate leaves
(Amaranthus) and small flowers in dense,
compound inflorescences; perianth segments
4–5, usually scarious; ovary superior; fruit
usually a circumscissile capsule.

252 **Purslane Family** *Portulacaceae*. Marsh or water
plants (Montia) with minute, white flowers and
opposite, oblong and fleshy leaves; petals 4–6;
stamens 3–5.

Pink Family *Caryophyllaceae*. Flowers
symmetric, without corolla, in clusters
(Scleranthus, Herniaria). See page 25.

88 **Spurge Family** *Euphorbiaceae*. Plants with milky
172 juice (Euphorbia), or without (Mercurialis);
360 leaves alternate (but opposite in Mercurialis);
444 flowers unisexual, greenish; perianth segments 5
or absent; ovary superior; fruit a capsule. In
Euphorbia, what appears to be a flower is in
fact a cyathium, consisting of a 5-cornered
involucre with 4 bean- or half-moon-shaped
glands, in the centre of which there are 10–20
male flowers, each of a single stamen, and a
single stalked female flower, consisting of just
one ovary; the cyathia are arranged in umbels
with many branches.

100 **Bindweed Family** *Convolvulaceae*. Twining, pale-
180 coloured parasitic plants with 4–5-merous,
symmetric flowers in clusters (Cuscuta). See
page 28.

114 **Plantain Family** *Plantaginaceae*. Plants with
188 undivided, alternate leaves with parallel veins,
216 often in rosettes; flowers 4-merous, in dense
282 terminal spikes or heads; corolla scarious;
446 stamens 4, filaments very long; ovary superior;
fruit a circumscissile capsule or an achene
(Plantago).

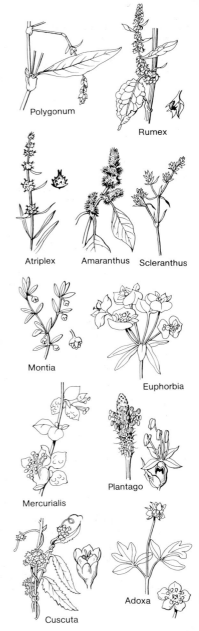

Polygonum

Rumex

Atriplex   Amaranthus   Scleranthus

Montia

Euphorbia

Mercurialis

Plantago

Cuscuta

Adoxa

400 **Moschatel Family** *Adoxaceae*. Small plants with opposite, trifoliate leaves and small greenish flowers in almost cube-shaped heads; corolla 4–5-merous; stamens 4–5, very deeply bifid (Adoxa).

Trees and shrubs over 1m tall.

Coniferous trees

290 **Yew Family** *Taxaceae*. Leaves needle-shaped, evergreen; plants dioecious, flowers unisexual; male flower with stamens arranged like a catkin; fruit berry-like, red (Taxus).

290 **Pine Family** *Pinaceae*. Leaves needle-shaped, evergreen, or deciduous (Larix); plants usually monoecious; male flowers with stamens arranged like a catkin; female flowers arranged in cones with two kinds of scales, one woody and protective and the other bearing the ovule, either falling off as a whole (Pinus, Picea), or the scales falling individually (Abies).

130 **Cypress Family** *Cupressaceae*. Leaves needle-shaped or scale-like (Juniperus); female flowers with 3–6 fruiting scales (sporophylls), ripening into a globose sphalerocarpium (Juniperus).

Deciduous trees and shrubs (see also dwarf shrubs, page 36).

252 **Willow Family** *Salicaceae*. Leaves alternate;
318 flowers in catkins, usually appearing before the
414 flowers, without perianth; stamens 2–5 (Salix) or 8–30 (Populus).

320 **Walnut Family** *Juglandaceae*. Leaves alternate, imparipinnate; flowers unisexual, plants monoecious; male flowers in pendant catkins, female flowers 2–3 together; fruit a nut with a fleshy skin (Juglans).

252 **Birch Family** *Betulaceae*. Plants monoecious,
322 flowers unisexual; male flowers in catkins, female flowers several together in the axil of each bract; the fruiting catkin scales are 3-lobed, falling on ripening with the winged fruit (Betula); or 4–5 lobed and becoming woody on ripening and remaining on the tree as a cone (Alnus).

322 **Hazel Family** *Corylaceae*. Plants monoecious, flowers unisexual; male flowers solitary in the axils of bracts of a drooping catkin; female flowers in loose or bud-like catkins; fruiting involucre 3-lobed (Carpinus), or cup-shaped and split (Corylus).

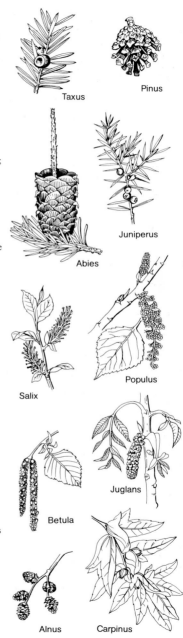

Taxus

Pinus

Juniperus

Abies

Salix

Populus

Juglans

Betula

Alnus

Carpinus

324 **Beech Family** *Fagaceae*. Plants monoecious; male flowers in globose (Fagus), elongated upright (Castanea) or hanging (Quercus) catkins; fruit solitary (Quercus), or several together, enclosed by a leathery or woody involucre (cupule); leaves entire (Fagus), lobed (Quercus), or with aristate teeth (Castanea).

326 **Elm Family** *Ulmaceae*. Leaves undivided, asymmetric at the base; flowers hermaphrodite, in capitate inflorescences; perianth segments 4–5; fruit a nut with a broad encircling wing (Ulmus).

338 **Barberry Family** *Berberidaceae*. Thorny shrubs; leaves alternate, simple; flowers actinomorphic, in racemes, usually with 6 petals; fruit a berry (Berberis).

340 **Currant Family** *Grossulariaceae*. Leaves alternate, 3–5-lobed; flowers actinomorphic, 5-merous, in racemes; petals free; ovary inferior; styles 2; fruit a berry (Ribes).

242 **Rose Family** *Rosaceae*. Trees, shrubs or herbs; leaves alternate, simple or compound, usually with stipules; flowers symmetric, usually 5-merous; styles 1–5 or numerous; receptacle and fruit very different in the various genera; e.g. Rosa, receptacle pitcher-shaped, with the petals and stamens attached at the neck; achenes numerous and enclosed in the fleshy hypanthium (hip); in Rubus, receptacle conical, achenes (drupelets) fleshy, combined into a compound fruit. Padus, Prunus: fruit a drupe, formed from 1 fleshy carpel. Pyrus, Malus, Sorbus, Amelanchier: carpels 2–5, walls cartilaginous in fruit and enclosed in the fleshy receptacle, fruit apple-like. Mespilus, Crataegus, Cotoneaster: carpels 2–5, partly united and making a fruit with a stony centre.

160
354 **Pea Family** *Papilionaceae* or *Fabaceae*. Shrubs (Sarothamnus, Colutea, Coronilla), or dwarf shrubs or herbs (see also page 30). Leaves alternate, pinnate or trifoliate, with stipules; flowers zygomorphic, usually in heads or racemes; petals 5, of which the upper is termed the standard, the two laterals the wings, and the two lower, which are sometimes combined, as the keel; stamens 10, all united into a tube or else 1 free; fruit a pod.

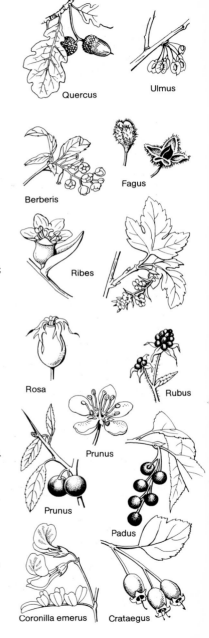

Quercus

Ulmus

Fagus

Berberis

Ribes

Rosa

Rubus

Prunus

Prunus

Padus

Coronilla emerus

Crataegus

362 **Rue Family** *Rutaceae*. Shrubs (Dictamnus) or
herbs (Ruta), with oil cells and strong-smelling;
leaves alternate, pinnate; flowers zygomorphic.

362 **Maple Family** *Aceraceae*. Leaves opposite,
palmately lobed; flowers symmetric, 4–5-
merous; fruit falling apart into two winged
halves (Acer).

364 **Bladder-nut Family** *Staphyleaceae*. Leaves
opposite, imparipinnate; flowers symmetric,
5-merous, in pendent panicles; fruit a capsule
(Staphylea).

362 **Holly Family** *Aquifoliaceae*. Usually evergreen
shrubs with alternate, leathery and spiny-
toothed leaves (Ilex); flowers symmetric, usually
4-merous, unisexual, small and white; fruit
berry-like and red.

364 **Spindle-tree Family** *Celastraceae*. Leaves
opposite, entire; flowers symmetric, 4-merous;
fruit a 3–5-angled capsule (Euonymus).

366 **Buckthorn Family** *Rhamnaceae*. Leaves opposite
or alternate, simple, with small stipules; flowers
inconspicuous, symmetric, 4–5-merous, axillary;
fruit a drupe (Rhamnus).

366 **Lime Family** *Tiliaceae*. Leaves alternate, simple;
flowers hermaphrodite, 5-merous; stamens 10
and more; inflorescence with wing-like bracts
(Tilia).

370 **Oleaster Family** *Elaeagnaceae*. Twigs
terminating in a thorn; leaves alternate; flowers
unisexual, small; plants dioecious; fruit orange,
drupe-like (Hippophae).

172 **Tamarisk Family** *Tamaricaceae*. Shrubs; leaves
alternate, needle- or scale-like, small; flowers
(4)5-merous, in dense racemes, small, pink
(Myricaria).

378 **Dogwood Family** *Cornaceae*. Trees or shrubs
with undivided, usually opposite leaves and
4-merous flowers; petals free; fruit a drupe
(Cornus).

384 **Olive Family** *Oleaceae*. Trees with pinnate
leaves (Fraxinus), or shrubs with undivided
leaves (Ligustrum); flowers symmetric,
4-merous; fruit a winged nut (Fraxinus) or a
berry (Ligustrum) or a capsule.

396 **Honeysuckle Family** *Caprifoliaceae*. Usually
shrubs, rarely herbs (Linnaea); leaves opposite,
simple or divided; flowers symmetric, 5-merous,
in corymbose panicles (Sambucus, Viburnum)
or zygomorphic, 2-lipped, in pairs on the same
stalk (Lonicera).

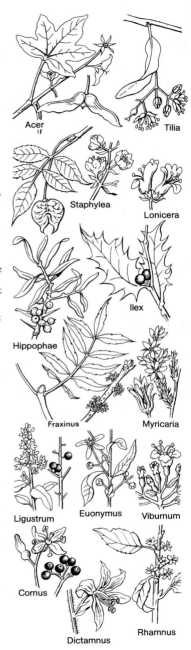

Acer

Tilia

Staphylea

Lonicera

Ilex

Hippophae

Fraxinus

Myricaria

Ligustrum

Euonymus

Viburnum

Cornus

Dictamnus

Rhamnus

Dwarf shrubs, low shrubs
under 1m in height.

414 **Willow Family** *Salicaceae*. Prostrate shrubs, creeping on the ground (Salix). See page 33.

252 **Birch Family** *Betulaceae*. See page 33.

326 **Mistletoe Family** *Loranthaceae*. Semi-parasitic half shrubs on trees; leaves opposite, evergreen (Viscum) or deciduous (Loranthus); flowers symmetric, unisexual or hermaphrodite, whitish, in sessile clusters; perianth segments 4 or 6; stamens 4; ovary inferior; fruit a sticky berry.

342 **Rose Family** *Rosaceae*. Dwarf shrubs (Cotoneaster); leaves simple; flowers hermaphrodite, small, in 2–10 flowered inflorescences on short side-shoots; see page 34.

172 **Daphne Family** *Thymelaeaceae*. Leaves
368 alternate, simple, without stipules; flowers symmetric, corolla absent, calyx coloured, 4–6 lobed; stamens 8–10; fruit a drupe (Daphne).

270 **Crowberry Family** *Empetraceae*. Evergreen dwarf shrubs; leaves needle-shaped, alternate; flowers solitary, 2–3-merous; stamens 2–4; fruit a berry-like drupe (Empetrum).

270 **Heath Family** *Ericaceae*. Low shrubs
380 (Rhododendron) or dwarf shrubs; leaves usually
424 evergreen, alternate or whorled (alternate in Andromeda); corolla rotate, 4-merous (Oxycoccus), bell-shaped or pitcher-shaped, 4-merous (Calluna, Erica, Vaccinium), bell-shaped, 5-merous (Loiseleuria), or rotate and funnel-like (Rhododendron, Rhodothamnus); anthers opening by apical pores, often with horn-like appendages.

Water plants
(submerged or floating in water)

440 **Tasselweed Family** *Ruppiaceae*. Submerged plants of brackish water with linear lvs which are sheathing at the base, and inconspicuous flowers without perianth in 2-flowered spikes (Ruppia).

440 **Sea-grass Family** *Zosteraceae*. Submerged plants of seawater with grass-like lvs and inconspicuous flowers which are arranged in 2 rows on the compressed axis of a spike.

224 **Naiad Family** *Najadaceae*. Submerged plants with linear, serrate lvs and inconspicuous unisexual flowers which are enclosed in a cup-shaped involucre.

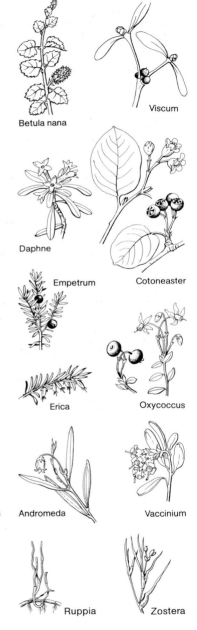

Betula nana

Viscum

Daphne

Empetrum

Cotoneaster

Erica

Oxycoccus

Andromeda

Vaccinium

Ruppia

Zostera

244 **Duckweed Family** *Lemnaceae*. Submerged or floating plants in which the stem and leaf are not distinct; shoot (thallus) rounded, leaf-like, small; flowers tiny, without perianth (Lemna, Spirodela).

254 **Water-lily Family** *Nymphaeaceae*. Plants with large, entire floating leaves and thick rhizome; flowers long-stalked and floating; petals and stamens graded from one to the other (Nuphar).

254 **Hornwort Family** *Ceratophyllaceae*. Plants submerged; leaves whorled, dichotomously forked, often with spiny teeth; flowers unisexual; perianth segments 8–12; rarely flowering (Ceratophyllum).

256 **Buttercup Family** *Ranunculaceae*. Plants with floating and submerged leaves (Ranunculus); flowers white; see page 25.

264 **Water-starwort Family** *Callitrichaceae*. Plants with thread-like stems; leaves entire, opposite; flowers without perianth, only with two bracts, 1 stamen, 1 ovary and 2 styles; fruit falling apart into 4 one-seeded mericarps (Callitriche).

266 **Water Chestnut Family** *Trapaceae*. Plants with rosettes of floating leaves with swollen petioles; flower 4-merous; fruit a 1-seeded drupe-like nut, enclosed in the enlarged fruiting involucre, with 2–4 thorns derived from the sepals (Trapa).

266 **Water-milfoil Family** *Haloragaceae*. Plants submerged, with whorled, comb-like pinnate leaves and 4-merous, inconspicuous flowers; stamens 2–6 (Myriophyllum).

266 **Mare's-tail Family** *Hippuridaceae*. Plants with numerous, linear, whorled leaves and inconspicuous axillary flowers; stamens 1; ovary superior (Hippuris).

276 **Bogbean Family** *Menyanthaceae*. Plants with roundish floating leaves (Nymphoides), or with trifoliate leaves (Menyanthes); flowers symmetric, 5-merous; ovary superior.

280 **Butterwort Family** *Lentibulariaceae*. Plants submerged with much-divided leaves with bladders which trap animals (Utricularia), or marsh plants with sticky-glandular, undivided rosette leaves (Pinguicula); insectivorous; corolla 2-lipped, spurred or pouched at the base; stamens 2; ovary superior.

442 **Goosefoot Family** *Chenopodiaceae*. On marine mud-flats (Salicornia), with thick, fleshy, leafless, articulated and much-branched stems; for further notes on the family see page 32.

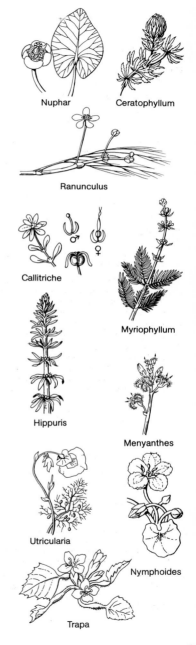

Nuphar

Ceratophyllum

Ranunculus

Callitriche

Myriophyllum

Hippuris

Menyanthes

Utricularia

Nymphoides

Trapa

Key
with special reference to flower colours

Organisation of the key
Flower colours

☐	white
▨	pink
▨	red, purple
▨	violet
▨	blue
▨	yellow
▨	greenish/brownish
▨	flowers inconspicuous

Symbols

●   no. of petals 4

●●   no. of petals *not* 4

✳   radially symmetric flowers

Δ   not radially symmetric flowers

Note: the numbers following the plant names refer to **plate** numbers.

Plants with radially symmetric flowers
Flowers over 10mm in diameter

Fields . . .
  e.g. Wild Radish 19
  e.g. Field Mouse-ear 15

Dry grassland . . .
  . . . . .
  . . . . .

Fields . . .
  e.g. Soapwort 8
  e.g. Field Bindweed 25

Dry grassland . . .
  etc.

Flowers under 10mm in diameter

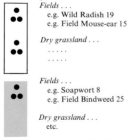

Fields . . .
  . . . . .
  . . . . .
  etc.

Plants with unsymmetric or zygomorphic flowers
Flowers over 10mm across

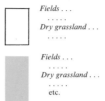

Fields . . .
  . . . . .
Dry grassland . . .
  . . . . .

Fields . . .
  . . . . .
Dry grassland . . .
  . . . . .
  etc.

Flowers under 10mm across

Fields . . .
  . . . . .
Dry grassland . . .
  . . . . .
  etc.

Plants not grass-like, flowers inconspicuous

Fields . . .
  . . . . .
Dry grassland . . .
  . . . . .

Plants grass-like, flowers inconspicuous

Fields . . .
  . . . . .
Dry grassland . . .
  . . . . .

Trees and shrubs over 1m high
With needle-shaped lvs and inconspicuous flowers

Dry grassland . . .
  . . . . .

With deciduous lvs and conspicuous flowers

Trees
  . . . . .
Shrubs
  . . . . .

Trees
  . . . . .
Shrubs
  . . . . .
  etc.

With deciduous lvs and conspicuous flowers

Dwarf shrubs and shrubs under 1m high

Δ	*Trees* . . . . . *Shrubs* . . . . .
	Trees . . . . . *Shrubs* . . . . .

✱	*Dry grassland . . .* . . . . . *Water . . .* . . . . . etc.
Δ	*Dry grassland and woods* . . . . . etc.

Example of an identification

In order to identify Meadow Buttercup we have to ask certain questions about its flowers. From the answers we find out which is the relevant group of plants that we have to look through next.

1. Question: What are the flowers like? *Answer:* radially symmetric

2. Question: What is the flower colour? *Answer:* yellow

3. Question: How large are the flowers? *Answer:* over 10mm diameter

4. Question: Where was the plant found? *Answer:* in a rich meadow

5. Question: How many petals are there? *Answer:* not 4

Now we have to look up the plants listed in that group and compare them with the plant that we want to name.

Plants with radially symmetric flowers
Flowers over 10mm in diameter

Fields . . .
- Wild Radish 13
- Dame's-violet 16
- Horse-radish 16
- Star-of-Bethlehem 3
- Bladder Campion 7
- White Campion 7
- Field Mouse-ear 8
- Umbellate Chickweed 8
- Large-flowered Orlaya 23
- Field Bindweed 25
- Thornapple 31
- Annual Fleabane 35
- Chamomiles 35, 36
- Scented Mayweed 35
- Scentless Mayweed 36
- Feverfew 36

Dry grassland . . .
- St. Bernard's Lily 44
- Branched St. Bernard's Lily 44
- White Asphodel 44
- St. Bruno's Lily 44
- Nottingham Catchfly 49
- Rock Campion 50
- White Stonecrop 53
- Navelwort 54
- Meadow Saxifrage 55
- Creamy Strawberry 56
- Stemless Carline Thistle 75

Rich meadows . . .
- Cuckooflower 80
- Pheasant's-eye Narcissus 79
- Purple Crocus 79
- Daisy 85
- Ox-eye Daisy 85

Water . . .
- Arrowhead 88
- Water-soldier 89
- Frogbit 89
- False Helleborine 99
- White Water-Lily 103
- River Water-crowfoot 104
- Common Water-crowfoot 104
- Fan-leaved Water-crowfoot 104
- Aconite-leaved Buttercup 105
- Grass-of-Parnassus 106
- Cloudberry 107
- Labrador Tea 111
- Bogbean 114
- Sneezewort 118

Woods . . .
- Ramsons 129
- Solomon's-seal 129
- Solomon's-seal species 129
- Snowdrop 130
- Spring Snowflake 130
- Wood Stitchwort 140
- Greater Stitchwort 140
- Berry Catchfly 140
- Christmas Rose 141
- Wood Anemone 142
- Snowdrop Windflower 142
- Erect Clematis 143

Flowers under 10mm in diameter

40

Flowers under 10mm in diameter

Flowers over 10mm in diameter

Flowers under 10mm in diameter

Cross-leaved Heath 111
Bell Heather 112
Blue Heath 112
Water Mint 115
Woods . . .
Greater Meadow-rue 141
Bittersweet 172
Wood Scabious 176
Alps
Purple Saxifrage 186
Dwarf Snowbell 189

Flowers over 10mm in diameter

Fields . . .
Borage 27
Alkanet 27
Creeping Bellflower 34
Large Venus'-looking-glass 34
Chicory 38
Cornflower 39
Dry grassland . . .
Fringed Gentian 65
Cross Gentian 65
Perennial Flax 61
Common Flax 61
Spring Gentian 65
Bladder Gentian 65
Devil's-bit Scabious 71
Sheep's-bit 72
Round-headed Rampion 72
Clustered Bellflower 72
Rampion Bellflower 72
Harebell 72
Blue Lettuce 76
Rich meadows . . .
Field Scabious 84
Spreading Bellflower 85
Water . . .
Siberian Iris 100
Marsh Gentian 114
Marsh Felwort 114
Woods . . .
Bluebell 127
Alpine Squill 128
Hepatica 142
Columbine 143
Lesser Periwinkle 169
Greater Periwinkle 169
Willow Gentian 169
Jacob's-ladder 170
Purple Cromwell 170
Lungwort 170
Nettle-leaved Bellflower 176
Peach-leaved Bellflower 176
European Michaelmas-daisy 177
Perennial Cornflower 178
Alpine Sow-thistle 180
Alps
Alpine Clematis 185
Sticky Primrose 189
Stemless & Trumpet Gentians 190
Bavarian Gentian 190
Alpine Gentian 190
Matted Globularia 192
Leafless-stemmed Globularia 192
Devil's-claw 192
Scheuchzer's Bellflower 192
Bearded Bellflower 192
Fairy's Thimble 192

Seashore . . .
Mertensia 198
Sea-holly 198
Common Sea-lavender 199

Flowers under 10mm in diameter

Fields . . .
Blue Woodruff 25
Grape Hyacinth 3
Scarlet Pimpernel 25
Bur Forget-me-not 26
Bugloss 27
Field Forget-me-not 27
Dry grassland . . .
Tassel Hyacinth 45
Water . . .
Water Forget-me-not 115
Woods . . .
Wood Forget-me-not 170
Alps
Sticky Primrose 189
Alpine Forget-me-not 191
King of the Alps 191

Flowers over 10mm in diameter

Fields . . .
Greater Celandine 11
Rape & Charlock 12
Hare's-ear Mustard 12
Perennial Wall-rocket 12
Bastard Cabbage 13
Woad 14
Buttercup species 10
Summer Pheasant's-eye 10
Creeping Cinquefoil 17
Silverweed 17
Common Evening-primrose 23
Henbane 31
Mulleins 31
Colt's-foot 34
Spring Groundsel 35
Chamomile 35
Tansy 37
Common Fleabane 37
Nipplewort 39
Hawkweed Oxtongue 39
Prickly Oxtongue 39
Wall Hawk's-beard 39
Beaked Hawk's-beard 40
Perennial & Prickly Sow-thistles 40
Greater Goat's-beard 40
Dry grassland . . .
Yellow Asphodel 44
Daffodil 45
Variegated Iris 45
Bulbous Buttercup 51
Yellow Pheasant's-eye 51
Reflexed Stonecrop 53
Orange Saxifrage 55
Hoary Cinquefoil 55
Spring Cinquefoil 55
Sulphur Cinquefoil 55
Spurge 62
St. John's-wort 63
Common Rock-rose 63
Common Fumana 63
Yellow-wort 66
Greater Golden Drop 67
Irish Fleabane 73

Woods . . .
 Herb Paris 130
 Stinking & Green Hellebores 141
 Deadly Nightshade 172
 Cape Gooseberry 173

Flowers under 10mm in diameter
(See also under: plants not grass-like, flowers
inconspicuous or under 3mm, p.47)

Fields . . .
 Narrow-leaved Pennywort 13
Dry grassland . . .
 Salad Burnet 56

Plants with unsymmetric or zygomorphic flowers
Flowers over 10mm across

Fields . . .
 Basil 28
 Cat-mint 28
 White Dead-nettle 29
Dry grassland . . .
 Late Spider-orchid 47
 Lizard Orchid 48
 Green-winged Orchid 48
 Shrubby Milkwort 61
 Teesdale Violet 63
 Mountain Germander 68
Water . . .
 Bog Arum 99
 Marsh Helleborine 100
 Water Lobelia 117
Woods . . .
 White Helleborine 131
 Narrow-leaved Helleborine 131
 Lesser Butterfly-orchid 133
 Greater Butterfly-orchid 133
 Common Spotted-orchid 134
 Bulbous Corydalis 145
 Wood Vetch 154
 Bastard Balm 171

Flowers under 10mm across

Fields . . .
 White Ramping-fumitory 10
 White Melilot 18
 Bird's-foot 18
 Hairy Tare 19
 Cat-mint 28
 Summer Savory 30
Dry grassland . . .
 Autumn Lady's-tresses 46
 Small-white Orchid 46
 Musk Orchid 46
 Mountain & Hare's-foot Clovers 58
 Fenugreek 59
 Mountain Germander 68
Rich meadows . . .
 White & Alsike Clovers 81
 Eyebright 83
Water . . .
 Summer Lady's-tresses 101
 Gipsywort 115
 Alpine Butterwort 116
Woods . . .
 Creeping Lady's-tresses 133
 Coralroot Orchid 133
 Greater Badassi 152

Flowers over 10mm across

Fields . . .
 Basil 28
 Motherwort 28
 Cat-mint 28
 Common Hemp-nettle 28
Dry grassland . . .
 'True' orchids 47
 Spiny Restharrow 57
 Crown Vetch 59
 Wall Germander 67
Water . . .
 Marsh Helleborine 100
Woods . . .
 Red Helleborine 131
 Round-leaved Restharrow 154
 Burning Bush 157
 Bastard Balm 171
 Toothwort 174
Alps
 Globe Orchid 182

Flowers under 10mm across

Fields . . .
 Common Fumitory 11
 Bird's-foot 18
 Motherwort 28
 Cat-mint 28
 Small Toadflax 32
Dry grassland . . .
 Burnt Orchid 47
 Hare's-foot Clover 58
 Fenugreek 59
Rich meadows . . .
 Alsike Clover 81
Water . . .
 Amphibious Bistort 102
 Water Mint 115
 Horse Mint 115
Woods . . .
 Greater Badassi 152
Seashore . . .
 Strawberry Clover 198

Flowers over 10mm across

Fields . . .
 Tuberous Pea 19
 Motherwort 28
 Large-flowered Hemp-nettle 29
 Red Hemp-nettle 29
 Common Hemp-nettle 29
 Spotted Dead-nettle 29
 Red Dead-nettle 29
 Black Horehound 30
 Marsh Woundwort 30
 Field Cow-wheat 32
Dry grassland . . .
 Fly Orchid 47
 Late Spider-orchid 47
 Early Spider-orchid 47
 Bee Orchid 47
 Burnt Orchid 47
 Green-winged Orchid 48
 Bug Orchid 48
 Pyramidal Orchid 48
 Spiny Restharrow 57
 Sainfoin 60
 Shrubby Milkwort 61

Water . . .
    Marsh Gladiolus 100
    Loose-flowered Orchid 101
    Early Marsh-orchid 101
    Irish Marsh-orchid 101
    Purple Loosestrife 108
    Marsh Lousewort 116
Woods . . .
    Dark-red Helleborine 132
    Early-purple Orchid 134
    Lady Orchid 134
    Bulbous Corydalis 145
    Solid-tubered Corydalis 145
    Round-leaved Restharrow 154
    Black Pea 155
    Spring Pea 155
    Bitter Vetch 155
    Narrow-leaved Everlasting Pea 155
    Indian Balsam 158
    Betony 172
    Wild Basil 172
    Foxglove 173
Alps
    Black Vanilla Orchid 182
    Alpine Sainfoin 188
    Beaked Lousewort 191

Flowers under 10mm across
Fields . . .
    Common Fumitory 11
    Motherwort 28
    Red Bartsia 33
Dry grassland . . .
    Fly Orchid 47
    Large Thyme 68
    Red Valerian 71
Rich meadows . . .
    Red Clover 81
Water . . .
    Purple Loosestrife 108
Woods . . .
    Mountain Zigzag Clover 152
    Red Trefoil 152
    Marjoram 172

Flowers over 10mm across
Fields . . .
    Common Vetch 19
    Wild Pansy 22
    Large-flowered Hemp-nettle 29
    Red Hemp-nettle 29
    Ground-ivy 30
    Marsh Woundwort 30
Dry grassland . . .
    Fragrant Orchid 46
    Bee Orchid 47
    Green-winged Orchid 48
    Lucerne 57
    Crown Vetch 59
    Teesdale Violet 63
    Large Self-heal 68
    Sage 69
    Whorled Clary 69
    Garden Lavender 69
    Purple Mullein 69
Rich meadows . . .
    Bugle 83
    Self-heal 83

Water . . .
    Loose-flowered & Early Marsh Orchids 101
    Marsh Pea 108
    Marsh Violet 108
    Alpine Bartsia 116
    Marsh Lousewort 116
    Water Lobelia 117
Woods . . .
    Red Helleborine 131
    Dark-red Helleborine 132
    Violet Limodore 132
    Ghost Orchid 133
    Calypso 133
    Common Spotted-orchid 134
    Bush Vetch 153
    Danzig Vetch 153
    Purple Wood Vetch 153
    Wood Vetch 154
    Black Pea 155
    Bitter Vetch 155
    Violets & Dog-violets 159, 160
    Hedge Woundwort 171
    Wood Cow-wheat 174
Alps
    Alpine Milk-vetch 187
    Purple Oxytropis 187
    Alpine Toadflax 191
    Alpine Calamint 191
    Pyramidal Bugle 191

Flowers under 10mm across
Fields . . .
    Hairy Tare 19
    Smooth Tare 19
    Fodder Vetch 19
    Corn Mint 30
    Vervain 30
    Small Toadflax 32
Dry grassland . . .
    Lucerne 57
    Tufted Milkwort 61
    Basil Thyme 67
    Ivy-leaved Toadflax 69
Rich meadows . . .
    Tufted Vetch 81
Water . . .
    Water Mint 115
    Common Butterwort 116
Woods . . .
    Heath Speedwell 173
    Nettle-leaved Speedwell 173
Alps
    Leafless-stemmed Speedwell 191
    Alpine Toadflax 191
    Pyramidal Bugle 191

Flowers over 10mm across
Fields . . .
    Forking Larkspur 10
    Wild Pansy 22
    Viper's-bugloss 26
Dry grassland . . .
    Lucerne 57
    Blue Bugle 67
    Large Self-heal 68
    Meadow Clary 68
    Garden Lavender 69
Rich meadows . . .
    Bugle 83
    Self-heal 83

Plants not grass-like, flowers inconspicuous or under 3mm

Plants grass-like, flowers inconspicuous

Trees and shrubs over 1m high
With needle-shaped lvs and inconspicuous flowers

With deciduous lvs and conspicuous flowers

Dwarf shrubs and shrubs under 1m high

Plate 1   Grass Family, Gramineae or Poaceae

1. Barren Brome, *Bromus sterilis* L., plant 30–80cm; lvs 3–5mm wide, hairy; ligule split and fringed; leaf sheaths softly hairy; stem glabrous above; panicle lax, spreading in all directions, with rough branches; spikelets 2–3cm long, becoming wider towards the tip; awn longer than the lemma; outer glume small, 1-nerved, upper larger, 3-nerved. Fl.6–7. Dry pathsides, rubble, walls; common. Almost all of Europe. Similar to it is **Drooping Brome**, *Bromus tectorum* L., but stem hairy above, panicle drooping to one side, upper branches softly hairy; paths, embankments, gravelpits; rare in Britain; S. Scandinavia, C. and S. Europe. The **Field Brome**, *B. arvensis* L., has its spikelets mostly tinged reddish-violet; both glumes about the same length, the lower 3–5 nerved and the upper 7–9 nerved; paths, rubbish, cornfields; Europe. See also *B. inermis*, Pl.41. The **Soft Brome**, *Bromus hordeaceus* L. (*B. mollis* L.), has softly hairy lvs and an erect panicle; spikelets usually hairy and glumes unequal. Waste places and meadows; common. Most of Europe. There are many similar species.

2. Annual Meadow-grass, *Poa annua* L., plant 5–25cm, forming a turf; lvs flat, with a double groove (tramline) in the centre on the upper side; ligule 2–4mm long, whitish; panicle 2–5cm long, lower branches 1–2 together, smooth, spreading horizontally; lemmas mostly green. Fl. all year round. Paths, trampled lawns, fields, gardens and in mountains; common. Europe. Similar species: **Prostrate Meadow-grass**, *Poa supina Schrad., but with lemmas usually tinged brownish-violet, ligules 0–2mm long; trampled lawns and pastures in the European mountains. See also *P. trivialis* Pl.77.

3. Bearded Finger-grass, *Dichanthium ischaemum* (L.)Roberty (*Bothriochloa i.*(L.)Keng), plant 15–60cm, ascending and bent at the nodes, with short stolons; leaves grey-green, sparsely ciliate, 3mm wide; ligule in the form of hairs; spikes 2–5, arranged digitately, narrow, 3–6cm long; spikelets linear, with violet, ciliate glumes and awns 1.5cm long. Fl.7–9. Dry or dryish grassland, banks of paths; likes warm places; rare. C. and S. Europe.

4. Common Couch, *Elymus repens* (L.)Gould (*Agropyron r.* (L.)P.B.)., plant 30–150cm, with long rhizomes; lvs green or blue-green, flat, 3–8mm wide, rough above, with auricles at the base which clasp the stem; spikelets flat, broadside on to the spike axis. Fl.6–8. Fields, waste places, paths, banks, riversides, water meadows; common. Europe.

5. Italian Rye-grass, *Lolium multiflorum* Lam. (*L. italicum* A.Br), plant 30–100cm; lvs light green, up to 10mm wide, shining below; leaf sheaths, upper part of culm and spike axis rough; spikelets flat, arranged in 2 rows, edgeways on to the main axis (compare with *Elymus*); lower glumes 7-nerved, half as long as the spikelet; lemmas 7–8mm long, awned. Fl.6–8. Pastures, paths, waste ground; cultivated; rather frequent. Europe. Similar species: **Perennial Rye-Grass**, *L. perenne* L., plant 20–60cm; lvs dark green, 3–4mm wide, finely ribbed above; ligule 1mm long, truncate; spike 10–20cm long; spikelets up to 2cm long; glumes about equal to the lemmas, lemma always awnless; lower glume 5-nerved, about $\frac{3}{4}$ length of spikelet; pastures, paths, common; Europe. In the now much less common **Darnel**, *L. temulentum* L., the glumes are 15–20mm long and 2–4 times as long as the usually awned lemma (seeds poisonous on account of alkaloid-containing fungi); cornfields; Europe.

1

3

4

5

Plate 2   Grass Family, Gramineae or Poaceae

1. Cockspur, *Echinochloa crus-galli* (L.)P.B., plant 30–90cm; stem smooth, with tufts of hairs at the nodes; lvs dark grey-green, undulate at the margins; ligules absent; panicle erect, up to 20cm long; spikelets clustered, sessile, ovate, pale green or violet tinged; glumes shortly ciliate, tapered or awned. Fl.7–10. Arable fields, waste places, ditches, vineyards; fairly common. Europe.

2. Hairy Finger-grass, *Digitaria sanguinalis* (L.)Scop. (*Panicum sanguinale* L.), plant 10–50cm; lvs and leaf sheaths hairy, lvs 5–10mm wide; ligules 1–2mm long; spikes narrow-linear, 4–8 together, digitately arranged; spikelets lanceolate, acute, 3mm long. Fl.7–10. Sandy arable fields, waysides; calcifuge; casual in Britain, common in Europe elsewhere. Similar to it is **Smooth Finger-grass**, *D. ischaemum* (Schreb.) Mühlenb. (*Panicum lineare* Krocher), but lvs and leaf sheaths glabrous; spikelets elliptic, blunt, 2mm long; spikes 2–4 together; acid, sandy fields; rather rare; almost all of Europe.

3. Loose Silky-bent, *Apera spica-venti* (L.)P.B., plant 30–100cm; lvs 3–4mm wide, rough on both sides, flat; ligules up to 6mm long; panicle up to 30cm long, lax-flowered, with long branches; spikelets 2–3mm long, 1-flowered, with an awn 5–7mm long. Fl.6–7. Cornfields, waste places; calcifuge; widespread. Almost all of Europe, but rare in Britain. The **Dense Silky-bent**, *A. interrupta* (L.)P.B., has lvs flat or inrolled, about 1mm wide with a ligule 1–2mm long; panicle narrowly contracted, irregularly interrupted, with short erect branches; awn 10–15mm long; plant 20–40cm; dry sandy soils, fallow land; rare. Britain, S. Scandinavia, Germany, S. Europe.

4. Green Bristle-grass, *Setaria viridis* (L.)P.B., plant 5–50cm, ascending to erect; lvs with whitish or violet central ribs; panicle 7–10mm thick, not interrupted; bristles green or violet tinged, 5mm long, with forward-directed teeth, so that the inflorescence feels rough when stroked downwards; spikelets 2mm long; fruit falling out when ripe. Fl.6–10. Arable fields, gardens, vineyards; fairly common, but casual in Britain. Europe. Similar species: **Yellow Bristle-grass**, *S. pumila* (Poir.)R.et Sch. (*S. lutescens* (Weigel) Hubb.), plant grey-green, bristles yellow, then rusty-red; lemmas wrinkled; arable fields, vineyards; widespread, but casual in Britain; almost all Europe. **Short-bristled Bristle-grass**, **S. decipiens* C.Schimp. (*S. ambigua* Guss.), is the hybrid between *S. viridis* and *S. verticillata*, spikes interrupted and raceme-like below; bristles 3mm long, green, with forward-directed teeth; arable fields, vineyards; likes warm places; rather rare. **Rough Bristle-grass**, *S. verticillata* (L.)P.B., bristles 3–5mm long, usually green, with backward-directed teeth, hence the inflorescence very rough when stroked upwards (opposite way round to other species); spike often interrupted and raceme-like below; arable fields; rather rare.

5. Wall Barley, *Hordeum murinum* L., plant 10–40cm, annual, mid-green, tufted; upper leaf sheaths rather inflated; spikes 4–12cm long; spikelets 1-flowered, arranged together in 3s, the central spikelet hermaphrodite, and the 2 lateral ones male or sterile; glumes bristle-like and long-awned, 15–20mm long, spreading ciliate on the margins; when ripe the spike axis breaks up under each triplet of spikelets. Fl.5–9. Dry waste places, waysides; likes warm places; common. Britain, S. and C. Europe, northwards to S. Scandinavia.

See *Dactylis glomerata* and *Holcus lanatus*, Pl.77; *Agrostis stolonifera*, Pl.91.

Plate 3   Rush Family, Juncaceae

1. Slender Rush, *Juncus tenuis* Willd., plant 15–40cm; stems closely tufted, with long grass-like, erect lvs at the base; inflorescence terminal, many-flowered, exceeded by 2 or 3 of the upper bracts; perianth segments 6, yellow-green, 3-veined, finely tapered, longer than the capsule. Fl.6–9. Tracks in woods, trampled grassland; common. Europe, introduced from N. America.

2. Hard Rush, *Juncus inflexus* L., plant 30–60cm; stem blue-green, dull, pruinose, apparently leafless; basal leaf sheaths shining blackish-brown; pith of stem and leaves interrupted with cross-walls; inflorescence lateral; stamens 6. Fl.6–8. Waysides, clearings, near water, on clayey or loamy soils; common. Most of Europe.

3. Bulbous Rush, *Juncus bulbous* L. (*J. supinus* Moench), plant 5–15cm, mat-forming; stem thickened into a bulb at base, rooting at the nodes, leafy to the top; flowers in clusters at the ends of the stems; clusters usually exceeded by the upper lvs; inflorescence often with leafy shoots; perianth segments 6, lanceolate, acute, greenish to brownish, 3–4mm long; capsule rather longer than the perianth segments. Fl.7–9. River banks, ditches, dried-up ponds, muddy soils, cart-ruts and bogs; widespread. Europe. Similar species: **Dwarf Rush**, *J. capitatus* Weigel, stem filiform, leafless; lvs channelled; flowers 4 to 8 together in terminal clusters; perianth segments whitish, later reddish-brown, long-pointed; plant 4–12cm; damp arable fields, heaths; rather rare.

4. Toad Rush, *Juncus bufonius* L., plant 10–25cm, much-branched from the base; lower leaf sheaths yellow-brown; flowers 3–7mm long, solitary and sessile on long, erect branches; perianth segments lanceolate, hyaline, with a green midrib, longer than the capsule. Fl.6–9. Damp arable fields, near water; widespread. Europe.

Lily Family, Liliaceae

5. Star-of-Bethlehem, *Ornithogalum umbellatum* L., plant 10–30cm, with spherical bulbs; lvs radical, linear, 2–5mm wide, with a white midrib; inflorescence a corymbose raceme, erect; perianth segments 6, 15–25mm long, 4–8mm wide, white, with a green band on the back. Fl.4–5. Waysides, vineyards, scrub and grassy places; widespread. Britain, C. and S. Europe. The **Drooping Star-of-Bethlehem**, *O. nutans* L., has a racemose, elongate inflorescence with 3–12 flowers; perianth segments white, greenish outside, 20–25mm long; vineyards, parks, grassy places; rare. Naturalised in Britain.

6. Grape Hyacinth, *Muscari neglectum* Guss. (*M. racemosum* (L.)Mill.), plant 10–20cm; lvs 3 to 6, narrow linear, semi-cylindrical, flaccid, channelled above, longer than the stem; inflorescence dense, 3–6cm long; flowers oblong-ovoid, scented, blue, with a white rim, 4–5mm long. Fl.4–5. Vineyards, gardens, sunny slopes, dryish grassland; widespread to rare. Britain, C. and S. Europe. Protected! Similar species: **Small Grape Hyacinth**, **M. botryoides* (L.)Mill., but the lvs are broadly lanceolate, widened from tip downwards, up to 8mm wide, stiffly erect, almost as long as stem; flowers globose-ovoid; scentless; rare. Germany, S. Europe. Protected!

7. Wild Onion, *Allium vineale* L., plant 30–60cm; lvs almost terete, glabrous, channelled above, 2–3mm wide, bluish-green; inflorescence globose, dense, with 1 acute bract, which soon falls away, few-flowered, often only with bulbils; perianth segments 4–5mm long, red, rarely greenish; stamens nearly twice as long as the perianth segments, inner filaments broadened, with a long sharp tooth on each side. Fl.6–8. Vineyards, waysides, fields; widespread. Britain, S. Scandinavia, C. and S. Europe.

Plate 4 Nettle Family, Urticaceae

1. Common Nettle, *Urtica dioica* L., plant dioecious with stinging hairs, 30–120cm; stem angled, upright; lvs opposite, narrow ovate with cordate base, coarsely dentate, dark green, 5–10cm long; petioles pubescent; inflorescence branches longer than the petiole. Fl.7–10. Paths, rubble, fens, woodland margins; common. Europe. Similar species: **Small Nettle**, *Urtica urens* L., but plant monoecious; lvs ovate, bluntish, light green, 2–4cm long; inflorescence branches shorter than the petioles; gardens, fields, weed communities; common. Europe. The **Roman Nettle**, *Urtica pilulifera* L., has spherical, stalked inflorescences; waste places; rare; C. and S. Germany, S. Europe.

2. Erect Pellitory, **Parietaria officinalis* L. (*P. erecta* Mert. et Koch), plant without stinging hairs, 20–100cm; lvs alternate, entire, ovate-lanceolate, acute, 3–12cm long; inflorescences globose; flowers inconspicuous, 4-merous, greenish. Fl.6–9. Shady walls and rocks, waste places, scrub margins; rare. C. and S. Europe. Similar species: **Pellitory-of-the-wall**, *P. diffusa* Mert. et Koch (*P. judaica* L.), but plant 5–30cm, decumbent and much branched; lvs roundish-ovate, acuminate, 2–4cm long; cracks in walls; local to rare; S. Germany, valleys of the southern Alps, W. and S.W. Europe.

Birthwort Family, Aristolochiaceae

3. Birthwort, *Aristolochia clematitis* L., plant 30–70cm; stem unbranched and flexuous; lvs ovate-cordate, light green, 6–10cm long; flowers 2–8 together in leaf axils, with tube curved and swollen at the base, yellowish; stamens 6. Fl.5–6. Vineyards, scrub, damp woods; likes warm places; poisonous! Medicinal; rather rare. Britain, C. and S. Europe.

Dock Family, Polygonaceae

4. Broad-leaved Dock, *Rumex obtusifolius* L., plant 50–120cm; lvs oblong-ovate, with cordate base, 10–30cm long; inner perianth segments with 3–8 teeth on each side, 1 or all 3 with tubercles. Fl.6–8. Weed communities, waste ground, paths and meadows; common. Europe. Similar species: **Clustered Dock**, *R. conglomeratus* Murray, but lvs long-stalked, oblong-ovate, 5–20cm long; plant 30–80cm with lvs all the way to the top, with lanceolate bracts between the whorls of the inflorescence; inner 3 perianth segments oblong, narrow, entire and without teeth, 2–3mm long, all 3 with tubercles; river banks, ditches, damp fields and sometimes woods; widespread; Britain, S. Scandinavia, C. and S. Europe. **Wood Dock**, *R. sanguineus* L., has inflorescences which are only leafy at the base, with only 1 tubercled perianth segment and all 3 perianth segments entire; grassy places, woodland clearings, fen carr; widespread; Britain, S. Scandinavia, C. and S. Europe.

5. Curled Dock, *Rumex crispus* L., plant 30–120cm; lvs oblong-lanceolate, up to 30cm long, margins crisped-undulate; inflorescence many-flowered; inner perianth segments broadly ovate with cordate bases, usually all 3 with tubercles. Fl.6–8. Weed communities, fields, ditches, river banks, meadows, pastures; common. Europe.

6. Sheep's Sorrel, *Rumex acetosella* L., plant 10–30cm; lvs acid-tasting, linear-lanceolate, with spreading lobes at the base; inflorescence lax-flowered; flowers unisexual; perianth segments without tubercles. Fl.5–8. Poor grassland, fields, paths, heaths, on sandy soils; calcifuge; widespread. Europe.

Plate 5   Dock Family, Polygonaceae

1. Golden Dock, *Rumex maritimus* L., plant 10–60cm, golden-yellow when in fruit; lvs linear-lanceolate, narrowed into the petiole; flowers clustered in whorls which crowd together above; inner perianth segments yellow-brown, ovate-triangular, with long narrow teeth. Fl.7–9. Ditches, ponds, often when water brackish, reservoirs, watering places for stock, muddy ground; rather rare. Britain, C. Europe, northwards to S. Scandinavia, N. Russia and Siberia. Similar species: **Marsh Dock**, *R. palustris* Sm., but plant brownish or reddish in fruit; inflorescence laxer; teeth of the inner perianth segments shorter than the width of the narrow tongue-shaped perianth segments; muddy soils, river banks, dried-up ponds; occasional to rare; Britain, S. Scandinavia, C. Europe.

2. Black-bindweed, *Bilderdykia convolvulus* (L.)Dumort (*Polygonum convolvulus* L.), plant 15–100cm; stem bending to and fro or twining, angled; lvs triangular to sagittate; flowers light green, 1–5 together in the axils or in few-flowered spike-like inflorescences; outer perianth segments keeled; fruit 4–5mm long, 3-cornered. Fl.7–10. Fields, gardens, waste places, widespread. Europe.

3. Knotgrass, *Polygonum aviculare* L., plant prostrate, 5–50cm; stem with dark stripes; lvs linear to ovate, shortly stalked, 0.5–3cm long, very variable in shape; flowers 1–5 together in the axils; perianth segments usually 5, 2–3mm long, greenish or pink. Fl.6–10. (Aggregate species with many microspecies.) Paths, waste places, ditches, gravelly places; common. Europe.

4. Redshank, *Polygonum persicaria* L., plant 10–80cm, similar to Pale Persicaria, but leaf sheaths closely appressed, shortly hairy, long ciliate on the upper margin; lvs shining; peduncles and flowers without glands; flowers pink or white. Fl.7–10. Fields, waste places and muddy soils by ponds; common. Europe.

5. Pale Persicaria, *Polygonum lapathifolium* L., plant usually erect, 20–80cm; lvs ovate lanceolate, broadest at about one-third the way up; petiole issuing from well out of the centre of the leaf sheath; sheaths loose, glabrous or arachnoid; flowers in dense curved spikes; peduncles and perianth segments glandular; flowers pink or white. Fl.7–10. (Aggregate species with many microspecies.) River banks, ditches, fields, muddy places; common. Europe.

6. Tasteless Water-pepper, *Polygonum mite* Schrank, plant 10–60cm; lvs lanceolate, gradually tapered at both ends, over 10mm wide; margin of leaf sheaths long ciliate; flowers in slender spikes, which are often drooping or curved; flowers pink or greenish. Fl.7–9. River banks, ditches, springs, damp tracks in woods; widespread. Europe. Similar species: **Water-pepper**, *P. hydropiper* L., but on chewing it has a burning taste; leaf sheaths shortly ciliate; perianth segments gland-dotted; poisonous! Ditches, river banks, damp tracks in woods; calcifuge; common. Europe. The **Small Water-pepper**, *P. minus* Huds., has leaves which are linear-lanceolate, under 8mm broad, rounded at the base and tasteless; inflorescence erect; damp fields and tracks in woods, beside ponds and lakes; calcifuge; widespread. Almost all Europe.

Plate 6 Dock Family, Polygonaceae

1. Buckwheat, *Fagopyrum esculentum* Moench (*F. sagittatum* Gilib.), plant 15–60cm; stem usually red, little branched; lvs spear- or arrow-shaped, lower long-stalked, the upper almost sessile, usually longer than broad; flowers 3–4mm long, several at a time in axillary inflorescences; perianth segments 5, pink or white; fruit 3-sided, 5–6mm long, smooth on the margins. Fl.7–10. Arable fields, waste places, waysides; calcifuge; widespread. Originally comes from C. Asia, nowadays in almost all of Europe. Similar species: **Green Buckwheat**, **F. tataricum* (L.)Gaertn., but stem stays green; lvs usually broader than long; flowers greenish; fruit with wavy-toothed margins; arable fields, paths; calcifuge; rare. Almost all Europe.

Goosefoot Family, Chenopodiaceae

2. Good-King-Henry, *Chenopodium bonus-henricus* L., plant 20–60cm; lvs triangular-sagittate, dull green, 5–10cm long; inflorescence compound with spherical clusters of flowers, leafless; flowers hermaphrodite, green. Fl.6–9. Weedy fields, hedges, paths, farmyards; local to rather common. Europe.

3. Many-seeded Goosefoot, *Chenopodium polyspermum* L., plant 10–50cm, prostrate to ascending; stem 4-angled, often flushed red; lvs oblong-ovate, long-stalked, entire, 2–5cm long; inflorescence leafy; seeds shiny black, 1mm broad, visible in the open fruiting perianth. Fl.7–9. Weedy fields, river banks, nitrogen-rich arable fields, vineyards, widespread. Almost all Europe. Similar species: **Stinking Goosefoot**, *C. vulvaria* L., but plant smelling of rotten fish; lvs ovate-rhomboid, 2–3cm long, entire, both sides mealy; flowers in axillary and terminal clusters; stem prostrate; weedy fields, margins of saltmarshes, waste places; likes warm places; rather rare. C. and S. Europe mainly. Protected!

4. Fat-hen, *Chenopodium album* L., plant 20–120cm; lvs and stem blue- or grey-green, stem sometimes flushed red; lvs variable in shape, ovate-rhomboid to lanceolate, acute, usually toothed, both sides mealy; inflorescence pyramidal in shape, almost leafless; perianth 5-merous, mealy, whitish-green. Fl.7–9. Weedy fields, waste places, river banks, woodland clearings; common. Europe. Similar species: **Oak-leaved Goosefoot**, *C. glaucum* L., which has lvs dark green above, dense-mealy pale bluish-green below, and oblong-ovate with sinuate-dentate margins; inflorescence dense and not mealy; weedy places; widespread but rare in Britain; almost all Europe.

5. Common Orache, *Atriplex patula* L., plants monoecious, very variable, 30–80cm; lower lvs oblong-rhomboid, with cuneate base, upper lvs oblong-lanceolate; flowers unisexual; male flowers with 5 perianth segments and 5 stamens; female flowers without perianth, but with 2 rhomboid, entire or toothed bracteoles, which are sometimes warty, and which become enlarged in fruit. Fl.7–10. (Genus with many very similar species.) Fields, gardens, waste places, and near the sea; common. Europe.

6. Spear-leaved Orache, *Atriplex hastata* L., plant 30–90cm, similar to *A. patula*, but lower lvs broadly triangular, light green; bracteoles (of the flowers without perianth) triangular, usually with small teeth, with truncate or broad-cuneate base. Fl.7–9. Nutrient-rich weedy fields, rubbish tips, river banks, ditches, muddy places and on salty ground near the coast; common. Europe.

Plate 7 Amaranth Family, Amaranthaceae

1. Common Amaranth, *Amaranthus retroflexus* L., plant 10–80cm; stem branched, densely short-hairy; lvs rhomboid-ovate, light green, hairy on the veins below, 5–15cm long; inflorescence terminal, dense, spike-like, or forming a conical panicle; flowers unisexual, small, crowded in clusters; perianth segments 5, greenish-white, 2–3mm long; bracteoles 3–6mm long, stiff, prickly. Fl.7–9. (Genus with many similar species.) Weedy fields, waste places, paths; rather likes warm places. Europe.

Pink Family, Caryophyllaceae

2. Corncockle, *Agrostemma githago* L., plant grey with shaggy hairs, 30–100cm; lvs opposite, linear; sepals 5, united into a bell-shaped tube, with long spreading lobes; corolla purple, 3–5cm wide, much exceeded by the free calyx teeth; seeds poisonous! Fl.6–7. Cornfields; formerly widespread, but now rare due to efficient seed cleaning. Almost all of Europe. Protected.

3. Bladder Campion, *Silene vulgaris* (Moench)Garcke (*S. cucubalus* Wib., *S. inflata* Sm.), plant 10–50cm; lvs ovate-lanceolate to broad ovate (very variable), bluish-green, usually glabrous; inflorescence a panicle; calyx globose-inflated, strongly net-veined, with 20 main veins; petals deeply bifid, white. Fl.5–9. (Aggregate species with many subspecies.) Paths, grassy slopes, quarries, scrubby places, poor grassland, up to 2500m in the Alps; common. Europe. Similar species: **Sea Campion**, *S. vulgaris* ssp. *maritima* (With.)A.et D. Löve (*S. maritima* With.), but is shorter, with prostrate, non-flowering shoots, fewer flowers, and petals with larger coronal scales. Maritime cliffs and shingle, locally common on Atlantic and North Sea coasts. **Sand Catchfly**, *S. conica* L., but plant glandular-hairy, 10–30cm, few-flowered; calyx with 30 veins, not net-veined, conical, finally weakly inflated, green, with glandular hairs; petals pink, bifid; lvs narrow-lanceolate; slopes, waste places, sandy fields, dunes, dry pastures; rather rare; mostly C. and S. Europe.

4. White Campion, *Silene alba* (Mill.)Krause (*Melandrium album* (Mill.)Garcke), plant glandular-hairy above, 30–100cm; lvs broad lanceolate; flowers unisexual, 2–3cm wide, first opening in the afternoon; calyx tube 18–25mm long; petals deeply bifid, white, with coronal scales at the throat; styles 5; teeth of fruiting capsule erect. Fl.6–9. Paths, waste places, arable fields, scrubby places; rather common. Europe. Similar species: the **Night-flowering Catchfly**, *S. noctiflora* L. (*Melandrium noctiflorum* (L.)Fries), but flowers hermaphrodite, scented; styles 3; petals pale pink; teeth of the fruiting capsule reflexed; plant sticky-glandular above, few-flowered; weedy fields, cornfields, paths; widespread; almost all of Europe.

Plate 8   Pink Family, Caryophyllaceae

1. Annual Gypsophila, *Gypsophila muralis* L., plant annual, 5–20cm; stem erect, branching from the base; leaves linear, blue-green, 5–20cm long; calyx tubular below, with hyaline stripes between the veins; petals pink, with dark veins, emarginate, 3–5mm long. Fl.6–10. Furrows in cultivated fields, or fallow land, river banks, damp paths; calcifuge; widespread. Almost all Europe, but not native in Britain. Protected!

2. Soapwort, *Saponaria officinalis* L., plant 30–80cm, with many upright finely pubescent stems; lvs elliptic, 3-nerved, 5–10cm long; flowers terminal, in compact corymbs; calyx tubular, glabrous, green or reddish, 2cm long; petals white to pink, slightly emarginate, with 2 coronal scales. Fl.6–9. Weedy fields, river banks, gravel heaps, hedgerows; rather common. Almost all of Europe.

3. Common Chickweed, *Stellaria media* (L)Cyr., plant mat-forming, 5–30cm; stem prostrate, with 1 line of hairs, terete; lvs ovate, acute; petals deeply bifid, white, as long as or shorter than the calyx; styles 3. Fl.3–10. Arable fields, gardens, vineyards, river banks; common. Europe.

4. Field Mouse-ear, *Cerastium arvense* L., plant loosely matted, hairy, 15–30cm; lvs oblong-lanceolate, commonly with axillary leaf clusters; bracts broadly hyaline at the tip; petals 11–15mm long, deeply bifid, white; calyx and pedicels glandular; styles 5. Fl.4–7. Waysides, walls, dry grassland, dunes; rather common. Almost all of Europe. See also *C. fontanum*, Pl.80.

5. Umbellate Chickweed, *Holosteum umbellatum* L., plant 5–25cm, bluish-green; lvs ovate, acute, in a basal rosette and opposite and sessile on the stem; inflorescence an umbel; pedicels finally deflexed; petals white, 4–6mm long, toothed at tip. Fl.3–5. Arable fields, fallows, waysides, walls, sandy soils; widespread. C. and S. Europe. Probably extinct in Britain.

6. Procumbent Pearlwort, *Sagina procumbens* L., plant procumbent, mat-forming, rooting at the nodes, 2–5cm; lvs linear, shortly awned, 5–12mm long; sepals usually 4, blunt, spreading in fruit; petals usually absent; pedicels recurved after flowering. Fl.5–9. Paths, fields on acid soils, lawns, verges, banks, streamsides; common. Europe.

7. Thyme-leaved Sandwort, *Arenaria serpyllifolia* L., plant 5–30cm, grey-green; leaves ovate, acute, 5–8mm broad, ciliate on the margins, sessile above and only the lowest stalked; flowers 5–7mm across, white; petals not emarginate, shorter than the ovate-lanceolate sepals. Fl.5–9. Arable fields, waysides, walls, dry grassland; widespread. Europe. Similar species: **Slender Sandwort**, *A. leptoclados* (Rchb.)Guss., but plant yellow-green, delicate; sepals lanceolate; flowers 3–5mm across; dry grassland.

8. Coral Necklace, *Illecebrum verticillatum* L., plant 5–25cm; stem prostrate, thread-like, rooting at the nodes; lvs obovate, 2–5mm long; flowers in axillary clusters; petals 5, white, filiform; sepals 5, white and spongy-thickened, 2mm long with an awned tip. Fl.7–9. Damp waysides, sandy fields and moist sandy places; calcifuge; rare. Britain, W. and C. Europe. Protected!

9. Smooth Rupturewort, *Herniaria glabra* L., plant prostrate, cushion-like, 5–20cm, annual or biennial; lvs ovate-elliptic, glabrous, sessile, yellow-green, 2–7mm long; flowers tiny, greenish, in dense subsessile axillary clusters; petals minute or absent. Fl.6–9. Paths, sandy and gravelly soils; widespread. Europe, rare in Britain.

Also in this family, see *Dianthus armeria*, Pl.50.

Plate 9   Pink Family, Caryophyllaceae

1. Annual Pearlwort, *Sagina apetala* Ard. (*S. ciliata* Fries), plant dark green, 3–10cm; sepals 4 (as in Procumbent Pearlwort, Pl.8), but the 2 outer acute (ssp. *apetala*), or all blunt (ssp. *erecta* (Hornem.)Herm.); petals absent or half as long as the calyx; flower stalks glabrous or glandular-hairy, reflexed after flowering; lvs 3–8mm long, glabrous or ciliate, with a long awn at tip. Fl.4–7. Arable fields, sandy dry grassland, cracks in pavements, paths, walls; calcifuge; common. Britain, C. and S. Europe.

2. Knotted Pearlwort, *Sagina nodosa* (L.) Fenzl, plant 5–15cm; upper stem lvs 1–2cm long, with short shoots in the leaf axils; sepals and petals 5; the petals white, twice as long as the sepals. Fl.6–8. Paths, moorland grassland, damp sandy places; widespread to rare. Principally N. and C. Europe. Likewise with 5-merous flowers, but with petals the same length as the sepals, are **Alpine Pearlwort**, *S. saginoides* (L.)Karst., with glabrous stem and lvs, on alpine and subalpine grassland and rock ledges, and **Heath Pearlwort**, *S. subulata* (Sw.)C.Presl., with glandular ciliate lvs and stem; in dry sandy, gravelly or rocky places; local; Britain, C. and S. Europe, also cultivated in rock gardens.

3. Annual Knawel, *Scleranthus annuus* L., plant annual or biennial, 5–15cm, prostrate to ascending; lvs subulate, blue-green; flowers in dense clusters; petals absent; sepals 5, acute, 3–4mm long, narrowly hyaline on the margins, 3–4 times longer than the 2–5 stamens. Fl.5–9. Sandy arable fields, paths, non-calcareous sandy soils; common. Europe. Similar species: **Perennial Knawel**, *S. perennis* L., but sepals obtuse, with broad hyaline border, scarcely longer than the 10 stamens; stem somewhat woody below; open grassland, dunes, paths, dry sandy fields; rare to local; C. and S. Europe, in isolated localities northwards to Britain, S. Scandinavia.

4. Corn Spurrey, *Spergula arvensis* L., plant 10–50cm; lvs linear, 2–3cm long, with longitudinal channels below, arranged in whorls; stem glandular-hairy; flowers 5–8mm across, in lax, dichotomously branched inflorescences; petals 5, undivided, rather blunt, white, 2–4mm long, almost as long as the glandular-hairy sepals; seeds with a narrow wing. (The Pearlwort Spurrey, *S. morisonii* Bor., has broadly winged seeds, see Pl.49.) Fl.6–10. Non-calcareous arable fields, sandy waysides, wood clearings; common; Europe.

5. Sand Spurrey, *Spergularia rubra* (L.)J.et C.Presl., plant prostrate to erect, glandular above, 5–25cm; lvs opposite, mucronate; stipules a shining silvery-white; flowers 5-merous, 6–8mm across, pink; stamens 10; seeds wrinkled, finely verrucose. Fl.5–9. Damp arable fields, river banks, sandy and gravelly soils; calcifuge; widespread. Almost all Europe. Similar species: **Prickly-fruited Spurrey**, **S.echinosperma* Čelak, which is distinguished by the prickly margin to its seeds; muddy soils, river banks; rare.

6 Small-flowered Catchfly, *Silene gallica* L., plant 10–45cm, annual, without non-flowering branches; lvs spathulate; stem hairy, glandular above; flowers in spike-like inflorescences, which are usually secund; petals 10–15mm long, pinkish-red, rarely white, more or less emarginate at tip; calyx roughly hairy, 7–10mm long, with 10 green veins; styles 3. Fl.6–7. Arable fields, waste places, weedy places; rather rare. Britain, C. and S. Europe.

Plate 10   Buttercup Family, Ranunculaceae

1. Forking Larkspur, *Consolida regalis* S.F.Gray (*Delphinium consolida* L.), plant 20–40cm, annual; lvs 1- to several times ternate, divided into narrow segments about 1mm wide; flowers dark blue, in few-flowered racemes, with a spur 25mm long, which is constructed out of 2 united nectaries; pedicels longer than the bracts; carpels 1; fruit glabrous. Fl.5–8. Arable fields, waysides; rare to local. C. and S. Europe, northwards to Britain and S. Scandinavia. Similar species: **Oriental Larkspur**, *C. orientalis* (Gay) Schrödinger, but flowers in many-flowered racemes; fruit softly hairy; arable fields, railway banks, waste places; rare.

2. Corn Buttercup, *Ranunculus arvensis* L., plant annual, 20–60cm; lvs deeply divided with 3-lobed, linear or lanceolate segments; flowers pale yellow, 4–12mm across; sepals spreading; achenes 3–8 together, about 5–6mm long, conspicuously prickly, bordered, with a straight beak. Fl.5–7. Loamy, clayey cornfields; local to rare. Europe. Protected!

3. Creeping Buttercup, *Ranunculus repens* L., plant 10–40cm, with long stolons which root at the nodes; basal lvs ternate, middle segment long-stalked, all segments 3-lobed, toothed; flowers 2–3cm across, egg-yellow; peduncle furrowed; beak of achene short, curved. Fl.5–8. Poisonous! Arable fields, fallow land, river banks, meadows, fen carr; common. Europe. See also *R. sardous*, Pl.51.

4. Summer Pheasant's-eye, *Adonis aestivalis* L., plant 25–60cm; lvs several times pinnate, with segments about 1mm wide; flowers solitary, terminal, 1–3cm across; sepals green, appressed to the petals; petals 5–8, red or pale yellow, often black at base; achenes in a dense head, with a pure green beak. Fl.6–8. Poisonous! Loamy, calcareous cornfields; widespread. C. and S. Europe (eastern); casual in Britain. Protected! Similar species: **Large Pheasant's-eye**, *A. flammea* Jacq., but sepals with scattered long hairs; achenes in a lax head, beak black at tip; arable fields; rare; poisonous! S. Germany, S. Europe (eastern). Protected!

5. Mousetail, *Myosurus minimus* L., plant 2–10cm, annual; lvs grass-like, narrowly linear, about 1mm wide, in a basal rosette; stem leafless; petals 5, pale green, 3–4mm long, ovate, with a short spur at the base which is appressed to the flower stem; nectaries filiform, yellowish green; receptacle elongating after flower to a length of 2–5cm, and hence resembling a mousetail. Fl.4–6. Damp waysides and arable fields, river banks, loamy, waterlogged, non-calcareous soils; rare. Almost all of Europe. Protected!

Fumitory Family, Fumariaceae

6. White Ramping-fumitory, *Fumaria capreolata* L., plant 30–100cm, glabrous; stem weak, climbing by the petioles; lvs blue-green, 2-pinnate; flowers 10–12mm long, whitish, with violet tip, in many-flowered, lax racemes; fruiting stalks recurved. Fl.5–9. Vineyards, gardens, waste places, hedgebanks; non-calcareous soils; local to rare. Britain, C. and S. Europe (western). The **Common Ramping-fumitory**, *F. muralis* Sonder, is likewise a climber, has flowers 5–7mm long, purple, with an almost black tip, in few-flowered, lax racemes; fruit stalks spreading. Fl.6–9. Arable fields, waste ground, hedgebanks, walls; common in the west, becoming rare in the east. W. Europe, western C. Europe.

Plate 11 Fumitory Family, Fumariaceae

1. Common Fumitory, *Fumaria officinalis* L., plant glabrous, branched, 10–30cm; lvs 2-pinnatisect, bluish-green, leaf segments 2–3mm broad; flowers apparently 2-lipped, upper petal spurred, 7–9mm long, dark red, in dense 20–40 flowered racemes; sepals 2–3mm long, caducous. Fl.5–9. Weedy arable fields, gardens, vineyards; common. Europe. There are other similar species, which have leaf segments 1–2mm wide and sepals 0·5–1mm long, such as **Few-flowered Fumitory**, *F. vaillantii* Loisel, leaf segments flat, flowers pale pink, 5–7mm long in 6–12 flowered racemes; fruiting pedicel under 3mm long; arable fields, vineyards; widespread; protected! **Schleicher's Fumitory**, **F. schleicheri* Soy.-Will., flowers dark pink, in 12–20 flowered racemes; fruiting pedicel 4mm long; leaf segments flat; arable fields, fallow land; rather rare. **Fine-leaved Fumitory**, *F. parviflora* Lam., flowers white, with dark purple tip, leaf segments channelled; arable fields; rare. Protected!

Poppy Family, Papaveraceae

2. Greater Celandine, *Chelidonium majus* L., plant with yellow juice, 30–70cm; lvs pinnate, bluish-green below, with oval blunt-toothed leaflets; flowers 1–2cm across, yellow, in 2–8 flowered umbels; sepals 2; stigma 2-lobed; fruit â siliqua 2–5cm long. Fl.5–9. Poisonous! Weedy places, wood and path edges, walls; widespread. Britain, S. Scandinavia, C. and S. Europe.

3. Common Poppy, *Papaver rhoeas* L., plant with white milky juice, 20–80cm; leaves pinnatisect, with toothed segments, hairy; flowers solitary, terminal; petals 2–4cm long, red, usually with a black spot at the base; filaments dark violet; stigma disc-shaped, 8–12 rayed; fruit a broadly obovoid capsule, rounded at the base. Fl.5–7. Cornfields, waste places, waysides; fairly common. Europe. The **Opium Poppy**, *P. somniferum* L., has ovate-oblong lvs, which are glabrous, blue-green and undulate, and entire to crenate or toothed; upper lvs clasp the stem; petals violet, red or white, with a dark blotch at base, 3–6cm long; fruit a capsule, up to 6cm long and globose; garden plant, sometimes naturalised in waste places; cultivated since the New Stone Age for the production of oil from the ripe seeds and for opium from the milky juice of the unripe capsule.

4. Long-headed Poppy, *Papaver dubium* L., plant 30–60cm, like Common Poppy, but stem leaves often 2-pinnatifid; stigmatic rays 5–8; capsule club-shaped, gradually tapered to the base, glabrous; filaments linear and not thickened; anthers bluish. Fl.5–7. Cornfields, waysides, waste places, quarries; widespread. Almost all of Europe.

5. Prickly Poppy, *Papaver argemone* L., plant 15–30cm; lvs 2-pinnatifid, almost to the central vein, with acute segments 2–3mm broad; petals 15–25mm long, dark red, with a black spot at the base, not overlapping each other at the margins; filaments thickened upwards and club-shaped; capsule oblong to club-shaped, bristly, with 4–5 stigmatic rays. Fl.4–6. Non-calcareous cornfields, on sandy soils; likes warm places; rather uncommon. Britain, C. and S. Europe. Similar species: the **Rough Poppy**, *P. hybridum* L., but the petals overlap at their margins; capsule globose to ovoid, bristly, with 5–9 stigmatic rays; cornfields, waste places; rare. Britain, C. and S. Europe. Protected!

Plate 12 Cabbage Family, Cruciferae or Brassicaceae

1. Rape, *Brassica napus* L., plant 60–120cm; all lvs blue-green, pruinose, almost glabrous, semi-amplexicaul, pinnatisect, with 1–4 oval lateral segments on each side and a much larger terminal segment; petals yellow, 10–14mm long, twice as long as the erect sepals; pedicels little longer than the flowers; open flowers not overtopping the buds; siliqua linear, 5–10mm long, each valve 1-veined. Fl.4–9. Oil-producing, fodder plant and vegetable, naturalised by roads and in waste places. Similar species: the **Wild Turnip**, *B. rapa* L., but rosette leaves grass-green, stem somewhat grey-green, all lvs amplexicaul, usually bristly; open flowers overtopping the buds; petals 7–11mm long; sepals spreading horizontally; cultivated since the New Stone Age as a source of oil, for fodder and as a vegetable. The **Black Mustard**, *B. nigra* (L.)Koch, has stalked pinnatifid lvs with irregularly toothed segments; upper lvs entire; petals 6–9mm long, yellow; siliquae 4-angled, 1–2cm long, with a beak of 2–3mm long on stalks of 2–3mm long, closely appressed to the stem; waste places, river banks, damp arable fields; cultivated since Roman times as a source of mustard.

2. Hairy Rocket, *Erucastrum gallicum* (Willd.)O.E.Schulz, plant 20–60cm; lvs pinnatifid, terminal lobe scarcely larger than the lateral ones; lower flowers with small bracts; petals 6–8mm long, pale yellow, green-veined; sepals erect, 3–5mm long; siliquae oblong, 2–4cm long, with a conical beak, on stalks of 4–10mm long. Fl.5–10. Arable fields, fallow land, waste places, river banks; widespread. C. Europe, casual in Britain. Similar species: **Watercress-leaved Rocket**, *E. nasturtiifolium* (Poir.)O.E.Schulz, but the flowers are without bracts, sepals spreading horizontally, petals 8–12mm long, yellow; river banks, damp waste places; rare. Casual in Britain, S.W. Europe, Lake Constance, upper Rhine valley.

3. Hare's-ear Mustard, *Conringia orientalis* (L.) Dum., plant glabrous, 10–50cm; lvs blue-green, alternate, ovate-elliptic, entire, amplexicaul; flowers yellowish or greenish-white, 15–20mm across; siliquae 6–10cm long, 4-angled, curving upward on spreading stalks 6–12mm long. Fl.5–7. Arable fields, fallow land, waste places; rare. Casual in Britain, C. and S.W. Europe. Protected!

4. Charlock, *Sinapis arvensis* L., plant 30–60cm; lvs sinuate-dentate, the lower almost lyrate, rough-hairy; petals sulphur-yellow, 8–12mm long; sepals 5–6mm long, spreading almost horizontally; siliquae 2–4cm long and 2–3mm wide, almost glabrous, with the beak 10–15mm long and scarcely flattened; valves 3–5 nerved; seeds black. Fl.6–10. Arable fields, fallow land; common. Almost all Europe. Similar species: **White Mustard**, *S. alba* L., but lvs are lyrate-pinnatifid; petals pale yellow, 7–10mm long; siliquae stiffly hairy, usually curved, beak strongly compressed; seeds yellowish; arable fields, waste places; widespread; cultivated for oil and mustard production.

5. Perennial Wall-rocket, *Diplotaxis tenuifolia* (L.)DC., plant 30–80cm; woody at the base; stem densely leafy; lvs deeply pinnatifid, with narrow entire or toothed lobes which slant forwards, lvs glabrous, somewhat blue-green, giving an unpleasant smell when bruised; petals yellow, 8–14mm long; pedicels 2–3 times longer than the flowers; siliquae 2–6cm long on stalk of the same length; seeds in 2 rows in each cell. Fl.5–8. Weed communities; widespread. Doubtfully native in Britain, wild in C. and S. Europe. Similar species: **Annual Wall-rocket**, *D. muralis* (L.)DC., but stem scarcely leafy; lvs basal, pinnatifid to sinuate-dentate; petals 4–8mm long, sulphur-yellow; pedicels as long as the flowers; stalks of the siliquae about half their length; arable fields, waste places; widespread; naturalised in Britain, from C. and S. Europe.

Plate 13   Cabbage Family, Cruciferae or Brassicaceae

1. Wild Radish, *Raphanus raphanistrum* L., plant rough-hairy, annual, 30–80cm; lower lvs with narrow lateral pinnate lobes and large terminal lobes, upper lvs entire; flowers 2–3cm across, white or pale yellow, often with violet veins, sepals erect; siliquae 2–10cm long, jointed like a string of pearls, with a long seedless beak. Fl.5–9. Arable and cornfields, waste places; common. Europe.

2. Bastard Cabbage, *Rapistrum rugosum* (L.)All., plant annual, 25–60cm, with a small taproot; lower leaves pinnately lobed in the shape of a fiddle, with toothed, roundish terminal lobes, upper lvs entire; flowers lemon-yellow, siliquae 2-jointed, bristly hairy at first, 5–10mm long, upper joint spherical, suddenly contracted into the narrow style, which is as long as the upper joint, the lower joint narrower, cylindrical. Fl.6–10. Loamy arable fields; widespread. S. Europe and naturalised elsewhere. Similar species: **Steppe Cabbage**, *R. perenne* (L.)All., but plant perennial, with a thicker taproot; lower lvs deeply pinnately-lobed, siliquae 7–10mm long, 2-jointed, upper joint ribbed, as broad as the lower, glabrous; style shortly conical, about half as long as the upper joint; dry arable fields, paths, dry grassland; rare; native in S.E. Europe, naturalised elsewhere.

3. Field Pepperwort, *Lepidium campestre* (L.)R.Br., plant grey-green, hairy, 20–60cm; stem usually simple; lvs sagittate amplexicaul, undivided, lanceolate, toothed; flowers white, 2–3mm across, in spike-like racemes; siliculae broad ovate, densely covered with scaly vesicles, flat or somewhat curved, winged, emarginate at the style, style not exceeding the notch. Fl.5–6. Paths, dry waste places, arable fields; widespread. Europe. Similar species: **Smith's Pepperwort**, *L. heterophyllum* (DC.)Bentham, but plant with several arched ascending stems; basal lvs pinnately lobed; siliculae glabrous or with a few scales; style exceeding the apical notch of the fruit; arable fields; local to rare. Protected!

4. Narrow-leaved Pepperwort, *Lepidium ruderale* L., plant with a foetid smell on bruising, 10–30cm; basal lvs 1–2 pinnately divided, with narrow lanceolate segments; stem lvs pinnately divided to undivided, not amplexicaul; flowers inconspicuous, greenish; petals shorter than the sepals or absent; siliculae broad ovate, 2–2.5mm long, narrowly winged. Fl.5–7. Weedy fields, trampled grassland, paths, railway sidings; widespread. Europe. The **Poor Man's Pepperwort**, *L. virginicum* L., has linear-lanceolate sharply toothed stem lvs and bristly, fiddle-shaped pinnately lobed basal lvs; petals longer than the sepals; sandy waste places; widespread; origin N. and C. America.

5. Perfoliate Pepperwort, *Lepidium perfoliatum* L., plant 20–40cm; basal lvs 2-pinnately divided, with linear segments less than 1mm wide; stem lvs undivided, entire, ovate, with a deeply cordate amplexicaul base; flowers pale yellow, 2–3mm across; siliculae broad ovate or rhomboid, Fl.5–6. Paths, river banks, waste places, fallow land; rare. Casual in Britain; C. and E. Europe.

6. Hoary Cress, *Cardaria draba* (L.)Desv. (*Lepidium draba* L.), plant 20–60cm; lvs oblong-ovate, irregularly toothed, upper lvs with cordate-sagittate amplexicaul bases; flowers 5–6mm across, white, in dense racemes; siliculae heart-shaped, 4–5mm wide, not winged. Fl.4–7. Weedy fields, vineyards, paths, banks, waste places; widespread. Introduced in Britain; C. and S. Europe (eastern).

Plate 14   Cabbage Family, Cruciferae or Brassicaceae

1. Woad, *Isatis tinctoria* L., plant 50–120cm, blue-green above, pruinose; lower stem lvs stalked, lanceolate, upper stem lvs fiddle-shaped and amplexicaul; flowers yellow, 3–6mm across in dense, branched inflorescences; siliculae pendent, flattened, wedge-shaped, winged, black when ripe, 10–25mm long, 3–7mm wide. Fl.5–6. Paths, waysides, banks, quarries, calcareous grassland; likes warm places; widespread. Almost all of Europe, but rare in Britain.

2. Field Penny-cress, *Thlaspi arvense* L., plant glabrous, with a smell of onion when bruised, 10–40cm; stem angled; stem lvs lanceolate, with a sagittate amplexicaul base, entire or toothed; flowers white, 4–6mm across, in terminal racemes; siliculae 10–15mm long, flattened, nearly circular, broadly winged, with a U-shaped notch above. Fl.4–9 Weedy fields, arable land, waste places; rather common. Europe. Similar species: **Garlic Penny-cress**, *T. alliaceum* L., but stem with scattered long hairs below; siliculae obovate, 6–10mm long, narrowly winged on the upper part only, on horizontally spreading stalks; plant with a smell of garlic; arable fields; rare. Introduced to Britain; N.E. and S. Alps, S.W. Europe.

3. Perfoliate Penny-cress, *Thlaspi perfoliatum* L., plant 5–20cm; stem terete; stem lvs sessile, cordate-amplexicaul, entire, blue-green; flowers white; siliculae obcordate, 4–8mm long, broadly winged above, style shorter than the apical notch. Fl.3–5. Vineyards, arable fields, paths, limestone grassland; rare or local. Britain, C. and S. Europe, northwards to Scandinavia.

4. Shepherd's Purse, *Capsella bursa-pastoris* (L.)Med., plant 5–60cm; lvs deeply pinnately cut in basal rosettes, upper stem lvs lanceolate, with amplexicaul sagittate bases; flowers white, 4–5mm across; siliculae triangular, flattened, in elongated, leafless racemes. Fl.1–12. Arable fields, paths, waste places, gardens; common. Europe.

5. Ball Mustard, *Neslia paniculata* (L.)Desv. ssp. *paniculata*, plant rough-hairy, 15–18cm; lower stem lvs lanceolate, stalked, upper with a sagittate amplexicaul base; flowers golden-yellow, 4–5mm across; siliculae spherical, 2–3mm broad, with a network of wrinkles, on stalks which are 3–5 times as long. Fl.5–7. Cornfields; widespread. Almost all of Europe, but rare casual in Britain. Similar species: **Sharp-fruited Ball Mustard**, *N. paniculata* ssp. *thracica* (Velen.) Bornm., but siliculae spherical with a short additional point; fruit stalk 1.5–3 times the length; S. Europe, northwards to C. France, N. Italy and Switzerland.

6. Warty Cabbage, *Bunias orientalis* L., plant much-branched above, 30–120cm; rosette lvs oblong, deeply pinnate-lobed; stem lvs lanceolate, pinnately lobed to irregularly and coarsely toothed; flowers yellow, 8mm across, in dense, many-flowered racemes; siliculae obliquely ovate, 6–10mm long, with irregular tubercles. Fl.5–8. Paths, waste places, banks, by rivers; widespread. E. and S.E. Europe, introduced in Britain and C. Europe. Similar species: **Wing-fruited Cabbage**, *B. erucago* L., but siliculae with toothed and crested wings, 8–12mm long; waste places; likes warm places; rare; casual in Britain; S. Germany, Switzerland, N. Italy, S. Europe.

Plate 15   Cabbage Family, Cruciferae or Brassicaceae

1. Hoary Alison, *Berteroa incana* (L.)DC., plant annual to biennial, 20–70cm, with a slender taproot; plant grey-green because of stellate hairs; lvs lanceolate, usually entire, narrowed into short stalks, 3–5cm long; petals deeply bifid, white, 5–6mm long; siliculae elliptic, stellate-hairy, grey, 7–10mm long in spike-like racemes. Fl.6–10. Paths, waste places, railway sidings, quarries; local to rare. Introduced in Britain; C. and S. Europe (eastern).

2. Winter-cress, *Barbarea vulgaris* R.Br., plant 30–90cm; lower lvs stalked, pinnately lobed, with 2–5 ovate-oblong lobes on each side and a larger, roundish terminal lobe, upper lvs sessile, deeply toothed; flowers yellow, 7–9mm across; petals twice as long as the sepals; siliquae standing almost erect, 15–25mm long on slender stalks 4–6mm long. Fl.5–7. Waysides, hedges, river banks, gravel beds, damp places; common. Europe. Similar species: **Small-flowered Winter-cress**, *B. stricta* Andrz., but lower lvs with 1–2 lateral lobes and ovate-oblong terminal lobe, siliquae stiffly adpressed erect; river banks, waste places; rare.

3. Creeping Yellow-cress, *Rorippa silvestris* (L.)Bess., plant with stolons, glabrous, 15–60cm; lvs pinnate, with 3–7 lobes on each side, the lobes again toothed or pinnately divided; flowers 4–6mm across, golden-yellow; siliquae narrow linear, 8–20mm long, their stalks horizontal to upright. Fl.6–9. River banks, paths, damp arable fields and waste places; rather common. Europe. See also *R. islandica*, Pl.105.

4. Treacle Mustard, *Erysimum cheiranthoides* L., plant 20–50cm; lvs lanceolate, entire or distantly toothed, with stellate hairs with 2–4 branches; flowers yellow, 5–8mm across; pedicels 2–3 times longer than the calyx of about 2mm; petals 2–5mm long; siliquae 2–3cm long, on stalks of $\frac{1}{3}$ to $\frac{1}{2}$ the length. Fl. 5–8. Arable fields, paths, damp waste places; widespread. Almost all of Europe. The **Spreading Treacle Mustard**, **E. repandum* L., has sinuate-toothed lvs, petals 7–10mm long, and the siliquae 4–10cm long on stalks which spread almost horizontally, the stalks nearly as thick and very short; weedy fields; rare; C. and S. Germany, S.E. Europe.

5. Garlic Mustard, *Alliaria petiolata* (Bieb.)Cavara et Grande (*A. officinalis* Andrz.), plant 20–100cm, smelling of garlic when bruised; basal lvs heart- to kidney-shaped, margins sinuate-toothed, long-stalked; flowers white, 5–8mm across; siliquae 2–3cm long, standing almost erect on spreading stalks. Fl.4–6. Shady weedy fields, wood margins, hedges, parkland; common. Almost all of Europe.

6. Hedge Mustard, *Sisymbrium officinale* (L.)Scop., plant 30–60cm; lower lvs pinnately divided, sometimes cut almost to the middle nerve, lobes triangular to ovate and toothed, upper lvs undivided or with two lateral lobes; flowers yellow, 3–6mm across; siliquae linear, narrowly pointed, 1–2cm long, adpressed to the stem. Fl.5–8. Paths, waste places, road and river banks; common. Europe. A genus with many similar species; but with spreading siliquae are: **Eastern Rocket**, *S. orientale* L., plant grey-hairy; upper lvs stalked, tripartite or undivided; siliquae 4–10cm long, their stalks 4–7mm long, almost as wide as the fruit; paths, waste places; widespread; likes warm places. **Tall Rocket**, *S. altissimum* L., plant green, with spreading stiff hairs below; upper lvs usually sessile, pinnately divided; basal lvs pinnately divided almost to the middle nerve, with segments 1mm broad; flowers whitish yellow; sepals spreading at flowering time, the outer 2 horned at the tip; siliquae 5–10cm long, on stalks 5–10mm long; waste places, paths; likes warm places; widespread; originally from E. Europe; nowadays widely distributed in warmer areas.

Plate 16   Cabbage Family, Cruciferae or Brassicaceae

1. **Flixweed**, *Descurainia sophia* (L.)Webb (*Sisymbrium sophia* L.), plant 20–80cm, branched in the upper part; lvs grey-green, 2–3 pinnate, with linear segments under 1mm broad; petals pale yellow, 2mm long, as long or shorter than the sepals; siliquae 15–25mm long. Fl.5–9. Sandy soils, paths, waste places, walls; widespread. Almost all of Europe (eastern).

2. **Thale Cress**, *Arabidopsis thaliana* (L.)Heynh., plant 5–30cm, with a basal leaf rosette, lvs ovate-oblong, toothed or entire, 2–3cm long; petals 2–4mm long, white; siliquae 1–2cm long, horizontal to upright. Fl.4–6. Arable fields, walls, dry soils; calcifuge, widespread. Originally from the Mediterranean region, nowadays almost all of Europe. Similar species: **Swedish Thale Cress**, **A. suecica* (Fries)Norrlin, but basal lvs toothed to pinnately lobed; stem lvs usually toothed; petals 4–6mm long; siliquae 2–3cm long; sandy dry grassland; Germany, N. Europe. See also *Cardaminopsis arenosa*, Pl.52.

3. **Gold-of-pleasure**, *Camelina sativa* (L.)Crantz, plant 30–90cm, glabrous, branched above, annual; lvs lanceolate, usually entire, sagittate amplexicaul at base; petals 4–5mm long, yellow; siliquae pear-shaped, 7–8mm long, valves of fruit with a distinct middle nerve. Fl.5–6. Cornfields, arable land, dry waste places; nowadays has become rare. The **Small-fruited Gold-of-pleasure**, *C. microcarpa* Andrz., has a roughly hairy stem and lvs and siliquae 5mm long; cornfields, paths; rare; likes warm places; casual in Britain, S. Germany, S.E. Europe.

4. **Dame's-violet**, *Hesperis matronalis* L., plant 40–100cm, with long bristly hairs; basal lvs ovate, up to 15cm long, usually toothed; stem lvs ovate-lanceolate, numerous, becoming smaller above; inflorescence dense; flowers violet, lilac or white, 2cm across, scented; siliquae spreading-erect, 3–10cm long and 2mm wide. Fl.5–7. Waste places, hedgerows, damp woods; widespread. C. and S.E. Europe.

5. **Horse-radish**, *Armoracia rusticana* G.M.Sch. (*A. lapathifolia* Usteri), plant 60–120cm; basal lvs ovate-oblong, up to 50cm long, long-stalked, toothed; lower stem lvs pinnately lobed, upper stem lvs undivided, toothed or entire; petals 5–7mm long, white; flowers in many-flowered racemes; siliculae cylindrical to spherical, 4–6mm long, rarely ripening in Britain. Fl.5–7. Cultivated for the condiment made from the roots, and for medicinal purposes, often naturalised in weedy places, common; rich in mustard-oil. Originally from E. Europe, now almost all Europe.

See also *Alyssum, Erophila*, Pl.52.

Plate 17   Cabbage Family, Cruciferae or Brassicaceae

1. Hairy Bitter-cress, *Cardamine hirsuta* L., plant with basal leaf rosette, 5–20cm; terminal lobes of rosette lvs roundish; stem glabrous or with scattered hairs, with 2–4 pinnate lvs; petals white, 2–4mm long, about twice as long as the sepals; stamens usually 4, with yellow anthers; siliquae about 20mm long and 1mm across. Fl.4–6. Scrub margins, vineyards, waysides, arable fields, gardens; rather damp, sandy soils; somewhat preferring warm places; common. Almost all Europe (western). Similar species: **Small-flowered Bitter-cress**, **C. parviflora* L., but basal lvs with cuneate terminal lobes; stem always glabrous; petals 2–2.5mm long, white; stamens usually 6; siliquae 10–20mm long; river banks, muddy soils, waste places; rare; almost all of Europe. **Wavy Bitter-cress**, *C. flexuosa* With., like *C. hirsuta* but taller, with few basal lvs and with 6 stamens. Common in similar places.

2. Wild Candytuft, *Iberis amara* L., plant 10–20cm; lvs cuneate, distantly toothed, blunt, 3–5cm long, ciliate on the margin; flowers in corymbose panicles, which elongate later on; petals white, rarely pale violet, the outwardly directed ones 6–8mm long and twice as long as the inner; siliculae 4–6mm long and almost as wide, with winged border, fruiting pedicel spreading almost horizontally. Fl.5–8. Cornfields, dry stony calcareous soils; local to rare. Britain, W. and C. Europe. The **Umbellate Candytuft**, **I. umbellata* L. has lanceolate, acute, usually entire lvs and pinkish-red or violet flowers; marginal petals 8–15mm long; cultivated plant and occasionally naturalised, particularly in S. Europe.

Rose Family, Rosaceae

3. Creeping Cinquefoil, *Potentilla reptans* L, plant with creeping stem, rooting at the nodes, up to 1m long; lvs long-stalked, with 5–7 digitate lobes, lobes coarsely toothed; flowers solitary in the leaf axils, golden-yellow. 1–2cm across; petals 5, emarginate. Fl.6–8. Waysides, slopes, arable fields, waste places, hedgebanks; often in company with the next species; common. Europe.

4. Silverweed, *Potentilla anserina* L., plant with stolons, often reddish coloured, and up to 80cm long, rooting at the nodes; lvs pinnate, green above, silvery-hairy below, 5–25cm long, with 6–10 pairs of oblong-ovate larger leaflets, alternating with pairs of smaller ones; leaflets coarsely toothed; flowers solitary, long-stalked, yellow, 2–3cm across; petals 5, roundish. Fl.5–7. Paths, river banks, waste places, dunes, fallow land, pastures; common. Almost all of Europe.

See also *P. recta*, Pl.55.

Plate 18   Rose Family, Rosaceae

1. Parsley-piert, *Aphanes arvensis* L., plant usually grey-green, 5–20cm, without basal leaf rosette; stem lvs shortly stalked below, sessile above, rhomboid, divided to $\frac{1}{3}$–$\frac{1}{2}$, 4–8mm wide and 6–10mm long; flowers 1.8 to 2.5mm long, in 10–20 flowered clusters; sepals usually spreading; fruit over 2mm long. Fl.5–9. Cornfields, bare places in grassland; on acid or basic soils; widespread. C. and S. Europe, northwards to S. Scandinavia, Britain. Very similar to it is **Slender Parsley-piert**, *A. microcarpa* (Bois et Reuter)Rothm., but plant usually pure green, 3–12cm; flowers under 1.5mm long; sepals connivent; fruit about 1mm long; sandy, acid fields; local to rare; mainly C. and S. Europe. Protected!

Mignonette Family, Resedaceae

2. Wild Mignonette, *Reseda lutea* L., plant 20–50cm; lvs 1–2-pinnate, with narrow, entire or toothed lobes, all with undulate margins; leaf stalks narrowly winged; flowers pale yellow, scentless, in spike-like racemes; petals 6, the 2 upper longest, 2–3mm long. Fl.6–9. Paths, banks, waste places, quarries, especially on calcareous soils; common. Britain, C. and S. Europe. Similar species: **Weld**, *R. luteola* L., but plant 60–120cm; lvs undivided, narrowly lanceolate, undulate; flowers 4-merous, in long slender spike-like racemes; weed communities, arable land, especially on calcareous soils; common. C. and S. Europe; Britain; rarer to the north, S. Scandinavia. The **Corn Mignonette**, *R. phyteuma* L., has linear-spathulate lvs, sometimes with a pair of lateral lobes; flowers whitish, in lax racemes; sepals 6, 3–4mm long, enlarging in fruit, 5–13mm long; petals 6, divided into narrow lobes, 3–5mm long; plant 20–50cm; waste places, waysides; likes warm places; principally S. Europe, casual in Britain.

Pea Family, Fabaceae or Papilionaceae

3. White Melilot, *Melilotus alba* Med., plant 30–120cm, like the following species, but flowers white, 4–5mm long, inflorescences racemose, 4–6cm long, 40–80 flowered; pod 3–5mm long, glabrous, reticulate-veined, wrinkled, black when ripe. Fl.6–9. Weedy fields, waste places, railway land; widespread. Almost all of Europe.

4. Ribbed Melilot, *Melilotus officinalis* (L.)Pall., plant 30–100cm; lvs trifoliate; flowers yellow, 5–7mm long, in lax racemes, 4–10cm long; wings of the flowers longer than the keel; fruit 3–4mm long, glabrous, with transverse wrinkles, brown when ripe. Fl.6–9. Sunny weedy fields, dunes, quarries, waste places; common. Europe. Similar species: **Tall Melilot**, *M. altissima* Thuill., but wings of flower as long as the keel; racemes 2–6cm long; flowers 5–7mm long; pod appressed-hairy, reticulate veined, 3–5mm long; river banks, waysides, waste places; local to rare. The **Small Melilot**, *M. indica* (L.)All. (*M. parviflora* Desf.), has yellow flowers 2–3mm long and almost spherical pods, 2–3mm long.

5. Bird's-foot, *Ornithopus perpusillus* L., plant prostrate, 5–30cm; lvs pinnate, with 15–27 leaflets 2–5mm long; flowers 3–4mm long, in 3–7 flowered umbels or heads; flowers white, with yellowish keel and red veins on the standard; pod 1–2cm long, curved, falling into 4–7 sections when ripe; fruiting inflorescence like a bird's foot. Fl.5–6. Sandy fields, dunes, fallow land; calcifuge; local to rather rare. C. and W. Europe, northwards to Ireland, Scotland. Protected!

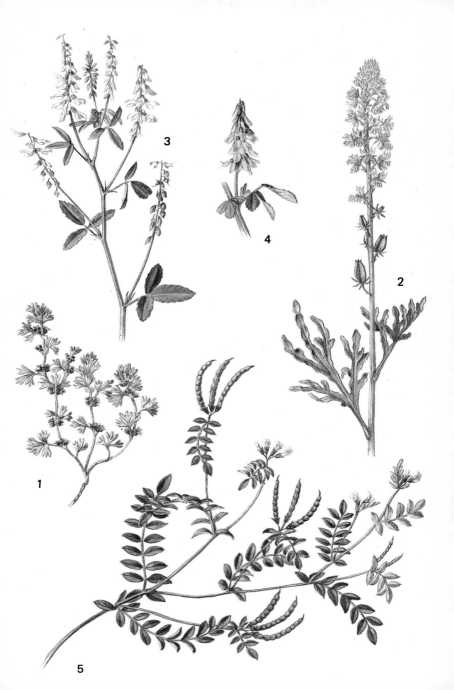

Plate 19   Pea Family, Fabaceae or Papilionaceae

1. Hairy Tare, *Vicia hirsuta* (L.)S.F.Gray, plant annual, 15–50cm; lvs pinnate, usually with 6–8 pairs of leaflets and a branched terminal tendril; leaflets emarginate or truncate at the tip, 5–12mm long; flowers pale violet or whitish, 3–6 together in long-stalked racemes; pod 2-seeded, softly hairy. Fl.6–8. Cornfields, paths, sandy grassland; common. Europe.

2. Smooth Tare, *Vicia tetrasperma* (L.)Schreb., plant 20–60cm, similar to *V. hirsuta*, but pods glabrous, 4-seeded; lvs with 6–8 leaflets; flowers 1–2 together, pale violet. Fl.6–7. Sandy grassland, neglected arable fields, also moorland grassland; calcifuge; widespread. Britain, S. Scandinavia, C. and S. Europe.

3. Common Vetch, *Vicia sativa* L., plants 30–80cm; lvs pinnate with 4–7 pairs of leaflets and branched terminal tendrils, leaflets ovate to linear; flowers 1–2 in leaf axils, short-stalked, with a bluish standard and reddish-violet wings; calyx teeth equal in length (in the similar Bush Vetch calyx teeth unequal, see Pl.153); pods upright, shortly hairy, brown when ripe, 3–8cm long. Fl.5–7. Fodder plant, naturalised in waste places; widespread. Europe.

4. Fodder Vetch, *Vicia villosa* Roth., plant 30–130cm, densely spreading soft-hairy; lvs with 14–18 linear, hairy, pinnate leaflets; racemes 10–20 flowered; flowers 15–20mm long; calyx teeth long-ciliate; calyx tube inflated at the base; pod 2–4cm long. Fl.6–8. Cornfields, waste places, waysides; widespread. Almost all of Europe (originally from S. Europe). See also Tufted Vetch, *V. cracca*, Pl.81.

5. Yellow Vetchling, *Lathyrus aphaca* L., plant 10–30cm; stem clambering or climbing, unwinged; lvs (which are in fact enlarged stipules) arranged in a row in pairs, leaf-stalk with tendril but without lvs, stipules 1–4cm long; flowers 1–2 together, 10–15mm long; calyx glabrous; pod 2–3cm long, glabrous. Fl.5–7. Dry places on sand, gravel or chalk, cornfields; likes warm places; widespread. W., C. and S. Europe. Protected! In the **Grass Vetchling**, *L. nissolia* L., the lvs consist only of leaf-like, broadened leaf-stalks, 4–10cm long, with an awn-like narrowed tip, and subulate stipules, 1–2mm long, at the base (often missing altogether); flowers 1–2 together, purple, with a dark-veined wing; pod 4–6cm long and 3–4mm broad; cornfields, grassy and bushy places; rare; Britain, France and Germany, S. Europe.

6. Tuberous Pea, *Lathyrus tuberosus* L., plant with root tubers, prostrate or climbing, 20–100cm; stem angled; lvs with 2 elliptic, mucronate leaflets and a terminal tendril; stipules half-arrow-shaped; flowers crimson, 12–16mm long, scented, 2–5 together; pod 2–4cm long, brown. Fl.6–8. Loamy arable fields, waysides, widespread. Almost all of Europe, rarely naturalised in Britain.

See also *Medicago*, Pls. 57 and 58, and *Trifolium*, Pls. 58 and 81.

Wood Sorrel Family, Oxalidaceae

7. Upright Yellow Oxalis, *Oxalis europaea* Jord. (*O. stricta* auct.), plant erect, 10–30cm; lvs trifoliate, like clover lvs, opposite or whorled; flowers pale yellow, 8–10mm across, 1–6 together in long-stalked, axillary, umbel-like inflorescences; petals rounded at tip; pedicels erect after flowering. Fl.6–9. Arable fields, gardens, paths; local to common. Europe. Similar species: **Yellow Oxalis**, *O. corniculata* L., but stem prostrate; lvs alternate; petals emarginate at tip; pedicels deflexed after flowering; gardens, paths; widespread.

Plate 20   Crane's-bill Family, Geraniaceae

1. **Hedgerow Crane's-bill**, *Geranium pyrenaicum* Burm., plant 20–60cm; stem softly hairy with spreading hairs; lvs roundish, divided rather more than halfway into 5–9 wedge-shaped lobes; flowers in pairs, in lax inflorescences; petals 7–10mm long, violet-red, deeply emarginate, twice as long as the sepals. Fl.5–10. Weedy arable fields, paths, scrub; fairly common. Almost all of Europe.

2. **Cut-leaved Crane's-bill**, *Geranium dissectum* L., plant 10–40cm; stem with spreading hairs; flowers in pairs; petals pinkish-red, emarginate, 4–6mm long, as long as the sepals; peduncle of the inflorescence shorter than its bract, glandular-hairy; lvs divided to the base, with linear lobes. Fl.5–9. Arable fields, gardens, paths; common. widespread to common. Europe. Similar species: **Long-stalked Crane's-bill**, *G. columbinum* L., but stem appressed-hairy, peduncles of the inflorescences longer than the bracts, not glandular; petals 7–9mm, emarginate; weedy fields; widespread.

3. **Round-leaved Crane's-bill**, *Geranium rotundifolium* L., plant softly hairy, 10–30cm; stem often reddish; lvs roundish, divided to halfway; petals 5–6mm long, pinkish-red, not emarginate. Fl.5–10. Sunny weedy fields, vineyards, walls; rather rare. Britain, C. and S. Europe.

4. **Dove's-foot Crane's-bill**, *Geranium molle* L., plant 10–30cm; stem with spreading, soft, shaggy hairs; lvs roundish, divided up to about the middle; petals 4–8mm long, pinkish-red, emarginate, scarcely longer than the sepals. Fl.5–9. Sunny weedy fields, dry grassland, waste places; common. Europe. Similar species: **Small-flowered Crane's-bill**, *G. pusillum* L., but petals 3–4mm long, dirty violet; stem shortly hairy; paths, vineyards, arable fields, dry grassland and waste places; widespread.

5. **Common Stork's-bill**, *Erodium cicutarium* (L.)L'Hér., plant coarsely hairy, 10–40cm; lvs pinnate, cut to the middle nerve, lobes further divided as well, with narrow acute segments; flowers 2–9 together in long-stalked inflorescences; petals 5–9mm long, pink or lilac, the two upper ones often larger; fruit coming apart when ripe into 5 mericarps, each with a beak which is twisted like a corkscrew. Fl.4–9. Paths, fallow land, vineyards, dunes; rather common. The **Long-fruited Stork's-bill**, *E. ciconium* (L.)L'Hér., has small extra lobes between the main leaf lobes; stem glandular-hairy; sepals 10–15mm long, with a distinct point 2–4mm long; petals blue-violet, almost as long as the sepals; weed communities; rare; S. Germany, S. Europe.

Spurge Family, Euphorbiaceae

6. **Trailing Spurge**, *Euphorbia humifusa* Willd., plant glabrous, blue-green, tinged red when older, 5–15cm; stem prostrate; lvs opposite, obovate, rather obtuse, 5–10mm long with small stipules; cyathia solitary. Fl.6–9. Gardens, paths, cracks in pavement, sandy waste places; rather rare. Originally from Asia, now widespread. Similar species: **Least Spurge**, *E. chamaesyce* L., stem and lvs spreading-hairy; stem internodes about as long as the corresponding pair of lvs; lvs usually uniform in colour; arable fields, railways; rare; likes warm places. The **Spotted Spurge**, *E. maculata* L., is pubescent, lvs usually red-spotted, stem internodes shorter than the corresponding pair of lvs; cracks in pavements, railways; likes warm places; rare; S. Germany, S. Europe.

Plate 21   Spurge Family, Euphorbiaceae

1. Sun Spurge, *Euphorbia helioscopia* L., plant 10–40cm; stem lvs alternate, obovate to wedge-shaped, pale green, shortly stalked, finely toothed at tip, 2–4cm long; flower umbel with 5 rays; bracts similar to the stem lvs; involucral glands yellow, roundish-oval; capsule smooth in fruit. Fl.4–11. Arable fields, gardens, vineyards, nutrient-rich soils; common. Europe.

2. Petty Spurge, *Euphorbia peplus* L., plant annual, 10–30cm, similar to previous species, but flower umbel with 3 rays, each ray branched 2–3 times more; bracts ovate, concave like a boat, shortly mucronate; involucral glands half-moon-shaped, running out into 2 long horns; lvs ovate, stalked, obtuse at tip; capsule in fruit trigonous, each valve with 2 narrow wings. Fl.5–10. Arable fields, gardens, waste places; common. Europe.

3. Dwarf Spurge, *Euphorbia exigua* L., plant annual, 5–20cm; lvs light green, alternate, linear, mucronate, sessile, 0.5–3cm long; flower umbels 3–5 rayed, each ray branched again dichotomously; bracts pale green, linear, sessile with cordate bases, not joined together; involucral glands half-moon-shaped, horned; capsule smooth in fruit. Fl.6–10. Loamy cornfields, fallow land, waste places; prefers warm places; widespread. Almost all of Europe; originally from S. Europe. Other similar annual species: **Corn Spurge**, **E. segetalis* L., plant glabrous, 10–30cm; lvs linear, acute, 0.5–4cm long, blue-green, entire, sessile; stem densely leafy; flower umbel 5-rayed; bracts triangular or rhomboid, with mucronate tips; involucral glands yellow, half-moon-shaped, running out into 2 very long horns (as in *E. peplus*); capsule finely warty in fruit; cornfields, waste places; likes warm places; rare; S. Germany, S. Europe. **Sickle Spurge**, **E. falcata* L., plant 10–30cm; lvs 0.5–2cm long, spathulate to lanceolate-cuneate, blue-green; flower umbel usually 3-rayed; bracts triangular-ovate, acute; involucral glands yellow, running out into 2 white, slender horns; capsule smooth in fruit; cornfields, waste places; likes warm places; rare; S. Germany, S. Europe. Protected!

4. Annual Mercury, *Mercurialis annua* L., plant 10–40cm; stem glabrous, 4-angled, with many opposite branches; lvs ovate-lanceolate, bluntly toothed, 3–8cm long; plant usually dioecious; female flowers almost sessile, in few-flowered axillary clusters; male flowers in many-flowered, spike-like clusters. Fl.5–10. Arable fields, vineyards, paths, waste places; likes warm places; widespread. Almost all of Europe, origin S. Europe.

Plate 22   Mallow Family, Malvaceae

1. Large-flowered Mallow, *Malva alcea* L., plant 50–100cm; upper part of stem, like lvs and calyx, with appressed stellate hairs; stem lvs palmately divided almost to the base; petals 2–4cm long, deeply emarginate, pink or lilac; epicalyx lobes ovate, broadened at the base, 3–5mm long. Fl.6–10. Paths, sunny slopes, banks; widespread. Almost all of Europe (similar to *M. moschata*, see Pl.63).

2. Common Mallow, *Malva sylvestris* L., plant 40–120cm; stem lvs roundish, with 5–7 toothed acute lobes; flowers 2–6 together in the axils; petals 2–3cm long, deeply emarginate, purple, with darker stripes, 3–4 times as long as the calyx, sepals united to halfway; epicalyx lobes 3, free, 4–8mm long. Fl.6–10. Paths, walls, waste places, on dry soils; common. Europe.

3. Dwarf Mallow, *Malva neglecta* Wallr., plant 10–40cm; lvs roundish, with 3–5 rounded lobes; petals 8–10mm long, deeply emarginate, white to pink, twice as long as the calyx; calyx lobes flat; mericarps smooth. Fl.6–9. Paths, walls, gardens, margins of arable fields; common. Almost all of Europe. Similar species: **Small Mallow**, *M. pusilla* Sm. (*M. borealis* Wallmann), but petals 3–5mm long, shallowly emarginate, whitish, about as long as the calyx; calyx lobes inflexed at tip; mericarps wrinkled; paths, waste places, vineyards; likes warm places; rather rare; principally C. and S. Europe.

St. John's-wort Family, Hypericaceae

4. Trailing St. John's-wort, *Hypericum humifusum* L., plant 5–15cm; stem threadlike, creeping, with 2 raised lines; lvs ovate, 10–15mm long, with black sessile glands on the margin beneath; petals 4–8mm long, yellow, with black glands at the margins; sepals oblong, entire; stamens 15–20. Fl.6–10. Damp arable fields, paths, clearings in woods, river banks, heaths and dry moors; widespread. Britain, W., C. and S. Europe. Likewise, **Marsh St. John's-wort**, *H. elodes* L., has a prostrate, creeping stem, but plant hairy, 10–30cm; stem terete; sepals ovate, glandular-ciliate; pools on heaths, wet peaty and sandy soils; rare; protected!

Violet Family, Violaceae

5. Field Pansy, *Viola arvensis* Murr., plant 5–20cm, very similar to the next species; stipules with ovate-lanceolate to ovate, crenate-serrate middle lobes; corolla cream-coloured and usually a little violet, the same length as the calyx; spur about as long as the sepal appendages. Fl.4–10. Arable fields, paths, waste places; common. Europe.

6. Wild Pansy, *Viola tricolor* L., plant very variable, 10–40cm; lvs heart-shaped, toothed; stipules pinnate, with lanceolate middle lobe; corolla longer than the calyx, blue-violet and yellow; spur nearly twice as long as the sepal appendages. Fl.5–10. Arable fields, fallow land, waysides; widespread. Almost all of Europe.

Plate 23　Gourd Family, Cucurbitaceae

1. Black-berried Bryony, *Bryonia alba* L., climbing plant, monoecious (but dioecious in S.E. Europe), 2–3m; tendrils coiled spirally; lvs divided into 5 triangular sharply toothed lobes; female flowers with ovary and staminodes; stigma glabrous; corolla greenish-white, about 1cm across; ripe berries black. Fl.6–7. Poisonous! Hedges, paths, scrub; rather rare. C. and S. Europe. Similar species: **White Bryony**, *B. cretica* L. ssp. *dioica* (Jacq.)Tutin, but plant always dioecious; flowers with ovaries only or stamens only; stigma hairy; ripe berries red; leaf lobes usually entire; poisonous! Hedges, fences; widespread; Britain, C. and S. Europe (western).

Willow-herb Family, Oenotheraceae or Onagraceae

2. Common Evening-primrose, *Oenothera biennis* L., plant erect, usually unbranched, 50–100cm; lvs ovate-lanceolate, 8–15cm long, toothed or entire; inflorescence erect; flowers saucer-shaped; petals 3–6cm long, yellow; 'calyx tube' (hypanthium) 2–5cm long. Fl.6–8. (Aggregate species with several microspecies.) Sandy, gravelly places, banks, quarries, sandy grassland; widespread. Originally from N. America, nowadays in almost all of Europe. Similar species: **Small-flowered Evening-primrose**, *O. parviflora* L., but inflorescence drooping; petals 1–2cm long; 'calyx tube' 1.5–3cm long; dry waste places; widespread. (Aggregate species with several microspecies.)

See also *Epilobium*, Pls. 109 and 161.

Carrot Family, Umbelliferae or Apiaceae

3. Dill, *Anethum graveolens* L., plant 50–120cm, glabrous; stem terete; lvs 2–4 pinnate, with linear, almost bristly segments; leaf sheaths short; umbel 15–40 rayed; bracts and bracteoles absent; petals yellow; fruit lens-shaped, 4mm long. Fl.7–9. Cultivated as a herb, rarely naturalised, weedy fields. Almost all of Europe. Similar species: **Fennel**, *Foeniculum vulgare* Mill., but plant blue-green, rootstock spindle-shaped (wild forms), or forming a large bulb; leaf sheaths long; cultivated as a herb, occasionally naturalised, often near the sea; likes warm places.

4. Small Bur-parsley, *Caucalis platycarpos* L. (*C. lappula* (Web.)Grande), plant 10–30cm, usually bristly hairy; lvs 2–3 pinnate; bracts 0–2; bracteoles 3–5; scarcely hyaline; flowers white; fruit 6–12mm long, with 4 rows of prickles. Fl.5–7. Cornfields, fallow land, likes warm places, on calcareous soils; rare. Casual in Britain, C. and S. Europe. Protected!

5. Large-flowered Orlaya, *Orlaya grandiflora* (L.)Hoffm., plant 10–30cm; stem glabrous, angled, branched; lvs 2–3 pinnate, with linear, finely pointed segments; umbels 5–12 rayed; bracts 5–8, lanceolate; bracteoles usually 5, broadly hyaline; the outwardly directed petals of the marginal flowers 10–15mm long, 7–8 times as long as the other petals; fruit 6–8mm long, with 2–3 rows of prickles. Fl.6–8. Cornfields, fallow land, dry grassland, on calcareous soils; rare. C., S. and S.W. Europe. Protected! The **Flat-fruited Orlaya**, *O. kochii* Heywood (*O. platycarpos* Koch), has 2–3 rayed umbels with 2–3 bracts; petals of the marginal flowers 5–7mm long; arable fields; S. Europe.

For other similar species see *Daucus*, Pl.64, *Anthriscus*, Pl.82, *Carum*, Pl.83 and *Aegopodium*, Pl.163.

1

2

3

4

5

Plate 24   Carrot Family, Umbelliferae or Apiaceae

1. Hemlock, *Conium maculatum* L., plant 50–200cm, smelling unpleasantly; stem terete, finely ridged, bluish-green, usually red-spotted below; lvs 2–3-pinnate, dark green above, grey-green below, segments oblong-lanceolate, coarsely toothed; umbel with 8–15 rays; bracts reflexed; fruit roundish, 3mm long, with wavy ribs. Fl.6–9. Highly poisonous! Ditches, rubbish tips, waysides; widespread. Britain, S. Scandinavia, C. and S. Europe.

2. Longleaf, *Falcaria vulgaris* Berh., plant blue-green, much branched, 30–80cm; lvs twice ternate, with sharply toothed, linear, slightly sickle-shaped segments, 10–15mm wide; umbels 10–20 rayed; bracts and bracteoles numerous, lanceolate; flowers white; fruit oblong-ovoid, 3–4mm long. Fl.7–9. Path and field margins, fallow land, sunny margins of scrub; on calcareous soils; widespread to rare. C. and S. Europe, occasionally to Britain and S. Scandinavia.

3. Wild Parsnip, *Pastinaca sativa* L., plant 40–120cm; stem shining, angled and furrowed; lvs yellow-green, once-pinnate, with oblong-ovate, lobed or coarsely toothed segments, about 5cm long; flowers yellow; bracts and bracteoles usually absent; umbel 5–15 rayed; fruit ovoid, 5–8mm long, broadly winged. Fl.6–9. Ssp. *urens* (Req.)Čelak is grey-hairy, dull; plant of squarrose habit, up to 250cm; stem round; umbel 5–7 rayed, and ssp. *sylvestris* (Mill.)Rouy et Cam. is likewise grey-hairy, but stem angled and furrowed and umbel 7–15 rayed. Weedy arable fields, quarries, paths, pastures; widespread. Originally from W. Asia, but cultivated since ancient times as a vegetable and a medicinal herb, nowadays almost all of Europe.

4. Upright Hedge-parsley, *Torilis japonica* (Houtt.)DC. (*T. anthriscus* Gmel.), plant 30–120cm, with appressed bristly hairs; lvs 2-pinnate, with pointed segments; umbels long-peduncled, 4–12 rayed; bracts 5–12; flowers white; fruit 2–3mm long, covered with rough, weakly curved spines. Fl.6–8. Weed communities, woodland paths, clearings, hedges and grassy places; common. Almost all of Europe. Similar species: **Spreading Hedge-parsley**, *T. arvensis* (Huds.)Link, plant annual, 30–80cm; umbel 3–9 rayed; bracts 0–2; fruit 3–5mm long, covered with hooked spines; arable fields, fallow land; rare. Britain, C. and S. Europe. Protected! **Knotted Hedge-parsley**, *T. nodosa* (L.)Gaertn., is distinguished by shortly stalked, clustered, leaf-opposed umbels without bracts; plant 15–30cm; waste places, arable fields, railway banks; likes warm places; local to rare. Britain, C. and S. Europe.

Also in this family, see *Bupleurum rotundifolium*, Pl.162.

Plate 25 Bedstraw Family, Rubiaceae

1. Cleavers, *Galium aparine* L., plant climbing, 60–200cm; stem hairy; lvs linear, 4–8mm broad, with backwards-directed tiny prickles on the margins, as on the stem, lvs 6–8 in a whorl; corolla 4-lobed, white, 2mm across; fruit spherical, with hooked bristles, 4–6mm broad. Fl.6–10. Arable fields, waste places, river banks, hedges, fen woods, sea shingle; common. Europe. The **False Cleavers**, *G. spurium* L., has greenish-white flowers, 1mm across, lvs 2–3mm broad and fruit 1.5–3mm across, finely wrinkled; stem glabrous but with prickles; cornfields; rather rare. Britain, C. and S. Europe. Protected!

2. Field Madder, *Sherardia arvensis* L., plant 5–20cm, annual; stem 4-angled, prostrate; lvs lanceolate, 1-nerved, rough on the margin, the lower 4 and the upper 6 to a whorl; flowers arranged in heads, subtended by 8–10 bracts giving a starry appearance, the bracts not ciliate; corolla 4-lobed, lilac, 2–3mm across. Fl.6–10. Cornfields, fallow land; widespread. Europe.

3. Blue Woodruff, *Asperula arvensis* L., plant erect, 5–25cm; lvs linear, the lower in 4 and the upper in 6–8 to a whorl; inflorescences in heads; bracts bristly ciliate; corolla 4–6mm across, blue. Fl.5–6. Loamy, calcareous arable fields; very rare. Casual in Britain, C. and S. Europe. Protected!

Primrose Family, Primulaceae

4. Scarlet Pimpernel, *Anagallis arvensis* L., plant 5–25cm; stem 4-angled, glabrous; lvs opposite, ovate, 5–20mm long; flowers long-stalked; petals usually scarlet-red, rarely blue or pink, overlapping, up to 7mm long and 6mm broad, entire or slightly crenate at tip, with 3-celled glandular hairs; sepals entire. Fl.6–10. Gardens, arable fields, vineyards, dunes; common. Europe. Similar species: **Blue Pimpernel**, *A. foemina* Mill. (*A. coerulea* L.), plant 5–25cm, but petals always blue, not overlapping, about 6mm long and 3.5mm wide, serrate at tip, with 4-celled glandular hairs; sepals finely serrate; cornfields; likes warm places; rather rare. Britain, C. and S. Europe.

Bindweed Family, Convolvulaceae

5. Hedge Bindweed, *Calystegia sepium* (L.)R.Br. (*Convolvulus sepium* L.), plant twining, 1–3m; lvs deeply cordate, 8–15cm long; flowers solitary, in the leaf axils, long-stalked; corolla funnel-shaped, white, 3–5cm long and wide; calyx enclosed by 2 cordate bracteoles; style lobes flat. Fl.6–9. Hedges, waste places, river banks, willow and alder scrub; widespread. Europe.

6. Field Bindweed, *Convolvulus arvensis* L., plant spread out on the ground or twining, 20–80cm; lvs cordate to sagittate, 3–4cm long; flowers in the leaf axils, long-stalked, with 2 filiform, short bracteoles in the middle of the stalk; corolla broadly funnel-shaped, 2–3cm long and broad, white or pink. Fl.6–9. Arable fields, gardens, paths; common. Europe.

Plate 26   Bindweed Family, Convolvulaceae

1. Greater Dodder, *Cuscuta europaea* L., reddish parasite, 20–100cm; stem twined about the host plant; lvs scale-like, colourless; flowers in small sessile clusters; corolla campanulate, 5-lobed, pinkish-white, 2mm across; corolla tube as long as the lobes, with erect coronal scales inside which do not close the tube. Fl.6–8. Parasitising Common Nettle, Hedge Bindweed and Mugwort; widespread. Almost all of Europe. (A genus with many species which are often hard to distinguish, and partly host-specific.) See also *C. epithymum*, Pl.66.

Borage Family, Boraginaceae

2. Lesser Honeywort, *Cerinthe minor L., plant blue-green, 15–60cm; basal lvs often white-flecked, upper stem lvs clasping the stem, spathulate; inflorescence with large herbaceous bracts; corolla 5-lobed, divided almost halfway into lanceolate segments, yellow, often red-spotted. Fl.5–7. Waysides and field margins, banks, fallow land; rather rare. C. and S. Europe. The **Smooth Honeywort**, **C. glabra* Mill., has a corolla with 5 short ovate lobes, which are reflexed at the tip; lvs never spotted; subalpine grassland, tall shrub communities, subalpine pastures; rather rare; Alps, Pyrenees, Carpathians, Abruzzi.

3. Bur Forget-me-not, **Lappula squarrosa* (Retz.)Dum. (*L. echinata* Gilib.), plant roughly hairy, branched above, 10–40cm; lvs lanceolate, sessile, appressed-hairy, 2–5cm long; flowers 2–4mm long, sky-blue; fruit stalks erect, nutlets with 2 rows of recurved hooked prickles on the angles. Fl.6–7. Fallow land, dry waste places, dry grassland; rare. Most of Europe. Protected!

4. Hound's-tongue, *Cynoglossum officinale* L., plant densely soft-hairy, grey, 20–80cm; stem densely leafy; lvs elliptic to lanceolate, semi-amplexicaul, grey-felted; inflorescence branched; flowers violet, then red-brown, 5–7mm across; corolla tube short, mouth closed with scales, lobes spreading; nutlets 4, with thickened borders, 6–8mm long, covered with recurved hooked prickles. Fl.5–7. Sunny weedy fields and borders of woods; widespread. Almost all of Europe. The **Green Hound's-tongue**, *C. germanicum* Jacq., has lvs which are glabrous and shining above and with scattered hairs beneath and red-violet flowers; nutlets without thickened borders; weedy mixed woodland, clearings, hedgebanks; rare; Britain, C. and S. Europe (distribution incompletely known).

5. Field Gromwell, *Buglossoides arvensis* (L.)Johnst. (*Lithospermum arvense* L.), plant roughly hairy, 10–50cm; lvs obovate to lanceolate, 1-nerved, 3–5cm long; stem with few leaves; corolla 3–4mm across, funnel-shaped, dirty white; nutlets brown, warty. Fl.4–6. Cornfields, widespread; Europe.

6. Viper's-bugloss, *Echium vulgare* L., plant with stiff bristly hairs, 30–80cm; lvs oblong-lanceolate, stiffly bristly; corolla irregular, 2-lipped, 15–20mm long, blue, with an open throat; flower buds red; stamens of unequal length. Fl.6–8. Sunny weedy fields, paths, quarries, dry grassland, dunes; local to common. Europe. The **Pale Bugloss**, **E. italicum* L., has a corolla which is white to pale violet, 8–12mm long, and a densely bristly long-hairy calyx; weed communities in the Mediterranean region.

Plate 27   Borage Family, Boraginaceae

1. Borage, *Borago officinalis* L., plant 20–60cm, with taproot; stem and lvs spreading hairy; lvs ovate-elliptic, narrowed into the winged stem, decurrent; inflorescence cymose; corolla sky-blue, 2–3cm across, with white coronal scales; sepals narrowly lanceolate; anthers dark. Fl.6–8. Herb and medicinal plant, often planted in gardens and often naturalised in areas with mild winters; waste places, vineyards. Originally from S. Europe.

2. Alkanet, *Anchusa officinalis* L., plant 30–80cm, softly hairy; lvs oblong-lanceolate, the lower ones narrowed into the stalk, the upper sessile with a rounded base, entire; flowers blue-violet, 10–15mm across; corolla tube straight, throat closed with ovate, velvety, coronal scales; inflorescence many-flowered. Fl.5–9. Dry sandy weedy fields, paths, banks, waste places; likes warm places; rare. Almost all of Europe; a garden escape in Britain. The **Large Blue Alkanet**, *A. azurea* Mill. (*A. italica* Retz.), has a sky-blue corolla, 12–15mm across and coronal scales bearded; lvs shining, undulate; waste places, railway tracks; likes warm places; rare. Casual in Britain; S. Germany, S. Europe. The **Yellow Alkanet**, *A. ochroleuca* M.Bieb., has a yellowish-white corolla, 7–10mm across; waste places, railway land; likes warmth; rare; casual in Britain; S. Germany, S. Europe (eastern).

3. Bugloss, *Anchusa arvensis* (L.)M.Bieb. (*Lycopsis arvensis* L.), plant densely bristly-hairy, 20–40cm; lvs oblong, 10–15cm long, sinuate-dentate, usually somewhat un-dulate; corolla pale blue, 7–10mm across, with a white, curved tube; nutlets finely warty, wrinkled. Fl.5–9. Vineyards, arable fields, sandy places; likes warm places; widespread. Europe.

4. Brown Nonea, *Nonea pulla* (L.)DC., plant 20–40cm, grey-hairy, glandular; lvs oblong-lanceolate, 8–12cm long, the upper amplexicaul; flowers red-brown, 10–15mm long; corolla funnel-shaped, with a cylindrical tube; calyx much enlarged at fruiting time; nutlets almost spherical. Fl.5–8. Cornfields, fallow land, waysides; likes warm places; rare. C. and S.E. Europe. Protected! The **Yellow Nonea**, *N. lutea* (Desr.)DC., has a yellow corolla 7–12mm long and 5–7mm wide, with hairy coronal scales; waste places, gravel pits; rare; likes warm places; S. Germany, S. Europe (eastern).

5. Field Forget-me-not, *Myosotis arvensis* (L.)Hill, plant densely grey-hairy, branched at the base, 10–40cm; lvs ovate-lanceolate, 2–8cm long, the lower gradually narrowed into the stalk, the upper sessile; inflorescence dense-flowered, without bracts; corolla funnel-shaped, 3–4mm across, blue; fruiting pedicels 2–3 times as long as the calyx. Fl.4–10. Arable fields, cornfields, woodland clearings, waysides, dunes; common. Europe.

Plate 28   Mint Family, Labiatae or Lamiaceae

1. Ground-pine, *Ajuga chamaepitys* (L.)Schreb., plant 5–15cm, hairy; lvs thickly set, 1–3cm long, tripartite, with linear lobes, upper lvs undivided; flowers solitary, shortly stalked, axillary; corolla 7–15mm long, yellow, with a very short upper lip and a 3-lobed lower lip with brownish markings. Fl.5–9. Cornfields, vineyards, fallow land, open places in chalk grassland; likes warm places; rare. Britain, S. Germany, S. Europe. Protected! See also *A. genevensis*, Pl.67.

2. Basil, *Ocimum basilicum* L., plant 20–40cm, glabrous, aromatically scented; lvs ovate to lanceolate, stalked, entire or toothed, 3–5cm long; flowers shortly stalked, in 6-flowered, superimposed whorls; corolla white or reddish, 10–15mm, upper lip with 4 obtuse lobes, lower lip undivided; stamens lying on the lower lip; calyx 2-lipped, with an undivided upper lip and a deeply 4-lobed lower lip. Fl.6–9. Herb and medicinal plant, rarely naturalised. Native of S.E. Asia.

3. Motherwort, *Leonurus cardiaca* L., plant erect, 30–100cm; lvs hairy, palmately 3–7 lobed, 6–12cm long, dark green above, pale green below; the upper lvs 3-lobed; flowers in a series of superimposed axillary whorls; corolla 2-lipped, 8–11mm long, pale purple; calyx campanulate, regularly 5-toothed, 5–8mm long; stamens much exceeding the corolla tube. Fl.6–9. Paths, hedges, waste places, nutrient-rich loamy soils; rather rare; an old medicinal plant. Almost all of Europe (eastern). Protected! Similar species: **False Motherwort**, *L. marrubiastrum* L., but lvs ovate-lanceolate, coarsely toothed, grey-felted below; corolla and calyx about equal in length; stamens scarcely exceeding the corolla tube; weed communities; rare; principally C. and E. Europe. Protected!

4. Cat-mint, *Nepeta cataria* L., plant erect, densely short-hairy, strongly aromatic, 40–100cm; lvs triangular-cordate, coarsely toothed, long-stalked, grey-felted below; flowers in a terminal inflorescence of many-flowered whorls, dense above; corolla 2-lipped, white or reddish, 1cm long; calyx 6–8mm long, often tinged violet, with 5 unequal, narrowly lanceolate teeth. Fl.6–9. Weed communities, paths, walls, hedge-banks; local to rare. Almost all Europe (eastern). Protected!

5. Common Hemp-nettle, *Galeopsis bifida* Boenn., plant 20–60cm; stem much thickened at the nodes and spreading bristly-hairy; lvs ovate-lanceolate, toothed; inflorescence with yellow-headed glandular hairs or glandless; corolla 10–15mm long, pale red; lower lip of corolla with an oblong central lobe, which is clearly emarginate and usually uniformly coloured. Fl.6–10. Arable fields, woodland clearings, wood margins, disturbed moorland; calcifuge; widespread. Europe.

Plate 29   Mint Family, Labiatae or Lamiaceae

1. Large-flowered Hemp-nettle, *Galeopsis speciosa* Mill., plant 30–80cm; stem stiffly hairy below the thickened nodes; corolla 25–40mm long, pale yellow, middle lobe of the lower lip violet. Fl.6–10. Paths, arable fields, especially on peaty soil, and wood clearings; widespread. Almost all of Europe. The **Hairy Hemp-nettle**, **G. pubescens* Bess., has slightly thickened, shortly and softly hairy nodes, with a few glandular hairs; lvs appressed-hairy above and spreading soft-hairy below; corolla 18–25mm long, dark red with a yellow throat, 2–3 times as long as the calyx; arable fields, wood clearings; widespread; S. Scandinavia, C. Europe, southwards to C. Italy and Bulgaria (eastern).

2. Red Hemp-nettle, *Galeopsis angustifolia* (Ehrb.)Hoffm., plant 10–40cm; stem not thickened at the nodes; lvs narrowly lanceolate, 2–6mm broad, entire or with 1–4 teeth on each side; corolla red, 1–2cm long, 3 times as long as the calyx, the latter often red-tinged; lower lip of the corolla with yellow and dark purple markings. Fl.6–10. Calcareous arable fields, stone and gravel quarries, banks; likes warm places; widespread. Britain, C. and S. Europe. The **Field Hemp-nettle**, *G. ladanum* Neck., has ovate-lanceolate lvs 7–15mm wide with 3–8 coarse teeth on each side; corolla red, twice as long as the calyx, which is densely glandular, with white, spreading hairs; quarries, dry arable fields, railway banks; widespread but rare in Britain. Also without thickened stem nodes is **Downy Hemp-nettle**, *G. segetum* Neck., with pale yellow flowers 2–3cm long, and ovate-lanceolate lvs; upper lvs and calyx glandular-pubescent; sandy arable fields, stony waste places; calcifuge; rare to local; Britain, W. C. and C. Europe. Protected !

3. Common Hemp-nettle, *Galeopsis tetrahit* L., plant 10–60cm; annual; stem 4-angled, much thickened at the nodes, spreading-hairy; lvs ovate-lanceolate, 3–10cm long, toothed; corolla 15–20mm long, red or white, scarcely twice as long as the bristly-hairy calyx; lower lip of corolla yellow and purple spotted, the middle lobe as broad as long, toothed at the tip; inflorescence usually with black glands. Fl.6–10. Arable fields, waste places, paths, woodland clearings; common. Europe.

4. Spotted Dead-nettle, *Lamium maculatum* L., plant 20–60cm; similar to White Dead-nettle, but lvs often white-spotted and corolla purple, 20–30mm long, with darker spots on the lower lip; corolla tube curved upwards with a transverse ring of hairs inside. Fl.4–9. Weedy fields, woodland margins, hedges and fen carr. A local garden escape in Britain, but common in C. and S. Europe.

5. Red Dead-nettle, *Lamium purpureum* L., plant often reddish-violet tinged, with an unpleasant smell, 10–30cm; lvs ovate-cordate, unequally crenate, softly hairy, shortly stalked; corolla 10–15mm long, pinkish-red. Fl.3–9. Arable fields, gardens, waste places, paths; common. Europe. Similar species: **Henbit Dead-nettle**, *L. amplexicaule* L., but upper lvs and bracts sessile and amplexicaul, roundish kidney-shaped, deeply crenate, 1–2cm long and 1–3cm wide, shorter than the stem internodes; flower whorls somewhat distant from each other; weedy fields; almost all of Europe.

6. White Dead-nettle, *Lamium album* L., plant hairy, 20–50cm; lvs long-pointed, like those of Common Nettle (but not stinging), sharply serrate, 3–7cm long; flowers in whorls; corolla 20–25mm long, white, 2-lipped; upper lip helmet-like; calyx tube curved upwards, with a ring of oblique hairs inside. Fl.4–10. Paths, hedges, walls, stockyards, waste places; nitrogen indicator; common. Europe.

2

3

6

4

1

5

Plate 30 Mint Family, Labiatae or Lamiaceae

1. Ground-ivy, *Glechoma hederacea* L., plant 20–40cm; stem creeping, rooting at the nodes; lvs roundish kidney-shaped; flowers 2–3 together in the axils; corolla blue-violet, 1–2cm long; upper lip straight, emarginate at tip, lower lip 3-lobed, with the central lobe the largest. Fl.4–6. Meadows, pastures, scrub margins, river banks, waste places; common. Europe.

2. Black Horehound, *Ballota nigra* L., plant smelling unpleasantly, dark green, 30–100cm; lvs softly hairy, coarsely crenate-serrate, 2–5cm long; flowers 4–10 in a whorl; corolla purple, 1–2cm long; lower lip 3-lobed; calyx funnel-shaped, 10-nerved, glandular-hairy. Fl.7–9. Paths, hedges, waste places rich in nitrogen; widespread. Principally C. and S. Europe (eastern).

3. Marsh Woundwort, *Stachys palustris* L., plant 30–100cm, with rhizomes; lvs oblong-lanceolate, with cordate base, sessile or very shortly stalked, shortly hairy, serrate, 3–12cm long; flowers in whorls of 6–10; corolla 14–18mm long, purple, twice as long as the shortly hairy calyx; upper lip of corolla entire, lower lip 3-lobed, with dark markings. Fl.6–9. River banks, pondsides, damp arable fields, pathsides; widespread. Almost all of Europe. The **Field Woundwort**, *S. arvensis* L., has broadly ovate, stalked lvs, 1–3cm long and bluntly serrate; plant prostrate to ascending, 10–30cm; stem with villous glandular hairs; corolla 6–9mm long, pale pink, scarcely longer than the calyx; arable fields, fallow land, non-calcareous damp soils; local to rather rare. Britain, C. and S.W. Europe.

4. Summer Savory, **Satureja hortensis* L., plant 10–30cm, aromatic; lvs narrowly lanceolate, shortly stalked, entire, obtuse, 1–3cm long; flowers very shortly stalked, 1–3 together in the axils of the upper lvs; corolla 4–6mm long, bluish or whitish; calyx campanulate, with 5 almost equal acute teeth, 3–5mm long. Fl.7–9. Herb and medicinal plant, and a forage plant for hive bees; rarely naturalised; likes warm places. Originally from S.E. Europe. The **Winter Savory**, **S. montana* L., is a smaller, woody dwarf shrub, 10–50cm, with the corolla white or pale violet, 6–10mm long; lvs acute, coriaceous, almost sessile; rocky slopes, calcareous, stony soils; likes warm places; pot herb; southern edge of the Alps, S. Europe.

5. Corn Mint, *Mentha arvensis* L., plant aromatic, 10–40cm; lvs ovate or elliptic, slightly toothed to crenate, 2–6cm long; whorls of flowers in the axils, stem ending in a tuft of lvs; corolla lilac, 5–8mm long; calyx campanulate, calyx teeth triangular-ovate. Fl.7–8. Damp arable fields, wet meadows; fairly common. Almost all of Europe.

For other species in this family, see also *Mentha*, Pl.115, *Teucrium*, Pl.67, *Stachys annua*, Pl.68 and *Salvia*, Pl.69.

Vervain Family, Verbenaceae

6. Vervain, *Verbena officinalis* L., plant 20–80cm; stem 4-angled, with almost leafless branches; lvs lanceolate, opposite, middle stem lvs 3-lobed, with a large terminal lobe; flowers in dense narrow spikes; corolla weakly 2-lipped, 2–5mm across, pale lilac, with a bent tube. Fl.7–9. Paths, walls, hedges, siliceous grassland; widespread. Europe.

Plate 31   Nightshade Family, Solanaceae

1. Henbane, *Hyoscyamus niger* L., plant sticky-villous, 30–60cm; lvs oblong, 10–20cm long, deeply lobed, the upper amplexicaul; flowers several together, axillary; corolla funnel-shaped, 2–3cm long, dirty yellow, with a network of violet veins; calyx with 5 spiny teeth; fruit a subglobose capsule. Fl.6–10. Poisonous! Pathsides, walls, waste places; rather rare. Almost all of Europe. Protected!

2. Thorn-apple, *Datura stramonium* L., plant glabrous, 20–100cm; lvs ovate, acute, deeply toothed, 10–20cm long; flowers solitary, axillary; corolla funnel-shaped, with 5 acute lobes, 6–8cm long, white, rarely violet; fruit ovoid, prickly, green, 5–7cm long. Fl.6–10. Poisonous! Waste places, paths, gardens; likes warm places; rather rare. Europe.

3. Black Nightshade, *Solanum nigrum* L., plant 10–60cm, dark green; lvs ovate or triangular, sinuate-dentate, more rarely entire, stalked; corolla with 5 spreading lobes, white, 6–10mm across; calyx teeth separated by acute sinuses; anthers longer than the filaments; berries black, rarely greenish-yellow, 6–10mm broad. Fl.6–10. Waste places, gardens, arable fields; widespread. Europe. Similar species: **Hairy Nightshade**, **S. luteum* Mill. ssp. *luteum*, but stem with long spreading hairs; calyx teeth separated by rounded sinuses; anthers as long as the filaments; berries yellow; stem bluntly angled, smooth; waste places, paths; likes warm places. The **Winged Nightshade**, **S. luteum* ssp. *alatum* (Moench) Dostál, has narrowly winged, rough tuberculate and thinly hairy stems and branches; berries red; weedy arable fields; rare. Both subspecies principally in S. Germany and S. Europe.

Figwort Family, Scrophulariaceae

4. Dark Mullein, *Verbascum nigrum* L., plant 50–150cm; stem angled above; lower lvs cordate at base, long-stalked, all lvs nearly glabrous above, grey-felted below; inflorescence elongate, rarely branched; flowers arranged in a spike of clusters; corolla 12–22mm across, yellow, with purple marks at the base; filaments clothed with purple hairs. Fl.6–8. Waste places, banks, woodland clearings, waysides; fairly common to rather rare. Almost all of Europe. Likewise with violet hairs on the filaments is **Moth Mullein**, *V. blattaria* L., but the flowers are long-stalked, pale yellow, reddish in bud, 1–2 together in the axils of a simple raceme; lvs ovate-oblong, narrowed into the stalk, glabrous on both sides; waysides, banks, waste places; rather rare; Britain, C. and S. Europe.

5. Great Mullein, *Verbascum thapsus* L., plant 20–200cm; lvs ovate-oblong, felted-hairy, up to 40cm long, decurrent down the stem, which is consequently winged; flowers shortly stalked in a spike of clusters; corolla broadly funnel-shaped, 15–22mm across, pale yellow, the 3 upper stamens white-woolly, the lower 2 glabrous. Fl.7–9. Waste places, banks, wood clearings; rather common. Europe. Similar species: **White Mullein**, *V. lychnitis* L., but lvs not decurrent, almost glabrous above, dusty grey-felted with stellate hairs below; flowers white or pale yellow, 10–15mm across; scrubby places, calcareous grassland, dry waste places; widespread; Britain, C. and S. Europe.

6. Large-flowered Mullein, *Verbascum densiflorum* Bertol. (*V. thapsiforme* Schrad.), plant 30–200cm; similar to *V. thapsus*; lvs yellow-green, woolly-hairy, crenate, long-pointed, decurrent down the stem; inflorescence shortly branched below; corolla flat, 30–35mm across, pale yellow. Fl.7–9. Sunny, weedy fields, waysides, woodland clearings; widespread. Rare casual in Britain; C. and S. Europe.

Plate 32   Figwort Family, Scrophulariaceae

1. Common Toadflax, *Linaria vulgaris* Mill., plant 20–60cm; lvs alternate, close toge-ther, linear to lanceolate, 3–8cm long; inflorescence many-flowered; corolla with a long straight spur, 15–30cm, sulphur-yellow, with an orange palate. Fl.6–10. Sunny, weedy fields, railway banks, quarries, woodland clearings, hedgebanks, waste places; common. Europe.

2. Small Toadflax, *Chaenorhinum minus* (L.)Lange, plant annual, glandular-hairy, 5–25cm; lvs narrowly lanceolate, the lower opposite; flowers in lax racemes; corolla with spur 5–10mm long, pale violet, with yellow palate, throat open. Fl.6–10. Arable fields, paths, banks, gravel pits, railway tracks; rather common. Almost all of Europe.

3. Field Cow-wheat, *Melampyrum arvense* L., plant 20–60cm; lvs linear-lanceolate; flowers in cylindrical spikes, not secund; corolla 20–25mm long, purple, with a pale yellow tube and yellow throat; upper lip helmet-shaped; bracts ovate-lanceolate, with awn-like teeth, flat, purple, rarely yellow-green. Fl.5–9. Cornfields, waysides, sunny . hedges; rare to local. Almost all Europe. Similar species: **Crested Cow-wheat**, *M. cristatum* L., but spike 4-sided, bracts roundish-cordate, folded, toothed like a comb, whitish or reddish; flowers yellowish-white, tinged red; sunny places, woodland mar-gins, dry oak-pine woods; rare; principally C. Europe.

4. Fingered Speedwell, *Veronica triphyllos* L., plant prostrate, densely glandular-hairy, 5–15cm; lvs deeply 3–5 lobed, sessile, crenate, dark green; stem lvs graded into small bracts, the upper more simple than the lower; flowers in lax racemes; corolla 6–9mm, dark blue; calyx-teeth 6–8mm long in fruit; capsule obcordate, densely glandular-hairy. Fl.3–5. Arable fields, dry sandy grassland; calcifuge; local to rare. Principally C. and S. Europe.

5. Common Field-speedwell, *Veronica persica* Poir., plant prostrate, 10–30cm; lvs triangular-ovate, coarsely toothed, shortly stalked; stem lvs scarcely diminishing upwards; flowers solitary, axillary; corolla 8–12mm across, blue, lowest lobe paler; calyx lobes ovate-lanceolate; capsule 8–10mm wide and 4–6mm long, emarginate with obtuse lobes. Fl.1–12. Arable fields, vineyards, gardens; common. Almost all of Europe.

6. Ivy-leaved Speedwell, *Veronica hederifolia* L., plant 8–30cm; lvs 3–7 lobed, like ivy leaves, roundish, long-stalked; corolla 6–9mm across, pale blue, dark-veined; calyx teeth broadly cordate, long-ciliate; capsule almost spherical. Fl.3–5. Arable fields, gardens, vineyards, paths, hedges, woodland clearings; common. Most of Europe.

7. Wall Speedwell, *Veronica arvensis* L., plant 5–20cm; stem with 2 rows of hairs; lvs undivided, ovate, crenate, 10–15mm long; bracts narrowly lanceolate, becoming smal-ler above; pedicel shorter than the calyx; corolla 3–4mm across, pale blue; capsule broadly cordate, ciliate, deeply emarginate with acute lobes. Fl.4–10. Arable fields, paths, walls, open places in grassland and heaths. Originally from S. Europe, now-adays spread worldwide.

8. Slender Speedwell, *V. filiformis* Sm., plant creeping, with filiform stems, 5–30cm, glandular-hairy; lvs roundish, weakly crenate, 5–12mm long; pedicels 2–4 times as long as the bracts; corolla bluish-violet, 8–12mm across; capsule 4–6mm across, with a style 3–4mm long, usually without seeds (capsule very rare in Britain); spreading by small bulbils in the leaf axils. Fl.4–5. Lawns, pastures, paths; widespread. Originally from the mountains of S. W. Asia, nowadays spread worldwide.

Plate 33 Figwort Family, Scrophulariaceae

1. Thyme-leaved Speedwell, *Veronica serpyllifolia* L., plant 5–25cm; lvs glabrous, roundish-ovate, shortly stalked or sessile, entire or crenate; flowers solitary in the axils of the upper lvs, forming a long, many-flowered inflorescence; corolla 5–6mm across, whitish, with blue veins. Fl.4–9. Pastures, paths, arable fields, river banks, heaths, damp places in mountains (up to 2400m in the Alps); common. Europe.

2. Red Bartsia, *Odontites verna* (Bell.)Dum. (*O. rubra* (Baumg.) Opiz), plant branching; lvs opposite, lanceolate, usually toothed; flowers on long secund racemes; corolla red, felted-pubescent, 8–12mm long; upper lip helmet-shaped, 3-lobed. Fl.8–10. Dry sandy grassland, trampled places, arable fields; fairly common. Europe.

3. Sharp-leaved Fluellen, *Kickxia elatine* (L.)Dum., plant annual, softly glandular-hairy, 10–40cm, with creeping, filiform stems and alternate, short-stalked, hastate to sagittate lvs; flowers on long stalks, which are usually glabrous, axillary; corolla 5–7mm long, pale yellow, with a straight spur of 6mm; upper lip of corolla purple inside. Fl.7–10. Cornfields, fallow land, loamy soils; local to rare. C. and S. Europe, Britain, S. Scandinavia. Similar species: **Round-leaved Fluellen**, *K. spuria* (L.)Dum., but all lvs ovate; pedicels roughly hairy; corolla with a curved spur.

Plantain Family, Plantaginaceae

4. Greater Plantain, *Plantago major* L., plant 5–30cm; lvs ovate-elliptic, 5–9 veined; blade clearly marked off from the leaf stalk, dark green, rough; flowers inconspicuous, in narrow spikes, up to 15cm long; corolla 4-merous, yellowish-white; anthers at first lilac, then yellowish. Fl.6–10. Paths, trampled places, pastures; common. Europe. Similar species: *P. major* ssp. *intermedia* (DC.)Arc. (*P. intermedia* Gilib.), but plant smaller, lvs gradually narrowed into the stalk, 3–5 veined, peduncles curved below, then erect, spikes short; damp arable fields, lake margins; widespread; Europe.

Valerian Family, Valerianaceae

5. Common Cornsalad, *Valerianella locusta* (L.)Laterr., plant 10–20cm; lvs oblong, the lower spathulate, 2–7cm long, upper lanceolate, acute; inflorescence terminal, dichotomously branched; with a small flower in the fork below the flower heads; corolla 5-lobed, bluish-white, 2mm across; fruit compressed, roundish, shortly pointed, smooth. Fl.4–6. Arable fields, paths, hedgebanks, rocky outcrops, dunes; common; cultivated. Europe. (The genus has several very similar species.)

Teasel Family, Dipsacaceae

6. Teasel, *Dipsacus fullonum* L. (*D. sylvestris* Huds.), plant 70–150cm; stem prickly; lvs ovate, usually undivided, glabrous on the margins, or with scattered prickles, up to 30cm long; stem lvs opposite, connate at the base, serrate; flower heads spherical, 3–8cm long, with long, curved, prickly bracts, and bristly receptacular scales, longer than flowers; corolla lilac, 1cm long. Fl.7–8. Paths, waste places, fallow land, river banks, copses, rough pasture; common. Britain, S. Scandinavia, Germany, C. and S. Europe. **Cut-leaved Teasel**, *D. laciniatus* L., has pinnate stem lvs and widely spreading, short bracts; flowers white; paths, wood margins; rare; Germany, S. Europe; protected! **Small Teasel**, *D. pilosus* L., has long-stalked basal lvs, shortly stalked stem lvs not connate at base, and spherical flower heads, 20–25mm across; bracts short, slightly longer than white or yellowish flowers; damp woods and paths, fen carr; rare.

5

6

4

2

3

1

Plate 34   Bellflower Family, Campanulaceae

1. Creeping Bellflower, *Campanula rapunculoides* L., plant 30–100cm, with rhizomes; stem shortly and roughly hairy; lower lvs stalked, shortly ovate, serrate, upper lvs lanceolate, sessile; flowers nodding, in secund racemes; corolla funnel-shaped, 2–3cm long, pale violet, usually ciliate on the margin. Fl.6–9. Wood margins and waysides, hedges, arable fields, open oak and pine woods; widespread. Almost all of Europe. Similar species: **Pale Bellflower**, **C. bononiensis* L., but plant without rhizomes; lvs densely pubescent beneath; all the bracts shorter than the flowers; corolla 1–2cm long; inflorescence not secund; dry bushy places, dryish grassland; rare; C. and S. Germany, valleys of the S. Alps, E. and S.E. Europe.

2. Large Venus's-looking-glass, **Legousia speculum-veneris* (L.)Chaix, plant 10–30cm; lvs oblong or obovate, margins weakly undulate, rough; inflorescence lax-paniculate; corolla rotate, 20–25mm across, violet; calyx teeth linear, as long as the ovary, scarcely longer than the corolla. Fl.6–8. Cornfields, calcareous arable fields; widespread. C. and S. Europe. Similar species: **Venus's-looking-glass**, *L. hybrida* (L.)Delarbre, but corolla 6–15mm across, broadly campanulate, purple; calyx teeth lanceolate, half length of ovary, longer than corolla; inflorescences few-flowered, terminal; cornfields; warm places; local to rare; Britain, C. and S. Germany, S. Europe. Protected!

Daisy Family, Compositae or Asteraceae

3. Canadian Fleabane, *Conyza canadensis* (L.)Cronq. (*Erigeron c.* L.), plant 30–100cm; lvs lanceolate, entire or finely serrate, 2–5cm long, hairy; flower heads numerous in dense panicles; heads 2–5mm long, 3–5mm across, ray florets whitish, scarcely longer than the yellowish disc florets; bracts linear, almost glabrous. Fl.7–10. Waste places, waysides; common. Almost all of Europe (originally from N. America). Similar species: **Horseweed**, **C. bonariensis* (L.)Cronq., but plant 15–50cm, ray florets absent; flower heads 7–8mm long, greenish, bracts often reddish-tinged; lvs densely hairy, grey-green, serrate; weed communities; warm places; C. and S. Europe (from N. America).

4. Colt's-foot, *Tussilago farfara* L., plant 5–20cm; flowers appearing before the lvs; flower stalk single-headed, with reddish scales; flower head 2–3cm across, yellow; bracts in 1 row; lvs all radical, in clumps, roundish cordate, 10–30cm broad, with black-edged teeth, grey-felted beneath; leaf stalk channelled above. Fl.2–4. Paths, gravel pits, waste places, arable fields, river banks, landslides; common. Europe.

5. Gallant Soldier, *Galinsoga parviflora* Cav., plant 20–70cm, sparingly appressed-hairy above; lvs ovate, acute, finely serrate, opposite; flower heads 3–5mm long and wide, usually with 5 short, white ray florets and a few yellow disc florets; receptacular scales broader above, usually 3-lobed. Fl.5–10. Arable fields, gardens, waste places; local to common. Europe. Similar species: **Shaggy Soldier**, *G. ciliata* (Raf.)Blake, but stem white villous-hairy; lvs coarsely and distantly toothed; receptacular scales lanceolate; arable fields, widespread. Europe. Both species are naturalised in S. America.

6. Groundsel, *Senecio vulgaris* L., plant 10–30cm; lvs pinnate, usually glabrous, with oblong, toothed and rather fleshy lobes; flower heads 10mm long and 4–5mm wide; outer involucral bracts 8–12, black-tipped, about $\frac{1}{4}$ as long as the 21 green, glabrous, acute inner bracts; ray florets usually absent. Fl.2–11. Arable fields, ditches, paths, waste places, wood clearings; common. Europe. Similar species: **Oxford Ragwort**, *S. squalidus* L., plant 20–30cm, with spreading yellow ray florets, outer bracts very short. Common in waste places. Britain, France, introduced from Italy.

Plate 35   Daisy Family, Compositae or Asteraceae

1. Spring Groundsel, *Senecio vernalis* W. et K., plant 15–50cm; stem and lvs arachnoid-woolly; lvs pinnatisect, with ovate, serrate segments, upper lvs sessile, broadly auriculate at the base; flower heads 2–3cm across; exterior bracts glabrous at the tip, black to nearly halfway; ray florets 8–10mm long; fruit densely white-hairy. Fl.5–11. Arable fields, fallow land, paths; widespread. C. and S.E. Europe.

2. Sticky Groundsel, *Senecio viscosus* L., plant 15–50cm, with an unpleasant smell, sticky with glandular hairs; lvs pinnately lobed, thinly arachnoid-hairy and densely glandular-hairy; flower heads 8–12mm across, with short rolled-back ray florets; outer bracts half as long as the inner, all glandular-hairy. Fl.6–10. Fallow land, railway tracks, wood clearings, stony soils, sea shores; likes warm places; local to common. S. Scandinavia, Britain, C. and S. Europe.

3. Annual Fleabane, *Erigeron annuus* (L.)Pers., plant 50–100cm; stem with scattered to dense spreading hairs, branched above; lower lvs broad lanceolate, coarsely toothed, suddenly narrowed into the stalk, upper lvs narrower, sessile; flower heads 15–20mm across, in lax corymbs; ray florets longer than the disc florets; in ssp. *septentrionalis* (Fern.et Wieg.) Wagtz., and in ssp. *strigosus* (Mühlb.)Wagtz., 4–6mm long, white, in ssp. *annuus* 7–10mm long, usually lilac; pappus of the fruit with 2 rows of hairs, the outer short and the inner long. Fl.6–10. River banks, waysides and banks, gravel pits, damp woods; common. Europe (originally from N. America).

4. Scented Mayweed, *Chamomilla recutita* (L.)Rauschert. (*Matricaria chamomilla* L.), plant aromatically scented, 15–40cm; lvs 2–3 pinnate, with narrow-linear segments; flower heads long-stalked, 10–25mm across, ray florets white, soon reflexed, disc florets yellow and 5-toothed; receptacle conical and hollow. Fl.5–7. Cornfields, paths, waste places; medicinal plant; rather common, but becoming rarer nowadays. Europe.

5. Pineappleweed, *Chamomilla suaveolens* (Pursh)Rydb. (*Matricaria matricarioides* auct.), plant 5–30cm, with a smell resembling pineapple; flower heads conical; shortly stalked, greenish-yellow, 5–8mm across, without ray florets; disc florets 4-toothed. Fl.5–8. Trampled grassy places, nitrogen-rich weedy fields, waste places; common. Europe (originally from N. America).

6. Austrian Chamomile, *Anthemis austriaca* Jacq., plant 30–50cm; lvs 2-pinnatisect, the segments like the teeth of a comb; flower heads hemispherical; receptacular scales with a distinct mucronate point; heads 2–4cm across; fruit compressed, 4-angled. Fl.7–9. Arable fields, paths, dry sunny calcareous places; rare. C. and S.E. Europe. See also *A. tinctoria*, Pl.75.

Plate 36   Daisy Family, Compositae or Asteraceae

1. Corn Chamomile, *Anthemis arvensis* L., plant aromatic, 10–50cm; lvs 2–3 pinnate, glabrous or weakly hairy; flower heads 2–3cm across, with white ray florets; receptacles conical, with scales, which are generally tapered to a sharp point; fruit bluntly 4-angled, border crenate. Fl.5–10. Cornfields, paths, locally common on calcareous soils. Europe. Similar species: **Stinking Chamomile**, *A. cotula* L., but plant foetid, 10–50cm; receptacle conical, scales linear, without a sharp point; ray florets sterile, without stigmas; fruit nearly terete, with tuberculate ribs; arable weed; widespread; Europe.

2. Scentless Mayweed, *Matricaria maritima* L. (*Tripleurospermum m.* (L.)Koch), plant scentless, 10–50cm; lvs 2–3 pinnate, with long slender segments; flower heads long-stalked, 2–4cm across, with 12–30 white ray florets; receptacle hemispherical, pithy, without scales. Fl.6–10. Waste places, arable fields, dunes, shingle; common. Europe.

3. Mugwort, *Artemisia vulgaris* L., plant much-branched, with unpleasant smell, 60–120cm; lvs 1–2 pinnate, with lanceolate segments 3–8mm broad, dark green above, white woolly beneath, auriculate; segments of upper stem lvs deeply serrate; flower heads ovoid, 3–4mm long, reddish-brown, in much-branched panicles; outer bracts ovate, felted, margins broad hyaline. Fl.7–9. Waste places, river banks, hedgerows; common; pot herb. Europe. Similar species: **Chinese Mugwort**, *A. verlotiorum* Lam., but plant aromatic, with long rhizomes; segments of the upper stem lvs entire; flower heads globose; bracts linear and nearly glabrous; waste places; rather rare; originally from E. Asia, spreading in Europe. **Wormwood**, *A. absinthium* L., plant 30–80cm high, aromatic, bitter and woody at the base; stem grey-felted; lvs 2–3 pinnate; leaf segments 2–3mm wide; flower heads 3–4mm across, drooping, with yellow florets; waste places, sunny walls; widespread; medicinal plant; Europe. **Field Mugwort**, *A. campestris* L., young leaves sparsely grey-felted, becoming glabrous later; lvs 2–3 pinnate, with lanceolate segments, 0.5–1mm wide; flower heads ovoid, 2–3mm wide, red-brown, with glabrous bracts; plant prostrate to ascending. 30–80cm; dry, sandy, poor grass-land, dunes, banks; Europe, rare in Britain.

4. Field Cudweed, *Logfia arvensis* (L.)Holub (*Filago arvensis* L.), plant 10–30cm, branched above, with almost simple, erect branches, grey-felted; lvs narrowly lanceolate, 1–2cm long; flower heads 4–5mm long, 5-angled, with receptacular scales, 2–7 together in dense clusters, florets tubular; bracts densely woolly-felted, obtuse, finally spreading like a star. Fl.7–9. Fallow fields, sandy soils, dry sandy grassland; calcifuge; rare. Casual in Britain; most of Europe. Similar species: **Common Cudweed**, *Filago vulgaris* Lam. (*F. germanica* L.), but flower heads 10–30 together; bracts tapered to awn-like points, erect; lvs lanceolate, often undulate at margins, overlapping; fallow land, heaths, waysides, acid sandy soils; local or rare; Britain, S. Germany, S. Europe.

5. Marsh Cudweed, *Filaginella uliginosa* (L.)Opiz. (*Gnaphalium u.* L.), plant felted, much-branched, 5–20cm; lvs narrowly lanceolate, both sides felted, sessile. 1–4cm long; flower heads 3–4mm long, yellowish, without receptacular scales, 3–10 together in dense clusters, exceeded by the spreading upper lvs; bracts glabrous, pale brown. Fl.7–9. Damp acid soils, paths, ditches; widespread. Most of Europe.

6. Feverfew, *Tanacetum parthenium* (L.)Schultz-Bip. (*Chrysanthemum p.* (L.)Bernh.), plant strong-smelling, 30–80cm; lvs delicate, pinnate, with 3–9 ovate, pinnatisect or serrate lobes; flower heads 1.5–2cm across, in lax corymbs; ray florets short, obovate, white; disc florets yellow; fruit with 10 ribs. Fl.5–8. Waysides, gardens, waste places; widespread; medicinal plant, most of Europe. See also *T. corymbosum*. Pl.178.

Plate 37   Daisy Family, Compositae or Asteraceae

1. **Tansy**, *Tanacetum vulgare* L. (*Chrysanthemum vulgare* (L.)Bernh.), plant 60–120cm; lvs pinnately divided, 5–25cm long, lobes 8–12 on each side, lanceolate, pinnately toothed; flower heads 7–12mm across, in corymbs, golden-yellow; ray florets absent; bracts pale green with hyaline margins, glabrous. Fl.7–9. Paths, waste places, road and river banks; common. Europe.

2. **Common Fleabane**, *Pulicaria dysenterica* (L.)Bernh., plant 30–60cm; lvs ovate, with a cordate amplexicaul base, often undulate, serrate, grey-felted beneath, 3–8cm long; flower heads 15–30mm across, in lax panicles; ray florets yellow, spreading, twice as long as the green and glandular-hairy involucre. Fl.7–8. Trampled grassland, paths, moorland pasture, ditches, wet meadows; common. Britain, C. and S. Europe. Similar species: **Small Fleabane**, *P. vulgaris* Gaertn., but plant 10–30cm, annual; ray florets erect, only about as long as the disc florets; flower heads about 10mm across, dirty yellow; upper stem lvs sessile with a rounded base; damp weedy fields, river banks, ditches; rare to local; Britain, S. Scandinavia, C. and S. Europe. Protected!

3. **Woolly Burdock**, *Arctium tomentosum* Mill., plant 60–120cm; lvs ovate-cordate, stalked, white-woolly below; flower heads globose, 1.5–3cm across, densely arachnoid-woolly; outer bracts green with a hooked tip, inner red, with a short straight spiny point; with disc florets only, purple. Fl.7–8. Waste places, by paths, river banks; widespread. Casual in Britain, but most of Europe. Similar species: **Greater Burdock**, *A. lappa* L., has bracts green to the tip, with hooked spines, glabrous; flower heads 3–4cm across, long-stalked in corymbose panicles; lvs whitish-green below, petiole furrowed with a pithy centre; weedy places; widespread; former medicinal plant (burdock oil); Britain, S. Scandinavia, C. and S. Europe. **Wood Burdock**, *A. nemorosum* Lej. et Court (*A. vulgare* (Hill)Ev.), has bracts, especially the inner ones, red, hooked; flower heads 3–4cm across, short-stalked; branches spreading almost horizontally; petiole furrowed, hollow; lvs almost glabrous below; wood clearings and woodland paths; widespread; distribution like last. The **Lesser Burdock**, *A. minus* (Hill) Bernh., has flower heads 1–2cm across, somewhat arachnoid; branches spreading-erect; lvs grey-green below; weedy places; common. Almost all of Europe.

4 **Welted Thistle**, *Carduus acanthoides* L., plant 30–100cm; stem and stalk of flower heads with spiny wings; lvs green on both sides, usually glabrous, deeply pinnately lobed, with whitish, strong spines, 6–7mm long; flower heads 1–2cm across, pale red, several together on short, wavy-winged stalks. Fl.6–9. Waysides, waste places, fallow land, farmyards; widespread. Principally C. and S. Europe. Similar species: **Curled Thistle**, *Carduus crispus* L., but leaves arachnoid-felted beneath, spines white; weedy fields, river banks, and wet woods; common; almost all of Europe, but casual in Britain.

5. **Musk Thistle**, *Carduus nutans* L., plant 30–100cm; stem white-felted; lvs lanceolate, crisped, deeply pinnate-lobed, with triangular, 2–5 lobed spiny segments; lvs running down the stem in broad spiny wings; flower heads solitary, 3–6cm across, nodding, purple; bracts overlapping at the base, lanceolate, acute, spreading and recurved; pappus of rough, simple hairs. Fl.7–9. (Has many varieties.) Waysides, waste places, poor pastures; local to common. Britain, S. Scandinavia, C. and S. Europe.

Plate 38 Daisy Family, Compositae or Asteraceae

1. Spear Thistle, *Cirsium vulgare* (Savi)Ten., plant 60–120cm; lvs pinnate, grey-felted beneath, rough from fine prickles above, decurrent down the stem; leaf segments tipped with spines; flower heads solitary or 2–3 together, 2–4cm across and 4–8cm long; bracts ovate. Fl.6–9. Waysides, waste places, river banks, woodland clearings; common. Europe.

2. Creeping Thistle, *Cirsium arvense* (L.)Scop., plant 60–120cm; stem without spiny wings; lvs sinuate-toothed or pinnately lobed, often crisped and undulate, shining, spiny on the margins, not decurrent; flower heads 1.5–3cm long, in corymbose panicles; bracts dark violet, arachnoid-hairy; florets 5-lobed to the base, lilac. Fl.7–9. Arable fields, paths, waste places, wood clearings, gardens; common. Europe.

3. Cotton Thistle, *Onopordum acanthium* L., plant 50–150cm; stem broadly winged; lvs pinnately lobed, decurrent, with broadly triangular, spiny toothed segments; flower heads usually solitary, 3–6cm broad and long; bracts narrow lanceolate, 2–4mm broad, running out into a long stiff point; corollas of the florets 2cm long, purple. Fl.6–8. Sunny weed communities in dry and warm places; widespread. Principally C. and S. Europe. Protected! The **Milk Thistle**, *Silybum marianum* (L.)Gaertn., is likewise a tall, robust plant, 50–150cm, with solitary purple flower heads 4–5cm across, but lvs white-spotted, oblong-ovate, deeply pinnate-lobed, narrowed into the winged stalk or sessile, with yellow spines; waste places, weedy fields, stony grassland; naturalised in Britain, native in S. Europe, cultivated in C. Europe and occasionally escaping.

4. Chicory, *Cichorium intybus* L., plant stiffly branched, 30–120cm; stem tough, furrowed; stem lvs broadly lanceolate, entire or distantly toothed, semi-amplexicaul; basal lvs pinnately toothed, like dandelion lvs, dark green, roughly hairy below, especially on the veins; flower heads 3–5cm across, blue, usually sessile. Fl.7–10. Path and field margins, waste places; common. Almost all of Europe. The **Endive**, *C. endivia* L., has broadly ovate, amplexicaul stem lvs, and glabrous, weakly serrate and pale green basal lvs, which are often undulate; flower heads 3–6cm across; cultivated in many places as a salad plant and occasionally naturalised; originally from S. Europe.

Plate 39   Daisy Family, Compositae or Asteraceae

1. Cornflower, *Centaurea cyanus* L., plant 30–80cm, white felted-hairy; stem much branched, angled; lvs lanceolate, 2–5mm wide, the lowest pinnately lobed; flower heads solitary, 2–3cm across; ray florets blue, spreading, much longer than the violet disc florets; bracts ovate, 10–15mm long. Fl.6–10. Cornfields, waste places; once common, but now becoming rare. Europe generally, originally from S.E. Europe.

2. Nipplewort, *Lapsana communis* L., plant 30–100cm; lvs triangular-ovate, sinuate-dentate, the lowest with 1–2 lanceolate lateral lobes on either side of the winged petiole; flower heads pale yellow, 1–2cm across, with 8–15 florets; bracts in 2 rows, the inner much longer than the outer; fruit 3–5mm long, with 20 longitudinal ribs, pappus absent. Fl.6–9. Arable fields, waysides, waste places, hedges; common. Almost all of Europe.

3. Hawkweed Oxtongue, *Picris hieracioides* L., plant 30–80cm, squarrosely branched; stem densely leafy; lvs oblong-lanceolate, sinuate-dentate or entire, the upper sessile, semi-amplexicaul; lvs and stem with stiff bristly hairs; flower heads yellow, in a lax, corymbose panicle; bracts of the flower heads overlapping, the outermost spreading, with a central stripe of bristly hairs; fruit 3–5mm long, shortly beaked, transversely wrinkled and with longitudinal ribs; pappus hairs feathery. Fl.7–10. (Highly variable.) Paths, banks, quarries, dryish grassland, especially on calcareous soils; likes warm places; widespread. C. and S. Europe, Britain, S. Scandinavia.

4. Prickly Oxtongue, *Picris echioides* L., plant annual, 30–60cm, with stiff bristly hairs; lower lvs obovate, narrowed into the narrowly winged stalk, the upper lvs sessile, cordate, amplexicaul; outer bracts 3–5, enlarged, cordate-ovate, 4–7mm across, forming an outer involucre; fruit long-beaked, 6–9mm long, without longitudinal ribs. Fl.7–9. Arable fields, gardens, waste places, roadsides; likes warm places; medicinal plant; local to rare. S. Europe, introduced to Britain and C. Europe.

5. Wall Hawk's-beard, **Crepis tectorum* L., plant annual, grey-green, 10–60cm, with a slender, whitish taproot; lvs narrowly lanceolate, entire or distantly pinnately lobed; stem lvs sessile with a sagittate base, upper lvs with rolled-up margins; involucre of the flower heads campanulate, hairy; receptacles shortly hairy; heads yellow, 11–13mm long; fruit 3–4mm long, with 10 ribs, narrowed above, but nevertheless unbeaked; pappus hairs white, soft, flexuous, not feathery. Fl.5–7. Fallow land, walls, paths, dry stony places which are warm in summer; widespread. Almost all Europe.

Plate 40 Daisy Family, Compositae or Asteraceae

1. Beaked Hawk's-beard, *Crepis. vesicaria* L. (*Crepis taraxacifolia* Thuill.), plant 20–80cm; stem leafy, many-headed; lvs sinuate-dentate to pinnately lobed, the lower narrowed at the base into the winged stalk; flower heads erect, in lax, corymbose inflorescences; florets yellow, the outer reddish beneath; bracts glabrous or sparsely bristly, 8–12mm long; outer bracts broadly hyaline on the margin; fruit 10-ribbed, usually narrowed into a long beak, 6–8mm long altogether. Fl.5–6. Waysides, walls, slopes, usually on calcareous soils; widespread. Britain, C. and S. Europe (western). (Genus with many species.) See also *Hieracium aurantiacum*, Pl.194.

2. Perennial Sow-thistle, *Sonchus arvensis* L., plant 50–150cm; stem simple, first branches in the flower umbel, glabrous below, golden-glandular above; lvs glabrous, lanceolate, sinuate-dentate or pinnately lobed; stem lvs with rounded, appressed auricles at the base; flower heads 4–5cm across, yellow, in corymbs; bracts and peduncles usually golden-glandular. Fl.7–10. Arable fields, river banks, ditches, dunes; common. Europe. Similar species: the **Marsh Thistle**, *S. palustris* L., but stem lvs with an acutely sagittate base; bracts and peduncles black-glandular; fruit (as in *S. arvensis*) with 5 longitudinal ribs; ditches, river banks, margins of fen carr, moorland pasture; rather rare; Britain, S. Sweden, C. Europe, southwards to N. Italy, Corsica, Serbia. The **Prickly Lettuce**, *Lactuca serriola* L., plant 30–150cm, stem stiffly erect, lvs pinnatifid and sagittate at base, spiny on the margins and below; heads 1–1.5cm across, pale yellow, peduncles not glandular. Waste places and sand dunes, widespread. Europe.

3. Prickly Sow-thistle, *Sonchus asper* (L.)Hill., plant 30–80cm; stem branched; lvs coarse, dark green, shining, often undivided, with spiny teeth, with appressed, rounded auricles at the base; flower heads dark yellow, in lax panicles; bracts almost glabrous; fruit with 3 longitudinal ribs, smooth between the ribs. Fl.6–10. Arable fields, gardens, waste places; common. Europe. Similar species: **Small Sow-thistle**, *S. oleraceus* L., but lvs blue-green, dull, unequally serrate, not spiny, with acute spreading auricles; fruit with 3 longitudinal ribs, transversely wrinkled between the ribs; weedy places; widespread; almost all of Europe, originally from S. Europe. The **Slender Sow-thistle**, **S. tenerrimus* L., has glabrous, dark green lvs which are pinnatisect to the midrib; segments again pinnately lobed or with obtuse teeth; upper stem lvs with 2 acute auricles, the lower lvs stalked; bracts glabrous; waste places, walls, rocks. S. Europe.

4. Greater Goat's-beard, **Tragopogon dubius* Scop., plant 20–60cm, usually unbranched, with remains of previous year's lvs at base; lvs narrowly lanceolate, long-pointed, entire; peduncle of the flower head thickened above in a club shape, hollow; flower heads 4–6cm across, pale yellow; bracts not constricted above the base; fruit (with beak) 20–35mm long, 5-sided. Fl.5–7. Sunny weed communities, dry grassland; rare. C. and S. Europe. **Salsify**, *T. porrifolius* L., has likewise thickened peduncles, but the heads are purple; waste places, waysides; likes warm places; origin S. Europe, but formerly cultivated as a root vegetable and rarely naturalised. See also Goat's-beard, *Tragopogon pratensis*, Pl.86.

Also in this family, see *Taraxacum*, Pl.86.

Plate 41 Cypress Family, Cupressaceae

1. Juniper, *Juniperus communis* L., shrub with erect, narrow habit, rarely a tree, 2–5m; lvs needle-shaped, prickly, 10–15mm long, grey-green; flowers in the leaf axils; female flowers in cones with 2–4 scales (sporophylls); fruit a globose berry, 5–9mm across, green at first, then turning bluish-black in the second year after flowering. Fl.5–6. Chalk downs, heaths, moors, stony slopes, light, dry woods; local to common. Almost all of Europe. Similar species: the **Dwarf Juniper**, *J. communis* ssp. *nana* Syme (*J. sibirica* Lodd.), is a procumbent shrub of mountains and N. Europe, 20–50cm; lvs 4–8mm long, scarcely prickly; up to over 3000m in the Alps. The **Savin**, **J. sabina* L., is a prostrate shrub, to 2m high, with lvs on the older branches which are scale-like, 1–2mm long, decussate, and needle-like on the young branches; poisonous! Dry grassland, dry, warm alpine slopes, especially the central Alps, Pyrenees, Apennines, Carpathians and the Balkan peninsula; occasionally planted in gardens.

Grass Family, Gramineae or Poaceae

2. Upright Brome, *Bromus erectus* Huds. (*Zerna e.* (Huds.) Gray), plant forming dense stands, 30–100cm; stem erect, lower lvs inrolled, bristle-like, upper lvs flat, 2–4mm wide, distantly ciliate; ligules 1–2mm long, lower leaf sheaths with scattered, spreading hairs; panicle erect, 10–15cm long, with long branches; spikelets many-flowered, somewhat compressed, flat, 2–4cm long; outer glumes 1-nerved, inner 3-nerved; lemma with a short awn. Fl.5–10. Dry grassland, preferably calcareous, sunny slopes, dry pastures; fairly common. Britain, S. Scandinavia, C. and S. Europe.

3. Hungarian Brome, *Bromus inermis* Leys., plant 30–100cm, with rhizomes; lvs and leaf sheaths glabrous, lvs flat, 6–10mm wide; ligules very short, up to 2mm long; panicle erect, not secund, 10–20cm long; branches up to 5cm long, with many spikelets; spikelets linear, obtuse, 2–3cm long; outer glume 1-nerved, inner 3-nerved; lemma without awn or with awn 1–2mm long. Fl.6–7. Waysides, fields, dry poor grassland; widespread. Rare casual in Britain, N. and C. Europe (eastern), southwards to the southern edge of the Alps, Balkan peninsula.

4. Quaking-grass, *Briza media* L., plant 20–50cm; lvs 2–5mm wide, rough on the margins; ligules 1mm long, rounded; spikelets broadly ovate, 3–12 flowered, 5–7mm long, drooping, without any awns; panicle branches slender, wavy. Fl.5–6. Dryish grassland, meadows. Almost all of Europe. The **Great Quaking-grass**, *B. maxima* L., has silvery-white spikelets, 1–2cm long, on very long, slender, drooping branches; dry slopes, S. Europe; cultivated plant, and escapes to waste places; casual in Britain.

5. Mat-grass, *Nardus stricta* L., plant 10–30cm, forming dense, grey-green tufts; rootstock growing out like a brush; lvs bristle-like, inrolled, rough, surrounded at the base by tufts of old dead sheaths; spikes 3–8cm long, secund; spikelets 1-flowered, 7–15mm long, awned. Fl.5–6. Poor pasture, heaths, moors; common. Europe, in the south only in the mountains.

6. Heath-grass, *Danthonia decumbens* (L.)DC. (*Sieglingia d.* (L.)Bernh.), plant tufted, 15–45cm; lvs grey-green above, 2mm wide, with scattered hairs on the margins; leaf sheaths ciliate, with a ring of hairs in place of the ligule; inflorescence an erect raceme with 4–12 erect-spreading spikelets with 2–5 flowers; glumes 8–10mm long, with an obtuse tip with 2 teeth; lemma 4–6mm long, with 3-toothed tip. Fl.6–7. Siliceous grassland, heaths; local. Europe.

Plate 42   Grass family, Gramineae or Poaceae

1. Sheep's-fescue, *Festuca ovina* L., plant 10–40cm, very variable, in dense tufts; basal and stem lvs bristle-like, rolled or folded, usually filiform; ligule very short; leaf sheaths open almost to the base; panicle 3–12cm long, narrow, with erect branches; spikelets 4–7mm long; lemma usually awned. Fl.5–8. (Aggregate species with many microspecies.) Heaths, acid and sandy grassland and on chalk or limestone, oak and pine woods, according to microspecies. See *F. rubra* and *F. pratensis*, Pl.77.

2. Hairy Melick, *Melica ciliata L., plant 20–70cm, grey-green; lvs 2–3mm wide, densely short-hairy above; ligules 2–4mm long, torn; lower leaf sheaths glabrous; spike 6–14cm long, stiff, erect, lax, usually secund; spikelets 5–7mm long; lemmas ciliate with stiff hairs 2–3mm long. Fl.5–6. Sunny rock outcrops, walls, dry grassland; likes warm places; rare, commoner in the south. S. Scandinavia, C. and S. Europe. The **Eastern Melick**, **M. transsilvanica* Schur, has pure green lvs which are keeled below; lower leaf sheaths villous-hairy; spike dense, not secund; dry grassland, rocks; rare; S. and C. Germany, Vosges, Valais, S.E. Europe.

3. Grey Hair-grass, *Corynephorus canescens* (L.)P.B., plant 15–30cm, forming tufts; lvs bristle-like, inrolled, stiff, erect, grey-green; leaf sheaths pink; ligule 2–3mm long; panicle finely branched, silver-grey, closed up after flowering; awn of the lemma club-shaped and concealed in the spikelet. Fl.6–7. Sandy grassland, dunes, dry pine-woods; calcifuge; local to rare. E. Britain, S. Scandinavia, C. and S. Europe.

4. Crested Hair-grass, *Koeleria pyramidata* (Lamk.)P.B. (*K. cristata* auct.), plant 30–80cm; lvs flat, 2–3mm wide, spreading-ciliate; panicle narrow, pyramidal, 5–12cm long; spikelets 3-flowered, 6–8mm long; glumes and lemma acuminate to aristate, with short bristly hairs; stem shortly hairy below the panicle. Fl.6–7. Calcareous grassland, waysides, open pinewoods; widespread. Britain, S. Scandinavia, C. Europe. Similar species: **Slender Hair-grass**, **K. macrantha* (Ledeb.)Schult. (*K. gracilis* Pers.), but the lvs of non-flowering shoots are inrolled, not ciliate; panicle almost cylindrical; spikelets 2-flowered, 4–5mm long; dry grassland; widespread. **Blue Hair-grass**, *K. glauca* (Schk.)DC., has blue-green, glabrous, rough lvs, a bulbous base to the stem and rounded lemmas; rare. Protected!

5. Common Bent, *Agrostis capillaris* L. (*A. tenuis* Sibth.), plant 20–40cm, in loose tufts; lvs 2–4mm wide, flat, rough; ligule obtuse, 1mm long; panicle 5–12cm long; spreading after flowering; spikelets sometimes reddish-violet, 1-flowered, 3–4mm long. Fl.6–7. Acid grassland, open pine and oak woods; common. Europe. See *A. canina*, Pl.91.

6. Purple-stem Cat's-tail, *Phleum phleoides* (L.)Karst. (*P. boehmeri* Wib.), plant 30–60cm; stem leafless above; lvs 2–4mm wide, rough, with white margins; ligules obtuse, 1–2mm long; spike slender, up to 18cm long; spikelets lyre-shaped, with 2 points, 1-flowered; glumes longer than the lemmas. Fl.6–7. Dry sandy grassland, rocks, open pinewoods; rare to local. Britain, S. Scandinavia, C. and S. Europe. **Panicled Cat's-tail**, **P. paniculatum* Huds., has the culm rough above, panicle up to 8cm long, lvs 4–10mm wide, ligules 2–4mm long and stem leafy almost to the top; vineyards, arable fields; rare; S. Germany, S. Europe; protected! **Sand Cat's-tail**, *P. arenarium* L., has a whitish-green panicle, 1–4cm long; upper leaf sheaths somewhat inflated; sandy soils; coastal; W. Europe and Mediterranean. Protected!

Also in this habitat, see *Poa pratensis* and *Cynosurus*, Pl.77; *Trisetum, Avenula* and *Anthoxanthum*, Pl.78; *Sesleria*, Pl.123, and *Brachypodium* and *Deschampsia*, Pl.124.

Plate 43   Grass Family, Gramineae or Poaceae

1. Feather Grass, *Stipa pennata* L., plant 30–120cm; lvs 2–3mm wide, flat or inrolled; panicle narrow, branches erect; spikelets 1-flowered; glumes 15–20mm long with awn of 3–5cm long; lemma with an awn 20–30cm long, spirally twisted below and feathery above. Fl.5–7. Protected! Dry grassland, sunny cliffs, steppes; rare, but widespread in the south. S. Sweden, C. and S. Europe. Similar species: **Esparto Grass**, *S. capillata* L., but awn rough and not feathery, 10–15cm long, often twisted and intertwined with the others; steppes, stony grassland. Protected! S. Germany, S. Europe (western).

Sedge Family, Cyperaceae

2. Oval Sedge, *Carex ovalis* Good (*C. leporina* L.), plant 20–60cm, forming dense tufts; lvs 2–3mm wide, grey-green, shorter than the trigonous stem, which is rough above; spikes 5–6, ovoid, close together in a raceme, male and female flowers similar; fruit winged, gradually narrowed into the beak; glumes pointed, brown; styles 2. Fl.6–7. Poor, acid grassland, wood clearings, damp meadows; calcifuge; common. Europe.

3. Spring Sedge, *Carex caryophyllea* Latourr., plant 10–30cm, with short rhizomes; lvs 2–4mm wide, stiff, grey-green, overwintering, shorter than the smooth stem; female spikes 2–3, shortly stalked, 5–10mm long, male spike 1, terminal; fruit scarcely beaked, pubescent, glumes yellowish- or rusty-brown, with green midribs, acute; styles 3. Fl.3–5. Dryish grassland, heaths; widespread. Almost all of Europe. Similar species: **Short-leaved Sedge**, *C. montana* L., but lower leaf sheaths blood-red; lvs 1–2mm wide, hairy above; plant forming a dense mat; fruit densely pubescent, gradually narrowed into the beak; calcareous grassland, light dry woods. **Pill Sedge**, *C. pilulifera* L., plant without rhizomes, densely tufted; lower leaf sheaths yellow-brown; female spikes almost spherical; fruits globose, finely hairy; glumes with a sharp point due to the excurrent midrib; stems curved downwards in fruit; roots smelling of valerian when rubbed; acid grassland, woods, heaths; calcifuge. Most of Europe.

4. Monte-Baldo Sedge, *Carex baldensis* L., plant 15–40cm; lvs 2–3mm wide, flat, grey-green; male and female spikes of similar appearance, arranged in heads, much exceeded by the long, leaf-like bracts; inflorescence white; fruits ovoid-globose, un-beaked, white to yellowish-brown; styles 3. Fl.6–7. Sunny, calcareous, stony grassland, open mountain pinewoods, river shingle; rare. Bavaria, S. Alps. Protected!

5. Pale Sedge, *Carex pallescens* L., plant tufted, 10–50cm; basal sheaths red-brown, shining; lvs 2–3mm wide, with scattered hairs, yellow-green; stem trigonous, rough; female spikes ovoid, many-flowered, 8–15mm long, green, shining; fruit quite beakless, yellow-green, shining; styles 3; male spike 1, terminal. Fl.5–6. Damp woods, poor grassland, pastures, paths; local to common. Almost all of Europe.

Rush Family, Juncaceae

6. Field Wood-rush, *Luzula campestris* (L.)DC., plant 5–20cm; lvs 2–4mm wide, flat, sparingly long-ciliate on the margin; spikes 2–5, ovoid, 3–10 flowered; branches later recurved; perianth segments 6, 3–4mm long, chestnut-brown, with scarious border; anthers longer than the filaments. Fl.3–5. Acid grassland, lawns, heaths, sandy soils; calcifuge; common. Europe. Similar species: **Heath Wood-rush**, *L. multiflora* (Retz.)Lej., but spikes 8–15 flowered, 5–10 together, on straight, erect-ascending branches; anthers about as long as the filaments; heaths, moorlands, on acid soils. Europe. See also *Juncus inflexus, J. bulbosus*, Pl.3.

Plate 44 Lily Family, Liliaceae

1. St. Bernard's Lily, *Anthericum liliago* L., plant 30–70cm; lvs radical, grass-like, 4–5mm wide, about as long as the flowering stem; flowers in a simple raceme; perianth segments 6, all alike, 15–20mm long, white, twice as long as the stamens; style hooked at tip; bracts lanceolate, half as long as the pedicels; fruit an ovoid, acute capsule. Fl.5–6. Dry grassland, wood margins, open oak and pine woods; rather rare. C. and S. Europe, northwards to S. Sweden.

2. Branched St. Bernard's Lily, *Anthericum ramosum* L., similar to *A. liliago*, but inflorescence a branched panicle; perianth segments 8–13mm long, the inner broader than the outer, as long as the stamens; style straight; bracts one-fifth the pedicel length; capsule almost globose; lvs much shorter than the flowering stem. Fl.6–8. Dryish grassland, wood margins, dry scrub; rare. Mainly C. and S. Europe.

3. White Asphodel, *Asphodelus albus* Mill., plant 60–120cm, root thickened like a turnip; basal lvs fleshy, rush-like, up to 70cm long and 2.5cm wide; flowering stem robust, leafless; flowers numerous, in a dense raceme, often with further racemose side-branches; bracts brown, lanceolate; perianth segments 6, free or united at the base, with a brown midrib; stamens 6, broader below. Fl.6–8. Dry pastures and meadows, margins of scrub. S. and W. Alps, S. Europe.

4. Yellow Asphodel, *Asphodeline lutea* (L.)Reichb., plant 60–100cm; stem leafy to the top; lvs linear, mucronate, densely set, triangular in cross-section; flowers yellow, 2–3cm long, in a dense raceme 10–15cm long; perianth segments with a green midrib, rather unequal, the lower longer and narrower; bracts ovate, acute, longer than the pedicels; fruit globose, 10–15mm long. Fl.4–5. Dry pastures, rocky slopes. S. and W. Europe; cultivated. Similar species: **Lesser Yellow Asphodel**, *A. liburnica* (Scop.) Reichb., but daintier, 20–60cm; lvs narrow, 1mm wide, rough, flowering stem leafless above; flowers yellow, with small, triangular bracts; inflorescence laxer; S.E. Europe.

5. St. Bruno's Lily, *Paradisia liliastrum* (L.)Bertol., plant 30–50cm; lvs radical, linear, grass-like, up to 40cm long; flower stem leafless; inflorescence 3–10 flowered, secund; flowers funnel-shaped, white; perianth segments 6, free above, 3–5 veined, up to 6cm long, united below into a narrow tube; bracts scarious, longer than the pedicels. Fl.6–7. Mountain meadows from 1000 to 2500m. C., W. and S. Alps, Swiss Jura, Pyrenees, Apennines, Abruzzi.

6. Field Garlic, *Allium oleraceum* L., plant 30–70cm; lvs flat, channelled, 5mm wide, not tubular, usually rough below; flower umbels with sessile reddish bulbils; pedicels very distinctly long, about 2–6 times as long as the reddish or greenish-white flowers; filaments without lateral teeth; stamens hardly longer than the perianth segments; bracts longer than the inflorescence. Fl.7–8. Wayside banks, vineyard walls, dry grassland; widespread. Europe.

7. Sand Leek, *Allium scorodoprasum* L., plant 30–60cm; lvs flat, narrow-linear, smooth on the margins, dark green; flowers in many-flowered umbels, with a torn, scarious 1-valved spathe; flowers purple; inner stamens with 2 long, filiform teeth (see illustration); inflorescence without bulbils. Fl.6–8. Patchy dryish grassland, vineyards, wayside banks; rather rare. S. Europe, northwards to S. and C. Germany. Protected! The above description applies to ssp. *rotundum* (L.)Stearn. Very similar to it is ssp. *scorodoprasum*, but spathe 2-valved, inflorescence with bulbils, 0–12 flowered. Britain and as above. See also *A. schoenoprasum*, Pl.99.

Plate 45 Lily Family, Liliaceae

1. Keeled Garlic, *Allium carinatum* L. ssp. *carinatum*, plant 30–60cm; lvs flat, 2–4mm wide, weakly furrowed; flower umbels with greenish sessile bulbils; flowers 5–7mm long, pinkish-purple; stamens much longer than the perianth segments; filaments broadened below, without lateral teeth; pedicels 4–6 times as long as the flowers. Fl.6–8. Dry grassy places, moorland pasture; rare. Europe. Protected! Similar species: **Pretty Garlic**, *A. carinatum* ssp. *pulchellum* Bonnier et Layens, but umbels without bulbils; flowers 4–5mm long on pedicels 2–4 times longer; lvs 1–2mm wide; dry grassland, gravelly soils; very rare. S. Germany, S. Europe. Protected! Both subspecies naturalised in Britain.

2. Tassel Hyacinth, *Muscari comosum* (L.)Mill., plant 30–70cm; lvs linear, 10–25mm wide, finely serrulate; inflorescence a lax raceme 10–25cm long, with a tuft of erect, blue-violet, long-stalked sterile flowers at the top; fertile flowers 5–7mm, dark blue, with white teeth. Fl.4–5. Protected! Calcareous grassland, waysides, arable fields; likes warm places; rare. Casual in Britain. S. Germany, S. Europe. See also *M. neglectum*, Pl.3.

Also in this family are *Fritillaria tubiformis*, Pl.100, and *Scilla*, Pl.128.

Daffodil Family, Amaryllidaceae

3. Daffodil, *Narcissus pseudonarcissus* L., plant 15–40cm; all lvs radical, linear, 7–15mm wide, fleshy; inflorescence usually 1-flowered; flower 5–10cm across; perianth segments 6, united below into a tube 5–15mm long, free lobes spreading, pale yellow; corona trumpet-shaped, 3–5cm across, with crisped margin, egg yellow. Fl.3–4. Poisonous! Protected! Damp woods, non-calcareous meadows, alpine pasture with Mat-grass; rare. W. Europe, eastward to the Rhine, Lake Constance; cultivated and naturalised in some places.

Iris Family, Iridaceae

4. Pygmy Iris, **Iris pumila* L., plant 5–15cm; lvs sword-shaped, grey-green, mucronate, 1–2cm wide; stem prostrate, 1-flowered; perianth segments 6, the outer 3 reflexed and bearded above with a longitudinal patch of dense hairs; flowers violet, more rarely pale blue, white or yellow. Fl.4–5. Dry grassland, stony plains (steppe). Principally E. Europe; lower Danube, Moravia.

5. Variegated Iris, **Iris variegata* L., plant 20–40cm; lvs 1–3cm wide, about as long as the stem; flowers 4–6cm long, 2–4 together; bracts green at flowering time; outer perianth segments yellow and veined reddish-violet, reflexed, inner segments erect, bearded within, pure yellow. Fl.5–6. Dry pastures, walls, rock outcrops; rare. Cultivated plant, occasionally naturalised. S. Germany, S.E. Europe. Protected! Similar species: **Garden Iris,** or **Flag**, *I. germanica* L., but perianth segments violet, only yellowish at the base, with a yellow beard, bracts scarious at flowering time; vineyard walls; cultivated plant and naturalised. Originally from S.E. Europe.

Plate 46   Orchid Family, Orchidaceae

1. Autumn Lady's-tresses, *Spiranthes spiralis* (L.)Chevall. (*S. autumnalis* Rich.), plant 10–30cm; stem with scale leaves only, with a lateral rosette of ovate lvs at the base; inflorescence dense, spirally twisted; flowers greenish-white; petals narrowly lanceolate, curved outwards at the tip; lip undivided, oval with a crisped margin. Fl.8–9. Protected! Siliceous grassland, poor sheep pasture, seasonally dry soils; rare. Almost all of Europe, northwards to Ireland, Denmark. Similar species: **Summer Lady's-tresses**, *S. aestivalis* (Poir.)Rich., see Pl.101.

2. Frog Orchid, *Coeloglossum viride* (L.)Hartm., plant 5–25cm; lvs ovate to lanceolate; flower spike dense; bracts lanceolate, green, the lower longer than the flowers; perianth segments connivent into a hood, lip 3-lobed at tip, spur short, wide. Fl.5–6. Protected! Pastures and grassy hillsides, especially on chalky soils; uncommon. Europe.

3. Fragrant Orchid, *Gymnadenia conopsea* (L.)R.Br., plant 20–60cm; lvs lanceolate, 5–15cm long; flower spike cylindrical, 5–10cm long, dense; flowers reddish-lilac, scented; the 2 lateral perianth segments oval, 5–6mm long, lip with 3 about equal lobes; spur slender, almost twice as long as the ovary. Fl.5–6. Protected! Calcareous grassland, open woods, moorland pastures; locally common. Almost all of Europe. Similar species: **Short-spurred Fragrant Orchid**, **G. odoratissima* (L.)Rich., but spur at most the length of the ovary; central lobe of lower lip longer, acute; flowers red-violet, strongly scented of vanilla; lvs narrowly lanceolate; pinewoods, moorland and mountain pastures, in the Alps to 2200m; rare; principally mountains of C. and S. Europe, S. Scandinavia. Protected!

4. Small-white Orchid, *Pseudorchis albida* (L.)A.et D.Löve (*Leucorchis a.* E.Mey.), plant 10–30cm; lvs oblong to obovate; flower spikes dense, 3–6cm long; flowers white to yellowish-white, weakly scented; perianth segments small, 3–4mm long, connivent, lip 3-lobed; spur cylindrical, half as long as the ovary. Fl.6–8. Protected! Acid grassland, hilly pastures; rather rare. Britain, European mountains.

5. Musk Orchid, *Herminium monorchis* (L.)R.Br., plant 8–25cm, with short stolons; lvs basal, 2–5, lanceolate, 5–10cm long; stem occasionally with 1–3 small lvs; spike narrow, secund, with small, greenish-yellow, honey-scented flowers; lip 3-lobed, 3–4mm long. Fl.5–6. Protected! Calcareous grassland, poor pasture; rare. England, S. Scandinavia, C. Europe, Alps, Pyrenees, Apennines, mountains of the Balkan peninsula.

Plate 47 Orchid Family, Orchidaceae

1. Fly Orchid, *Ophrys insectifera* L. (*O. muscifera* Huds.), plant 15–30cm; lvs lanceolate, enfolding the stems like sheaths; inflorescence 2–10 (20) flowered; flowers about 2cm long; outer 3 perianth segments oval, greenish, 5–8mm long, the 2 inner lateral ones filiform, red-brown; lip 3-lobed, without appendage, twice as long as wide, purplish-brown, velvety, with a large grey spot. Fl.5–6. Protected! Calcareous grassland, dry open pinewoods and deciduous woodland on chalk or limestone; rather rare. Almost all of Europe, northwards to Scotland, Ireland, southwards to N. Spain, C. Italy.

2. Late Spider-orchid, *Ophrys fuciflora* (Crantz) Sw. (*O. arachnites* Lam.), plant 15–30cm; flowers 1–10; outer perianth segments white or pink, with green midribs, oval, 10–15mm long, inner perianth segments triangular; lip undivided, roundish, almost quadrangular, red-brown, with yellowish markings, with small bosses at the base on each side, and with an incurved appendage at the tip. Fl.5–6. Protected! Calcareous grassland, dry scrub margins; rare. Britain, C. and S. Europe.

3. Early Spider-orchid, *Ophrys sphegodes* Mill. (*O. aranifera* Huds.), plant 15–30cm; outer perianth segments oval, greenish, 8–12mm long, the 2 inner ones narrowly lanceolate; lip undivided, convex, red-brown, with a bluish marking like the letter H, without appendage at tip. Fl.4–5. Protected! Calcareous grassland, wood clearings, on calcareous soils; rare. England, C. and S. Europe.

4. Bee Orchid, *Ophrys apifera* Huds., plant 15–40cm; lvs ovate-lanceolate; outer perianth segments oval, inner 2 filiform; lip longer than wide, 3-lobed, strongly convex, brown, with yellow markings, with a yellowish-green, narrowly lanceolate and incurved appendage at the tip. Fl.5–6. Protected! Calcareous grassland, open oak and pine woods, banks and copses on chalky soil or clay; local to rare. Britain, C. and S. Europe.

5. Burnt Orchid, *Orchis ustulata* L., plant 20–30cm; lvs lanceolate; inflorescence 3–6cm, dense, cylindrical, so blackened before flowering as to have a burnt appearance; flowers small, up to 1cm long, scented; perianth segments forming a hemispherical hood, dark purple outside; lip 3-lobed, 5mm long, white with red spots, central lobe bifid. Protected! Fl.5–6. Calcareous grassland, dry pastures, edges of scrub; rare. Britain and almost all of Europe, northwards to S. Sweden.

6. Military Orchid, *Orchis militaris* L., plant 25–50cm; lvs narrowly oval, 5–15cm long, upper lvs enclosing the stem-like sheaths; inflorescence 5–10cm long; all 5 perianth segments connivent into a hood, lanceolate, acute, pale pink outside, with dark veins; lip 10–15mm long, lilac to white, with minutely hairy dark spots, the two segments of the middle lobe blunt, short, wider than the side lobes. Fl.5–6. Protected! Calcareous grassland, slopes, scrub margins. Rare in Britain, but commoner elsewhere. Almost all of Europe, northwards to S. Sweden.

7. Monkey Orchid, *Orchis simia* Lam., plant 30–40cm, very similar to the Military Orchid, but inflorescence globose to shortly cylindrical, flowering from the top downwards (the other species flower from the bottom upwards), all 4 segments of the lip narrow-linear, pointed, about equal in length, curved upwards, lip without hairy spots. Fl.5–6. Protected! Sunny calcareous grassland; rare. Principally S. Europe.

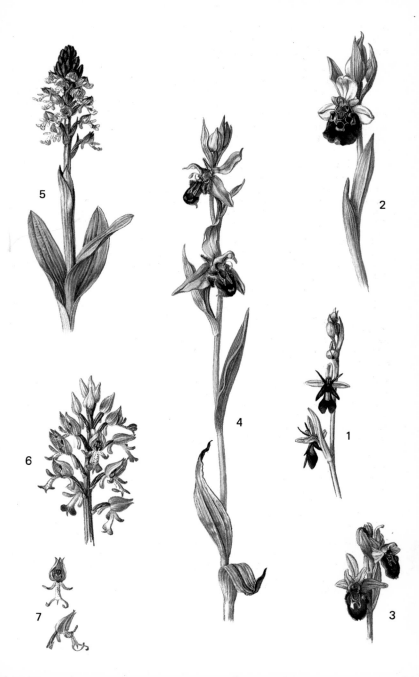

Plate 48   Orchid Family, Orchidaceae

1. Green-winged Orchid, *Orchis morio* L., plant 10–30cm; stem leafy, angled; lvs narrowly oval, broadest in the middle, blunt, not spotted, 3–8cm long; inflorescence 5–10cm long, lax-flowered; bracts scarious, about as long as the ovary, often coloured; all 5 perianth segments connivent into a hood, red (rarely white), with green stripes lengthwise; lip at least as broad as long, 3-lobed, the middle lobe often emarginate; spur shorter than the ovary, more or less horizontal. Fl.4–6. Protected! Poor grassland, unfertilised meadows, especially on calcareous soils; local to rare. Almost all Europe, northwards to S. Scandinavia, Britain. See also *O. mascula*, Pl.134.

2. Bug Orchid, **Orchis coriophora* L., plant 10–40cm; stem terete; lvs linear-lanceolate, 6–10cm long, channelled and pleated; inflorescence 3–6cm long, cylindrical, dense-flowered; flowers smelling of bugs; all perianth segments, apart from the lip, connivent into an oblong, acute hood, brownish-red, with green veins; lip 4–6mm long, longer than wide, the middle lobe entire, longer than the side lobes; spur conical, directed downwards;.bracts scarious. Fl.6–7. Protected! Poor grassland, moorland pasture; rare. C. and S. Europe, northwards to C. Germany.

3. Pyramidal Orchid, *Anacamptis pyramidalis* (L.)Rich., plant 25–50cm; lvs lanceolate; inflorescence 4–8cm long, at first conical, then cylindrical, dense-flowered; flowers a shining reddish-purple; outer 3 perianth segments spreading, lanceolate, 5–7mm long; lip as long as broad, 6–9mm, 3-lobed, middle lobe somewhat smaller; spur slender, as long as the ovary. Fl.6–7. Protected! Calcareous grassland, seasonally dry meadows, open woods, calcareous dunes; local to rare. Britain, C. and S. Europe (western), northwards to S. Scandinavia.

4. Lizard Orchid, *Himantoglossum hircinum* (L.)Spreng., plant 30–80cm, lvs ovate, upper lanceolate, 5–15cm long and 3–5cm wide, blue-green, enclosing the stem like sheaths; inflorescence 10–20cm, lax; bracts filiform, yellowish; flowers greenish-white, with red veins, smelling of goats; all 5 perianth segments forming a hemispherical hood; lip olive-green, with the middle lobe tape-like, twisted, 2–5cm long and 2–3mm wide with an emarginate tip, and with 2 short side lobes at the base, 5–15mm long and undulate; spur conical, 3–4mm long. Fl.5–6. Protected! Calcareous grassland, sunny scrub margins; rare. Britain, C. and S. Germany, S. Europe (western).

See also *Epipactis atrorubens*, Pl.132, and *Platanthera bifolia*, Pl.133.

Willow Family, Salicaceae

See Creeping Willow, *Salix repens*, Pl.102.

Dock Family, Polygonaceae

See Sheep's Sorrel, *Rumex acetosella*, Pl.4.

Plate 49   Sandalwood Family, Santalaceae

1. Pyrenean Bastard Toadflax, *Thesium pyrenaicum* Pourr. (*T. pratense* Ehrh.), plant 10–40cm, similar to *T. alpinum*, but lvs weakly 3-nerved; inflorescence not secund; flowers 5-merous, perianth segments after flowering inrolled at the tips only; each flower with 3 minute bracts; branches spreading almost horizontally when fruiting. Fl.6–7. Acid grassland, dryish grassland; rather rare. C. and S. Europe. Protected!

2. Beaked Bastard Toadflax, *Thesium rostratum* Mert.et Koch, plant 20–30cm; stems many, erect or arched-ascending; ending with a tuft of lvs without flowers; lvs very narrow lanceolate, 1-nerved; inflorescence racemose; each flower with only 1 bract (only the case with this species and *T. ebracteatum*), 5-merous, white; perianth in fruit about twice as long as fruit, which is consequently beaked. Fl.5–7. Stony soils, open, calcareous pinewoods; rather rare. The Alps and their foothills, S. Germany.

3. Alpine Bastard Toadflax, *Thesium alpinum* L., plant 10–30cm, arched-ascending; scale lvs at stem base distant from each other; lvs narrow-lanceolate, 1–4mm wide, 1-nerved; inflorescence usually secund; each flower with 3 bracts; flowers usually 4-merous; perianth segments at fruiting time inrolled at the tip only, shorter than the fruit. Fl.6–7. Subalpine and alpine poor grassland, in the Alps up to about 2400m; rather rare. The mountains of Europe, northwards to S. Sweden. See also *T. bavarum*, Pl.139. Similar species: **Bastard Toadflax**, *Thesium humifusum* DC., stem 5–45cm, prostrate to spreading; inflorescence terminal, flowers 5-merous. Calcareous grassland; local. Britain, W. Europe. **Flax-leaved Bastard Toadflax**, *T. linophyllum* L., has likewise 1-nerved, linear-lanceolate, yellow-green lvs, 1–4mm wide, but perianth segments after flowering inrolled right to the base; flowers 5-merous; dry and stony grassland, dunes, rock outcrops; rare; C. and S. Europe (eastern). Protected!

Pink Family, Caryophyllaceae

4. Sticky Catchfly, *Lychnis viscaria* L. (*Viscaria vulgaris* Bernh.), plant 30–60cm, with basal rosette; stem very sticky between the nodes; lvs narrow lanceolate, dark green; inflorescence an interrupted panicle, almost whorled; petals obtuse or emarginate at tip, red, with coronal scales; styles 5, calyx glabrous, 10-ribbed. Fl.5–7. Poor grassland, heaths, open woods, on rocks and cliffs; calcifuge; local to rare. Britain, and almost all Europe, in the south only in the mountains.

5. Pearlwort Spurrey, *Spergula morisonii* Bor. (*S. vernalis* Willd.), lvs narrow-linear, 1–2cm long, not furrowed beneath (as is *S. arvensis*, Pl.9); petals broadly ovate, rather blunt, touching one another; stamens 10; seed with broad scarious wing. Fl.4–6. Sandy dry grassland, dry pinewoods; calcifuge; widespread, but a rare introduction in Britain. C. and S. Europe. See also *Stellaria graminea*, Pl.102.

6. Nottingham Catchfly, *Silene nutans* L., plant 30–60cm, with a non-flowering leaf rosette; lower lvs spathulate, upper lanceolate; inflorescence lax, many-flowered, secund, nodding; calyx with glandular hairs, 10-ribbed, calyx teeth $\frac{1}{3}$ as long as the calyx tube; petals 5, deeply bifid, white, with coronal scales 1–3mm long. Fl.5–8. Calcareous grassland, rocks, open oakwoods; rare in Britain but fairly common in most of Europe. See also *S. vulgaris*, Pl.7.

Plate 50   Pink Family, Caryophyllaceae

1. Rock Campion, *Silene rupestris* L., plant 10–25cm, glabrous, branched; lvs ovate-lanceolate, acute, sessile, bluish-green; inflorescence lax, dichotomously branched, with long-stalked, white to pinkish-red flowers; petals emarginate, twice as long as pale green, bluntly serrate sepals. Fl.7–8. Non-calcareous slopes; rare. European mountains.

2. Tunic Flower, *Petrorhagia saxifraga* (L.)Link (*Tunica s.* (L.)Scop.), plant 10–25cm, densely matted; lvs subulate, 1cm long; flowers solitary in lax, paniculate inflorescences; calyx narrowly campanulate, 4–6mm long, with an epicalyx of 4 hyaline scales; petals pale lilac to pink, with 3 dark veins. Fl.6–9. Dry grassland, stony slopes; rare. S. Germany, valleys of the C. and S. Alps, S. Europe. Rare introduction in Britain. Similar species: **Proliferous Pink**, *P. prolifera* (L.)Ball et Heyw., but flowers several together and enclosed in 6–8 hyaline bracts; calyx tubular, 10–13mm long, enclosed in 2 hyaline scales; calcareous grassland, dunes, gravelly places; scattered to rare; C. and S. Europe, isolated localities as far as S. Sweden, Britain.

3. Deptford Pink, *Dianthus armeria* L., plant 15–40cm, annual-biennial, shortly hairy above, rough; lvs narrowly lanceolate; flowers up to 1cm across, 2–20 together, clustered in a head-like inflorescence; calyx scales (bracts) hairy, green, almost as long as the calyx; petals a dirty red, toothed at tip. Fl.6–7. Dry grassland, paths, broom scrub, rather rare. Britain, C. and S. Europe.

4. Large Pink, *Dianthus superbus* L., plant 30–60cm, glabrous; lvs narrow-lanceolate, grass-green, 3–5mm wide; inflorescence lax, few-flowered; calyx scales oval, suddenly pointed, $\frac{1}{4}$–$\frac{1}{3}$ as long as the calyx; petals 15–35mm long, irregularly split to halfway or more, purple. Fl.6–10. Moorland pasture, acid, damp oakwoods; scattered to rare. Almost all of Europe. Protected! The **Sand Pink**, *D. arenarius* L., has likewise its petals split to halfway, but stems usually 1-flowered, petals white, lvs 1mm wide.

5. Carthusian Pink, *Dianthus carthusianorum* L., plant 15–40cm, glabrous; lvs narrowly lanceolate, 2–4mm wide; leaf sheaths about 4 times as long as the width of the lvs; flowers 4–10 together, 20–25mm across, surrounded by long-awned epicalyx scales, which are glabrous, hyaline, and shorter than the calyx; petals dark purple, toothed at tip. Fl.6–9. Calcareous grassland, sunny wood margins; fairly common. Principally C. Europe, but a rare escape in Britain.

6. Cheddar Pink, *Dianthus gratianopolitanus* Vill., plant 10–25cm, in cushions or densely tufted; lvs blue-green, rather blunt; stem often 1-flowered; calyx scales 4–6, about half as long as the calyx; petals hairy at the throat, pale purple. Fl.5–6. Protected! Rock communities, dry rocky outcrops; rare. Principally C. Europe.

7. Wood Pink, *Dianthus sylvestris* Wulf., plant 10–40cm, in loose cushiony mats; lvs dark green, channelled, 1–2mm wide; stem 1–4 flowered; calyx scales 2, oval, suddenly contracted into a short point; petals 8–15mm long, pink, toothed at tip. Fl.6–7. Stony slopes, dry grassland, alpine pastures; rare. Mountains of C. and S. Europe. Similar species: **Maiden Pink**, *D. deltoides* L., lvs and stem shortly pubescent; flowers 1–3 together, red, with white spots and a darker ring; calyx scales 2, oval, with short awns; lvs narrowly spathulate, blunt; siliceous grassland, sandy soils, heaths; local. Britain, N. and C. Europe, in the south only in the mountains.

8. Rock Soapwort, *Saponaria ocymoides*, see p.150.

For other species of this family, see Pls. 8 and 9.

Plate 51   Buttercup Family, Ranunculaceae

1. Bulbous Buttercup, *Ranunculus bulbosus* L., plant 10–40cm, usually spreading-hairy; stem with a bulbous base; basal lvs tripartite, segments 3-lobed and toothed again, almost to the middle; flowers 2–3cm across, yellow; sepals reflexed, long-hairy; peduncles furrowed; achenes with a short curved beak. Fl.5–7. Calcareous grassland, poor pasture, slopes; rather common. Europe. Likewise the **Woolly Buttercup**, **R. illyricus* L., has reflexed sepals and bulbous based stems, but peduncles terete, not furrowed; flowers 2–3cm across; lvs tripartite, with linear-lanceolate segments, both sides with dense, long, white hairs; achenes compressed and winged; dry grassland; rare; C. and S.E. Europe. The **Hairy Buttercup**, *R. sardous* Crantz, also has reflexed sepals, but is distinguished by the spreading-hairy stem, pale yellow flowers and very short, scarcely curved beaks to the achenes; loamy, damp fields, wet meadows; local to rare; Britain, S. Germany, S. Europe. Protected!

2. Yellow Pheasant's-eye, *Adonis vernalis L., plant 10–30cm; lvs almost sessile, repeatedly pinnate, densely set, with segments 1mm wide; flowers 4–8cm across; petals 10–20, golden-yellow. Fl.4–5. Protected! Dry grassland, stony pasture (steppes), pinewoods; rare. C., S. and E. Europe.

3. Eastern Pasqueflower, **Pulsatilla patens* (L.)Mill., plant 5–30cm; basal lvs tripartite, with 3-lobed or digitate-pinnate leaflets, white-villous when young; flowers blue-violet, finally spreading like a star, villous outside; petals much longer than the stamens; style enlarged after flowering, feathery. Fl.4–5. Protected! Poor pastures, heaths, pinewoods; rare. Sweden, Germany, E. Europe. Very similar to it is **Pasqueflower**, *P. vulgaris* Mill. (*Anemone pulsatilla* L.), plant 5–40cm, stem villous white-hairy, basal lvs repeatedly pinnate (but not digitately) with segments 100–150 per leaf, about 4mm wide (in ssp. *grandis* (Wend.)Zam., segments 50–75 per leaf and 7mm wide); flowers campanulate, pale violet; petals 4–5cm long, villous outside; dry grassland, pinewoods, on calcareous soils. Britain, Scandinavia, W., C. and E. Europe. Protected! The **Small Pasqueflower**, **P. pratensis* (L.)Mill., has nodding, campanulate flowers with purple or blackish-violet petals which are little longer than the stamens; dry sandy grassland, pinewoods; rare. Protected! C. and E. Europe. See also *P. vernalis*, Pl.184.

Also in this family, see *Thalictrum minus*, Pl.141.

Peony Family, Paeoniaceae

4. Peony, *Paeonia officinalis* L., plant 50–100cm, with tuberous fleshy roots; basal lvs absent at flowering time; stem lvs 2–3 pinnately cut to the base, 20–40cm long, with ovate-lanceolate segments, dark green above, grey-green below; petals 5–8, roundish, 4–8cm long, red; sepals broadly ovate to lanceolate, green to red; carpels 2–3, white-felted when ripe, 3–5cm long. Fl.5–6. Stony dry slopes, dry scrub. Very rarely naturalised in Britain; S. Alps, S. Europe. Cultivated plant.

---

8. Rock Soapwort, *Saponaria ocymoides* L. (Pl.50), plant 10–30cm; stem prostrate or ascending, shortly glandular-hairy; lvs spathulate, 1–3cm long; flowers shortly stalked, clustered at the ends of the branches; calyx not glandular-hairy; petals 12–18mm long, light red. Fl.5–6. Calcareous rubble, open pinewoods; local to rare. Mountains of C. and S. Europe (western); garden escape in Britain.

Plate 52 Fumitory Family, Fumariaceae

1. Yellow Corydalis, *Corydalis lutea* (L.)DC., plant 15–30cm, much-branched, densely leafy; lvs 2–3 pinnate, leaflets oval; flowers 1–2cm long, in 5–15 flowered racemes, yellow, spurred; sepals 4–6mm long, toothed. Fl.5–9. Cracks in walls, rocks, scree; cultivated. From S. Alps, otherwise naturalised.

Cabbage Family, Cruciferae or Brassicaceae

2. Burnt Candytuft, *Aethionema saxatile* (L.)R.Br., plant 5–20cm, ascending to erect; lvs ovate to lanceolate, 1–2cm long, with bluish pruina, entire; petals flesh-red or white, 2–4mm long; sepals with hyaline margins; siliculae lens-shaped, compressed, broadly winged, on curved spreading stalks, but sometimes with 1-seeded fruits on straight, erect stalks. Fl.4–6. Sunny screes, river shingle, rocks; rather rare. Alps (up to 1900m) and foothills, S. Europe.

3. Buckler Mustard, *Biscutella laevigata* L., plant 15–30cm; basal lvs cuneate, oblong, narrowed into the petiole, entire or serrate, stiffly hairy or glabrous; stem lvs few; flowers 4-merous; sepals greenish-yellow; petals yellow, 4–8mm long; siliculae shaped like spectacles. Fl.5–8. Stony rubble, dry sandy grassland, stony slopes, river shingle of alpine rivers, moorland flushes; widespread. C. and S. Europe.

4. Small Alison, *Alyssum alyssoides* (L.)Nath. (*A. calycinum* L.), plant 5–20cm, annual; lvs oblong-ovate, narrowed into a short stalk, 2cm long, grey with stellate hairs on both sides; flowers small; sepals 1.5–2.5mm long, stellate-hairy; petals 2–4mm long, yellow, white after flowering; siliculae almost circular, 3–6mm, flat, hairy, 4-seeded. Fl.4–9. Sunny calcareous grassland, stone quarries, sandy fallow fields; widespread. Rare in Britain, principally C. and S. Europe.

5. Golden Alison, *Alyssum saxatile* L., plant 10–30cm, woody at the base; lvs broadly lanceolate, distantly toothed, grey-felted, up to 5cm long; inflorescence a panicle; sepals soon falling; petals golden-yellow, 4–6mm long, emarginate; siliculae almost circular, flat, 4–5mm, on a stalk 6–10mm long. Fl.3–5. Sunny rock outcrops, either acidic or basic, stony soils; rather rare. Cultivated and sometimes escaping, but wild in S.E. Europe, northwards to Saxony, Jura, Austria.

6. Common Whitlow-grass, *Erophila verna* (L.)Bess., plant 3–15cm; lvs elliptic, cuneate, entire or distantly toothed, stellate-hairy; stem leafless; sepals with scarious margins; petals white or reddish, deeply emarginate; siliculae oblong-elliptic, 5–10mm long. Fl.3–5. Open, poor grassland, sandy arable fields, paths, gravelly soils; common. Europe. The **Round-podded Whitlow-grass**, *E. spathulata* Lang, has almost circular siliculae, and the **Early Whitlow-grass**, *E. praecox* (Stev.)DC., broadly elliptic, weakly pointed siliculae; both kinds on sandy grassland and rather rare.

7. Tall Rock-cress, *Cardaminopsis arenosa* (L.)Hayek, plant 10–40cm, with spreading hairs below, with a basal leaf rosette; lvs usually pinnately lobed or coarsely toothed, upper stem lvs lanceolate, serrate or entire; petals 6–8mm long, white, sometimes pale lilac; siliculae 2–4cm long, 1mm wide, distinctly compressed, on stalks 1–2cm long. Fl.4–8. Dry sandy grassland, railway banks, drained moorland, rubble; widespread. Originally from E. Europe, but spreading in Europe.

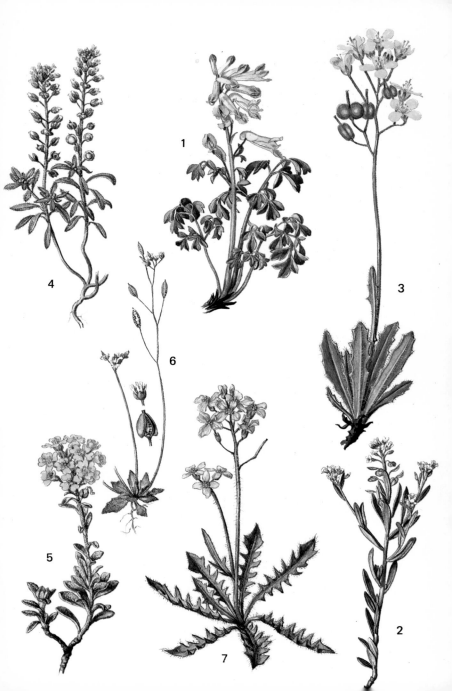

Plate 53   Cabbage Family, Cruciferae or Brassicaceae

1. Hairy Rock-cress, *Arabis hirsuta* L., plant 10–80cm; stem densely leafy; lvs of basal rosette ovate, narrowed into the short stalk, toothed or entire; stem lvs ovate-lanceolate, with cordate base, sessile, or sagittate and amplexicaul, usually very hairy; inflorescence many-flowered, elongated after flowering; petals 4–6mm long, white; siliquae 15–35mm long, erect, crowded, valves with midrib. Fl.5–7. (Variable species with several microspecies.) Calcareous grassland, limestone rocks, walls, open pinewoods; widespread. Almost all of Europe. The **Annual Rock-cress**, **A. recta* Vill. (*A. auriculata* auct.), has distant, spreading siliquae, and ovate, amplexicaul stem lvs; dry grassland, stony slopes; likes warm places; rather rare; S. and C. Germany, S. Europe.

For other species of this family, see Pls. 12, 14, 16 and 17.

Stonecrop Family, Crassulaceae

2. House-leek, *Sempervivum tectorum* L., plant 15–50cm, with a basal rosette 8–12cm across; lvs sharply pointed, bluish-green, often reddish-tinged, glabrous, ciliate on the margins; flowers pale red, 2–3cm across, with 10–16 petals. Fl.7–8. Protected! Walltops, old roofs, quarries; likes warm places; widespread. Britain, C. and S. Europe.

3. Reflexed Stonecrop, *Sedum reflexum* L. (*S. rupestre* L.), plant 15–35cm, woody at base, decumbent-ascending, non-flowering shoots densely leafy; lvs terete, 1–2cm long, apiculate, spurred at base; flowering shoots not so leafy; inflorescence cymose; petals 6–7mm long, lemon-yellow; sepals 3–4mm long, glabrous. Fl.6–8. Dry sandy grassland, walls, rock outcrops; widespread. Almost all of Europe. Similar species: **Creamy Stonecrop**, **S. ochroleucum* Chaix, but petals yellowish-white, 8–10mm long and sepals 5–7mm long, glandular-hairy; limestone rocks, vineyards, rare; mainly S. Europe.

4. White Stonecrop, *Sedum album* L., plant 8–20cm, loosely matted, with erect, flowering stems and creeping non-flowering stems; lvs green or reddish, cylindrical; inflorescence usually glabrous; petals 3–5mm long, blunt, white, often pink outside. Fl.6–7. Sunny, stony grassland, on rocks, walls, dunes and shingle; common. Britain, S. Scandinavia, C. and S. Europe. The **Thick-leaved Stonecrop**, *S. dasyphyllum* L., has bluish-green, ovoid lvs, 3–7mm long, flat above and curved beneath, and a glandular-hairy inflorescence; sunny rock outcrops, walls, rubble; rather rare; naturalised in Britain, native in the Alps, S. Europe.

5. Biting Stonecrop, *Sedum acre* L., plant 5–15cm, much branched, decumbent to erect, forming mats; lvs thick, ovoid, rounded at the base, without spur, 4mm long and 3mm wide, obtuse, in 6 rows on the stems, usually acrid-tasting; petals 6–7mm long, finely pointed, golden-yellow. Fl.6–7. Dry, sandy grassland, rock outcrops, walls, dunes, shingle, dry pinewoods; common. Europe. Similar species: **Tasteless Stonecrop**, *S. sexangulare* L., but lvs cylindrical, linear, spurred at base, never acrid-tasting; petals 3–5mm long; dry sandy grassland, walls; scattered; Britain, S. Scandinavia, C. and S. Europe.

Plate 54 Stonecrop Family, Crassulaceae

1. Roseroot, *Rhodiola rosea* L. (*Sedum rosea* (L.) Scop.), plant 10–35cm; stock with a scent of roses; stem erect, thick, fleshy, with lvs alternate, flat, lanceolate, bluish-green and toothed at the tip; flowers in dense cymes; males and female flowers on different plants; male flowers with 8 stamens and 4 yellowish or reddish petals; female flowers with rudimentary petals. Fl.6–8. Crevices of non-calcareous rocks, cliffs, mountain grassland communities; local to rare. Britain, Black Forest, Vosges, the Alps, Pyrenees, Apennines, mountains of the Balkan peninsula, Scandinavia.

2. Orpine, *Sedum telephium* L., plant glabrous, 20–40cm, with carrot-like rootstock; stem erect, thick, not rooting; lvs usually alternate, oblong-ovate, toothed, with a cuneate base, upper lvs sessile with a rounded base; flowers 5-merous, hermaphrodite, in dense, many-flowered cymose inflorescences; petals 4–5mm long, red, spreading. Fl.7–9. Stony banks, scrub margins, hedgebanks, paths, arable fields; scattered. Britain, N. and C. Europe, southwards to N. Italy and N. Spain. Similar species: **Greater Orpine**, **S. telephium* ssp. *maximum* (L.)Krocker (*S. maximum* (L.)Hoffm.), but lvs opposite or in whorls of 3, sessile with a cordate or rounded, not cuneate base; flowers pale greenish-yellow; stony soils, walls, waste places, sandy dry grassland; scattered; C. Europe, S. Scandinavia, southwards to the Pyrenees.

3. English Stonecrop, *Sedum anglicum* Huds., plant 5–15cm, grey-green, often reddish-tinged, loosely matted, with erect flowering stems and creeping non-flowering stems; lvs terete, 3–5mm long, spurred at the base, alternate; flowers 5-merous; petals 3–5mm long, pointed, pink or white; stamens 10, with blackish anthers. Fl.6–9. Siliceous rocks, non-calcareous sandy soils, dunes and shingle. Britain, W. Europe, S. Scandinavia.

4. Annual Stonecrop, *Sedum annuum L., plant biennial, glabrous, only with flowering stems, 5–15cm; lvs linear, thick but flattened on both surfaces, shortly spurred at base, not dense on the stem; flowers 5-merous, yellow, in spike-like inflorescences; petals twice as long as the sepals, keeled, acute. Fl.6–7. On rocks, walls, screes; colonist of bare places; rare. Principally in the European mountains. Protected!

5. Navelwort, *Umbilicus rupestris* (Salisb.) Dandy (*U. pendulinus* DC.), plant glabrous, 10–40cm, with tuberous, thickened roots; lvs fleshy, roundish to ovate, 2–6cm across, distantly toothed; leaf stalks attached in about the centre of the blade; corolla tubular, drooping, white or yellowish, rarely reddish, in long, many-flowered racemes; petals 7–10mm long, about $\frac{3}{4}$ united; tube 3–4 times as long as the calyx, whose lobes are united at the base. Fl.5–8. Cracks in rocks, walls, shady slopes; fairly common. Britain, S. and W. Europe, W. Alps.

Plate 55   Saxifrage Family, Saxifragaceae

1. Orange Saxifrage, *Saxifraga mutata* L., plant 15–30cm, with a large basal rosette; lvs tongue-shaped, 3–5cm long, almost entire; stem lvs alternate, glandular; inflorescence paniculate-branched, many-flowered; flowers reddish-yellow. Fl.6–7. Protected! Streambanks, damp gravel, rock faces with trickling water; rather rare. Alps and hinterlands.

2. Meadow Saxifrage, *Saxifraga granulata* L., plant 15–40cm, annual, with underground bulbils at the base; basal lvs in a rosette, roundish kidney-shaped, deeply crenate, shrivelled up by flowering time; stem branched, glandular-hairy; inflorescence a lax panicle; petals 10–15mm long, white. Fl.3–5. Paths, banks, gravel pits, dry grassland; local. Almost all of Europe. Similar species: **Bulbous Saxifrage**, *S. bulbifera* L., but stem lvs with bulbils in the axils; stem only branched above; inflorescence with short branches, narrowly contracted; petals 6–10mm long; dry, sandy soils, woods of Downy Oak; mainly S.E. Europe, S. Alps. The **Rue-leaved Saxifrage**, *S. tridactylites* L., has 3-lobed, spathulate lvs, usually withered by flowering time, and white petals, 3–4mm; plant 3–15cm; dry sandy grassland, walls, paths, rooftops; almost all of Europe.

Rose Family, Rosaceae

3. Hoary Cinquefoil, *Potentilla argentea* L., plant 10–40cm, erect or prostrate; lvs digitately 5–7 lobed, white-felted below, lobes inrolled at the margins, coarsely and deeply toothed; inflorescence paniculate; flowers 1–1.5cm across, petals yellow, obtuse or emarginate, slightly longer than the lanceolate sepals. Fl.6–8. Siliceous grassland, sandy soils, paths, banks; rather common, or local. Almost all of Europe.

4. Spring Cinquefoil, *Potentilla tabernaemontani* Aschers. (*P. verna* auct.), plant 5–15cm, with a thick, branched stock and forming a mat; lvs digitately 5–7 lobed, green on both sides, hairy, stipules of the basal lvs linear, those of the stem lvs ovate-lanceolate; flowers yellow, 1–2cm across; style thickened above, club-shaped. Fl.4–6. Dry, sandy grassland, stony slopes, rock outcrops, dry pinewoods, local. C. Europe, northwards to Britain, S. Scandinavia and southwards to S. France.

5. Sulphur Cinquefoil, *Potentilla recta* L., plant 30–70cm, erect, branched above; basal lvs palmately 5–7 lobed, green on both sides, often with long hairs, deeply serrate, on stalks 5–20cm long; stem and leaf stalks with long, soft hairs and short, bristly hairs; flowers 20–25mm across. Fl.6–7. Dry grassland, paths, gravel pits, sandy grassland, waste places; scattered to rare. Sometimes naturalised in Britain; native in C. and S.E. Europe.

Plate 56   Rose Family, Rosaceae

1. Shrubby Cinquefoil, *Potentilla fruticosa* L., dwarf shrub, 20–100cm; lvs 3–7 (usually 5) lobed, lobes ovate-oblong, entire, densely silky-hairy beneath; petals 5, golden yellow, roundish, 10–12mm long. Fl.6–8. Mountain crags and damp, stony places liable to flooding; rare as a wild plant, often planted in gardens and sometimes naturalised; Britain, Scandinavia, Pyrenees, S. Europe.

2. Creamy Strawberry, *Fragaria viridis* Duch., plant 5–15cm; lvs trifoliate, serrate, middle leaflet stalked, oval, the lateral sessile, green above, sparsely hairy, densely silvery-hairy below; leaf stalks and stem with erect to spreading hairs; flowers 1–1.5cm across, yellowish-white; calyx appressed to the berry, which is yellowish, rarely red, when ripe; calyx comes away with the fruit when picked (not so in the Wild Strawberry, which has a spreading calyx). Fl.5–6. Dryish grassland, scrub margins, open oak and pine woods; scattered. Mainly C. and S. Europe, northwards to S. Scandinavia.

3. Salad Burnet, *Sanguisorba minor* Scop. (*Poterium sanguisorba* L.), plant 30–60cm, with a basal leaf rosette; lvs pinnate, with 11–31 oval leaflets, which have 4–8 acute teeth on each side; flower head spherical, greenish, 1–1.5cm across, the lower flowers male, the middle hermaphrodite, and the upper female; filaments loosely hanging, 3–5 times as long as the calyx. Fl.5–8. Calcareous grassland, paths, slopes, sometimes on neutral soil; rather common. Almost all of Europe, northwards to S. Sweden, Britain.

4. Dropwort, *Filipendula vulgaris* Moench (*F. hexapetala* Gilib.), plant 30–80cm; root tuberous, thickened; lvs pinnate, leaflets 10–40 on each side, oblong, 1–2cm long, coarsely double-serrate or pinnatisect; inflorescence many-flowered, panicled; petals usually 6, oval, 5–10mm long, white to pale pink. Fl.5–7. Calcareous grassland, scrub margins, woodland edges, open oak and pine woods; scattered. Almost all of Europe, northwards to Britain, S. Scandinavia. See also *Agrimonia*, Pl.149.

Pea Family, Fabaceae or Papilionaceae

5. Eastern Dwarf Broom, *Chamaecytisus ratisbonensis* (Schaef.)Rothm. (*Cytisus ratisbonensis* Schaef.), plant 10–30cm; branches prostrate, ascending at tip, appressed silky-hairy; lvs trifoliate, stalked; flowers usually in pairs, axillary, shortly stalked; calyx tubular, twice as long as wide; corolla 1–2cm long, golden-yellow, standard red-brown. Fl.5–6. Dry grassland, wood margins, open, dry pinewoods; rather rare. C. and S. Europe (eastern). Similar species: **Clustered Broom**, *C. supinus* (L.)Link (*Cytisus capitatus* Scop.), but plant 20–50cm, villous-hairy; flowers in terminal heads and solitary, laterally; standard of the flower usually red-brown; sunny wood margins, dry grassland; rare. C. and S. Europe, northwards to C. France and S. Germany.

6. German Greenweed, *Genista germanica* L., plant 20–50cm; older branches leafless, thorny, prostrate to ascending; young twigs thornless, leafy, erect, spreading-hairy, with many-flowered, terminal racemes; lvs simple, lanceolate, 1–1.5cm long, grass-green, without stipules; calyx hairy; corolla 7–11mm long, standard glabrous, shorter than the shortly hairy keel; pod 7–12mm long, hairy. Fl.5–6. Heaths, wood margins, slopes; scattered. C. and E. Europe, northwards to S. Sweden, southwards to N. Italy. The **Petty Whin**, *G. anglica* L., has few-flowered, terminal and lateral racemes, glabrous flowering branches, calyx and pod; lvs grey-green; twigs thorny; dry acid grassland, heaths, moors, pinewoods; calcifuge; scattered to rare; principally W. Europe, Britain, N.W. Germany, Black Forest, northwards to S. Sweden.

Plate 57 Pea Family, Fabaceae or Papilionaceae

1. Dyer's Greenweed, *Genista tinctoria* L., plant 30–60cm, erect, thornless; lvs lanceolate, 1–4cm long, with 2 small, narrow lanceolate stipules at the base; flowers in terminal racemes, yellow; corolla 8–16mm long, glabrous; pod 2–3cm long, glabrous. Fl.6–8. Rough pastures, wood margins, waysides, open oakwoods; local to fairly common. C. and S. Europe, northwards to Britain, S. Sweden. The **Hairy Greenweed**, *G. pilosa* L., is likewise thornless, but flowers 1–2 together, axillary; underside of the lvs, pod and outside of corolla silky-hairy; plant 10–30cm, prostrate; heaths, dry, sandy grassland, clifftops; calcifuge; scattered to rare; W. and C. Europe, northwards to Britain, S. Sweden, southwards to N. Spain, C. Italy.

2. Winged Broom, **Chamaespartium sagittale* (L.)P.Gibbs (*Genista sagittalis* L.), plant 10–30cm, with creeping, woody stem, and ascending, usually simple, young branches; stem and branches broadly winged, spreading-hairy; lvs small, falling away; flowers in short, dense racemes; calyx spreading-hairy, with 2-lobed upper lip and 3-lobed lower lip; corolla 10–15mm long, yellow. Fl.5–6. Poor grassland, wood margins, waysides, stony slopes; widespread. C. and S. Europe.

3. Spiny Restharrow, *Ononis spinosa* L., plant 20–50cm, woody at the base, with thorny twigs; stem with 1 or 2 rows of hairs; lvs usually trifoliate, glabrous, leaflets ovate, serrate, the terminal leaflet longer-stalked; flowers shortly stalked, in the axils of the upper stem lvs; corolla 8–25mm long, pink to violet-red; pod as long as the calyx or longer. Fl.6–8. Calcareous grassland, paths, moorland pastures; local to common. C. Europe, northwards to Britain, S. Scandinavia, southwards to the Pyrenees, C. Italy. Similar species: **Common Restharrow**, *O. repens* L., but stem glandular-villous all the way round, often thornless; pod shorter than the calyx; leaflets glandular-hairy, rounded at tip; poor grassland and pasture, moorland grazing; widespread; almost all of Europe (western). Very rare is the **Large Yellow Restharrow**, **O. natrix* L., with stalked, yellow flowers; plant 20–50cm, glandular-hairy; sunny calcareous grasslands. S. and W. Europe, S. Alps. Protected!

4. Lucerne, *Medicago sativa* L., plant 30–80cm, erect, almost glabrous; lvs trifoliate, leaflets 2–3cm long, serrate towards the tip and mucronate; flowers in head-like racemes, blue or violet, 8–12mm long; pod in a lax spiral of 2–3 turns. Fl.6–9. Dry grassland, paths, slopes, cultivated as fodder plant, naturalised; common. Europe.

5. Sickle Medick, *Medicago sativa* ssp. *falcata* (L.)Arc. (*M. falcata* L.), plant 20–50cm, prostrate to ascending; lvs trifoliate; leaflets mucronate in the emarginate tip; flowers yellow, about 1cm long, in almost spherical racemes; pod sickle-shaped, 8–15mm long. Fl.6–9. Dry grassland, calcareous grassland, dry wood margins and waysides; scattered or rare. Almost all of Europe, northwards to Britain, S. Scandinavia.

6. Black Medick, *Medicago lupulina* L., plant 10–40cm; lvs trifoliate, appressed-hairy below; leaflets with mucronate tip (in contrast to *Trifolium dubium*, Pl.81); flower heads 10–50 flowered; flowers 3–5mm long; pod kidney-shaped, without spines, 2–3mm long. Fl.5–10. Dry grassland, meadows, waysides, arable fields; widespread to common. Europe. Similar species: **Spotted Medick**, *Medicago arabica* (L.) Huds., but leaflets usually black-blotched; flowers 1–4 to a head, 4–6mm long; pod in a spiral, with hooked spines. Waste or grassy places, likes sandy or gravelly soil; widespread. Britain, France, Germany and S. Europe.

Plate 58   Pea Family, Fabaceae or Papilionaceae

1. Bur Medick, *Medicago minima* (L.) Burtal., plant 5–25cm, prostrate to ascending; lvs trifoliate, silky-hairy; stipules entire or with short teeth; flowers 3–5mm long, yellow, in 3–8 flowered heads; pods coiled up like snails' shells, very spiny, with 2–6 narrow turns, 3–5mm across (without spines). Fl.5–6. Calcareous grassland, sand dunes, sheep pasture, paths, heaths; rather rare. Britain, C. and S. Europe.

2. Kidney Vetch, *Anthyllis vulneraria* L., plant 15–30cm, with taproot; lvs pinnate, terminal leaflet larger, lateral oval to lanceolate; basal lvs often only with terminal leaflet; inflorescence dense, head-like; corolla 1–2cm long, pale yellow to orange-yellow. Fl.5–8. Many subspecies: in ssp. **alpestris* (Kit.)A.et Gr. (2a), calyx 13–16mm long, villous-hairy; corolla golden-yellow; basal lvs with only a large, elliptic terminal leaflet, glabrous, fleshy. In ssp. **maritima* Schweigg., calyx 8–11mm long, corolla orange-red, stem much-branched, densely silky-hairy; lvs glabrous above, grey silky-hairy to felted below; stem lvs with 3–5 pairs of linear-lanceolate leaflets. Ssp. *vulneraria* (2b), stem sparingly branched, shortly appressed-hairy; lvs hairy on both sides, stem and basal lvs with an enlarged terminal leaflet and 3–5 pairs of elliptic lateral leaflets; corolla shining yellow. Calcareous grassland, paths, quarries; stony alpine grassland (ssp. *alpestris*); sand dunes by the North Sea and the Baltic (ssp. *maritima*); fairly common.

3. Common Bird's-foot Trefoil, *Lotus corniculatus* L., plant 5–30cm, curved-ascending; stem pithy or with a narrow hollow; lvs with 5 oval leaflets, the lowest leaflet pair close to the stem and distant from the other 3; flowers 6–14mm long, in 3–8 flowered, cymose heads, yellow; keel with a right-angled bend at the tip, often red-tipped. Fl.5–8. Meadows, pastures, dry grassland, quarries; widespread. Europe. Similar species: **Greater Bird's-foot Trefoil**, *Lotus uliginosus* Schk.(*L. pedunculatus* Cav.), but stem with a broad hollow; cymes 8–12 flowered; keel with an obtuse-angled bend; wet meadows, river banks, ditches; widespread. Europe.

4. Hop Trefoil, *Trifolium campestre* Schreb., plant 10–20cm; lvs trifoliate, middle leaflet with a longer stalk than the side ones; stipules ovate, widened at the base, shorter than the petiole; flowering head dense, 20–40 flowered, 7–10mm across; flowers 5mm long, pale yellow, withering to pale brown. Fl.6–9. Calcareous grassland, sandy fields, paths, arable land, roadsides; rather common. Europe. Similar species: **Large Hop Trefoil**, *T. aureum* Poll., but flower heads 10–15mm across; all 3 leaflets with very short stalks of the same length; stipules oblong-lanceolate, as long as the petiole, not widened at the base; non-calcareous dry grassland, waysides; waste places; scattered. Europe, naturalised in Britain.

5. Mountain Clover, **Trifolium montanum* L., plant 15–40cm, erect or ascending, densely hairy; lvs trifoliate, leaflets finely serrate, softly hairy beneath; inflorescence ovoid to spherical, 1–2cm long; flowers shortly stalked; calyx tube 10-nerved, hairy, with narrow, equal teeth; corolla 7–10mm long, white. Fl.5–7. Calcareous grassland, wood margins; scattered. Europe, north to S. Scandinavia, in mountains in the south.

6. Hare's-foot Clover, *Trifolium arvense* L., plant 5–30cm, erect, densely hairy; lvs trifoliate, leaflets oblong, grey-green, often reddish-tinged; flower heads ovoid to cylindrical, densely villous; calyx longer than the corolla, which is whitish at first, then reddish. Fl.6–9. Thin dry grassland, sandy fields, fallow land, dunes; fairly common. Originally from S. Europe, nowadays spread over almost all of Europe.

See also *Trifolium* and *Dorycnium*, Pl.152.

Plate 59   Pea Family, Fabaceae or Papilionaceae

1. Large Brown Clover, *Trifolium spadiceum* L., plant erect or ascending, 10–30cm; upper stem lvs almost opposite; lvs trifoliate, leaflets sessile or equally short-stalked; stipules oblong-lanceolate; flower heads becoming cylindrical, 6–25mm long and 10–13mm across, yellow to yellow-brown, withering from the top to blackish-brown; flower pedicel much shorter than the calyx tube; corolla 5–6mm long. Fl.6–8. Moorland pastures, paths, edges of ditches, poor alpine meadows; scattered. Almost all of Europe, principally in the mountains. Similar species: **Brown Clover**, *T. badium* Schreb., but plant 10–20cm; stipules ovate-lanceolate; flower heads hemispherical to ovoid, 13–20mm across, withering from the top to a tan colour; corolla 6–9mm long; pedicel as long as the calyx tube; lush pasture in the Alps and the other higher mountains of C. and S. Europe.

2. Sulphur Clover, *Trifolium ochroleucon* Huds., plant ascending, 20–40cm; stem and lvs densely hairy; lvs trifoliate; leaflets oblong-elliptic, 1–3cm long; stipules with a long subulate tip; flower heads solitary, globose or ovoid, 15–30mm long, shortly stalked; calyx tube 10-nerved, hairy; calyx teeth unequal; corolla 13–18mm long, yellowish white to pale yellow. Fl.6–7. Poor grassland, dry moorland pasture, wood margins; rare. Protected! C. and S. Europe, northwards to England, S. Poland and S. Russia.

3. Fenugreek, *Trifolium ornithopodioides* L. (*Trigonella o*. (L.)DC.), plant prostrate to ascending, glabrous, 5–10cm; lvs trifoliate, leaflets obovate or cordate, 4–10mm long, shortly stalked, finely serrate; flower heads 2–4 flowered; corolla 6–8mm long, pink or white; calyx teeth somewhat unequal, longer than the calyx tube; pods 5–10 seeded, 6–8mm long. Fl.6–7. Dry, sandy grassland, dunes. N.W. Germany, England, Ireland, Holland, Italy, W. Europe.

4. Gorse, *Ulex europaeus* L., densely spiny shrub, 60–120cm; twigs spreading-squarrose, densely spiny, spreading-hairy; lvs spine-like, prickly, subulate, 5–10mm long; flowers in the axils of thorn-leaves; corolla yellow, 15–20mm long, much longer than the calyx, which is densely spreading-hairy; pod 15–20mm long, densely hairy. Fl.4–6. Fallow land, wood margins, paths, heaths, non-calcareous sandy soils with a mild winter climate. W. Europe, Britain, S. Scandinavia, in C. Europe only in protected spots and usually planted and naturalised. Similar species: **Dwarf Gorse**, *U. minor* Roth, but shrub only 20–70cm; corolla 6–9mm long, about as long as the calyx; pods 8–10mm long; Britain, W. Europe.

5. Crown Vetch, *Coronilla varia* L., plant 30–80cm, prostrate to ascending; lvs pinnate, shortly stalked, with 11–23 oval leaflets; flowers in 10–20 flowered umbels; corolla 8–15mm long, white with a reddish standard and a violet tip to the keel; pod spreading-erect, straight, 2–8cm long. Fl.6–8. Dryish grassland, wood and scrub margins, slopes; scattered. C. and S. Europe (eastern), naturalised in Britain.

For more species of this family, see Pl.19.

Plate 60   Pea Family, Fabaceae or Papilionaceae

1. Dragon's-teeth, *Tetragonolobus maritimus* (L.)Roth (*Lotus siliquosus* L.), plant 10–25cm, prostrate; lvs with 5 leaflets, of which the lower 2 are close to the stem and somewhat smaller; leaflets grey-green, rather fleshy; flowers 1 or 2 together in the leaf axils, long-stalked, pale yellow, 2–3cm long; pod 4-angled or 4-winged, 4–5cm long and 3mm wide. Fl.5–6. Calcareous grassland, open pinewoods, upland pasture; salt tolerant; scattered. C. and S. Europe, occasionally northwards to Britain, S. Scandinavia.

2. Stemless Milk-vetch, **Astragalus exscapus* L., plant 3–8cm, villous-hairy, almost stemless; lvs and inflorescence basal; lvs with 25–39 ovate, rounded or emarginate, densely hairy leaflets; flowers 3–9 together; corolla 20–25mm long, yellow; pod ovoid. Fl.6–7. Dry grassland, stony steppes, on calcareous soil or gypsum; rare. Thuringia, C. and S. Alps, E. and S. Europe. Likewise with a very much shortened stem is **Sprawling Milk-vetch**, **A. depressus* L., with lvs glabrous above and appressed-hairy below; leaflets 17–25; inflorescence 6–14 flowered, with a peduncle of about ⅓ the length of the subtending leaf; corolla 8–12mm long, yellowish; Alps (Switzerland, Italy), Pyrenees, Apennines, mountains of the Balkan peninsula. See also *Astragalus glycophyllus*, Pl.154.

3. Hairy Milk-vetch, **Oxytropis pilosa* (L.)DC., plant 10–30cm, spreading villous-hairy; lvs with 19–27 lanceolate leaflets, 5–20mm long, glabrous or appressed-hairy below; flowers 5–25 together in erect racemes; calyx teeth almost as long as the calyx tube; corolla 10–12mm long, pale yellow, keel with a tooth at the tip; pods linear, shortly hairy. Fl.6–7. Sunny steppes, stony slopes, open Scots Pine woods; rare. S. Germany, Thuringia, S. Sweden, Alps, Apennines, E. Europe. Protected!

4. Horseshoe Vetch, *Hippocrepis comosa* L., plant 5–20cm; lvs long-stalked, with 5–15 oval to narrow-lanceolate leaflets; umbel 4–10 flowered; corolla yellow, 8–12mm long; pod 1–3cm long, spreading or somewhat pendent, with horseshoe-shaped segments. Fl.5–7. Sunny calcareous grassland, paths, quarries, open pinewoods, cliffs; local to fairly common. Britain, C. and S. Europe, Alps.

5. Sainfoin, *Onobrychis viciifolia* Scop., plant 30–60cm, curved ascending; lvs pinnate, with 13–27 ovate-oblong leaflets, 3–8mm wide; inflorescence oblong-ovoid before flowering; pedicels 1mm long; calyx 5–8mm long, teeth 2–4 times as long as the calyx tube; corolla 10–14mm long, red; keel almost as long as the standard; pod with 6–8 teeth of 1mm long, net-veined and tubercled. Fl.5–7. Calcareous grassland, paths; cultivated plant; local to common. Originally from S.E. Europe, nowadays in almost all of Europe. Similar species: **Small Sainfoin**, **O. arenaria* (Kit.)DC., but plant 10–30cm, leaflets 2–5mm wide; corolla 8–10mm long, flesh-coloured; pod with 4–5 slender spines; dry grassland, pinewoods; rare. C. Germany, S. and E. Europe.

Plate 61 Flax Family, Linaceae

1. Fairy Flax, *Linum catharticum* L., plant 5–30cm, erect, dichotomously branched; lvs opposite, ovate-lanceolate, 1cm long, very finely serrate; flower clusters drooping; sepals 4–5, 2–3mm long, glandular-hairy on the margin, 1-nerved; petals 4–5, 3–6mm long, white, yellow at base. Fl.6–7. Calcareous grassland, upland pasture, heaths, rock ledges and dunes; common. Europe, in the south only in the mountains.

2. Sticky Flax, **Linum viscosum* L., plant 30–60cm; stem spreading-hairy; lvs alternate, ovate-lanceolate, 4–9mm wide, glandular-ciliate; petals 1–2cm long, pink; sepals glandular on the margins. Fl.5–7. Calcareous grassland, open pinewoods, wood margins; rare. S. Germany, Alps and subalpine region, S. Europe. Similar species: **Narrow-leaved Flax**, **L. tenuifolium* L., but lvs 1–2mm wide, rough, not glandular; lvs and stem glabrous; calcareous grassland, likes warm places; S. and C. Germany, S. Europe; protected! The **Yellow Flax**, **L. flavum* L., is easily distinguished by its yellow flowers; stem sharp-angled, plant glabrous; dry grassland, on calcareous soils; rare; S. Germany, S.E. Europe. Protected!

3. Perennial Flax, *Linum perenne* L., plant 20–80cm, with several stems; lvs lanceolate, 1–2mm wide, 1–3 nerved, rough on the margin; flowers pale blue; petals 1–2cm long, imbricate (in ssp. *alpinum* (Jacq.)Ockenden, only overlapping at the base), inner sepals longer and wider than the outer. Fl.6–8. Dry grassland, dry pinewoods; rare. Britain, S. Germany, S. Europe. Protected!

4. Common Flax, *Linum usitatissimum* L., plant 30–60cm, annual, with only 1 stem; lvs narrow lanceolate, 3–4mm wide; petals 12–15mm long, blue; sepals 5–7mm long, finely pointed, finely ciliate, but not glandular. Fl.6–7. Cultivated for linen and linseed oil since the New Stone Age, occasionally naturalised.

Milkwort Family, Polygalaceae

5. Shrubby Milkwort, **Polygala chamaebuxus* L., plant 10–20cm, prostrate to ascending, woody below; lvs leathery, evergreen, pale green below; flowers 13–15mm long, like pea flowers, 1–2 together in the axils of the upper lvs, yellow, often red-tinged; sepals coloured, petal-like, the 2 inner (wings) large, 10–15mm long; petals 5, the lower (keel) large, with a fringed appendage. Fl.4–6. Calcareous grassland, dwarf shrub communities, dry pinewoods; scattered. The mountains of C. and S. Europe.

6. Tufted Milkwort, **Polygala comosa* Schk., plant 10–25cm, lvs herbaceous, alternate, ovate-oblong, 10–25mm long; inflorescence 5–20 flowered; flowers lilac or reddish, wings faintly net-veined, 4–7mm long; bracts longer than the pedicels, exceeding the flower buds, hence the plant appears tufted. Fl.5–6. Sunny calcareous grassland; scattered. C. and S. Europe (eastern), northwards to S. Sweden.

7. Common Milkwort, *Polygala vulgaris* L., plant 10–30cm; lower lvs alternate, not forming a basal rosette; inflorescence 10–20 flowered; bracts usually shorter than the pedicels and so not exceeding the flower buds; flowers blue (less often pink or white), wings 6–10mm long, net-veined. Fl.5–6. Siliceous grassland, heaths, dunes, waysides; widespread to common. Britain; Europe, in the south only in the mountains. Very similar to it is **Heath Milkwort**, *P. serpyllifolia* Hose, but the lower lvs are opposite.

Plate 62   Milkwort Family, Polygalaceae

1. Dwarf Milkwort, *Polygala amara* L., plant 10–20cm, prostrate, then ascending; lvs bitter-tasting, arranged in a rosette at the base, smaller above; inflorescence lax; flowers blue to reddish-blue, 3–7mm long, wings 4–8mm long, keel (lowest petal) with a markedly fringed, lobed appendage, keel much constricted at the juncture with the appendage. Fl.5–6. Subalpine and alpine, stony grassland, damp upland pasture, flushes; scattered to rare. Britain; mountains of C. and S. Europe (eastern).

Crane's-bill Family, Geraniaceae. See *Geranium, Erodium*, Pls.20 and 156.

Spurge Family, Euphorbiaceae

2. Warty Spurge, **Euphorbia brittingeri* Opiz (*E. verrucosa* L.), plant 20–50cm, curved-ascending; lvs alternate, oval, sessile, finely serrate, pale green; umbel usually 5-rayed; bracts under the individual flower heads oval, not connate, shortly stalked; involucral glands oval, yellow; fruit 3–4mm long, with hemispherical or cylindrical warts. Fl.5–6. Calcareous grassland, sunny slopes; scattered. C. and S. Europe.

3. Cypress Spurge, *Euphorbia cyparissias* L., plant 15–35cm; lvs alternate, narrowly linear, 2–3mm wide, pale green; non-flowering shoots densely leafy, like a twig of a fir tree; umbel 9–15 rayed; bracts of the individual flower heads pale yellow, becoming red, not connate; involucral glands half-moon-shaped, with 2 horns, yellow; fruit warty, and so finely punctate. Fl.4–6. Calcareous grassland, scrub, paths, slopes; local to common. Often attacked by rust fungi (host to a generation of the rust of peas, *Uromyces pisi*). Europe, but often an escape in Britain.

4. Steppe Spurge, **Euphorbia seguieriana* Neck., plant 15–30cm, with a thick, woody rhizome, often creeping horizontally; lvs alternate, lanceolate, 4–6mm wide, mucronate, blue-green; umbel 8–15 rayed, rays sometimes further branched 1–2 times dichotomously; involucral glands oval, yellow; fruit smooth, glabrous, 3mm long. Fl.4–6. Dry grassland, calcareous dunes, steppes; rare. C. and S. Europe. Protected!

Daphne Family, Thymeleaceae

5. Garland Flower, **Daphne cneorum* L., plant 5–30cm; lvs linear-spathulate, dark green, leathery, equally distributed along the twigs; inflorescence umbel-like at the ends of the twigs; flowers dark pink, hairy outside; calyx tube and twigs appressed-hairy. Fl.5–8. Protected! Dryish grassland, dry wood margins, open pinewoods, rock outcrops; rare. C. and S. Europe. Similar species: **Striped Spurge-laurel**, **D. striata* Tratt., but lvs bunched together at the ends of the twigs, linear-cuneate, bluish-green; flowers pink, finely striped, glabrous, strongly scented; calyx tube and twigs glabrous. Protected! Mountain Pine scrub, subalpine shrub communities; rather rare; Alps.

Tamarisk Family, Tamaricaceae

6. Tamarisk, *Tamarix gallica* L., grey-green shrub, up to 3m, with scale-like, imbricate lvs and small, 5-merous, pinkish-red flowers, 1.5–2mm across, flowers in spike-like inflorescences. Fl.5–8. River banks, coasts; cultivated plant. S.W. Europe, often planted.

Plate 63   Tamarisk Family, Tamaricaceae

1. Myricaria, *Myricaria germanica* (L.)Desv., shrub to 2m, with brush-like branches; lvs scale-like, grey-green, 2–3mm long; flowers reddish or white, 3–5mm across in terminal, dense racemes, which are often branched; sepals and petals 4–5; stamens 10. Fl.6–8. Shingle flats of alpine rivers; rare. Scandinavia, Alps and hinterlands as far as the Danube, Black Forest, Pyrenees, Abruzzi, mountains of the Balkan peninsula. Protected!

Mallow Family, Malvaceae

2. Musk Mallow, *Malva moschata* L., plant 20–80cm; lvs roundish in outline, palmately deeply 5–7 lobed, segments of the upper lvs narrowly strap-shaped; stem and lvs with simple hairs; flowers solitary, axillary, upper flowers often bunched in heads; epicalyx of 3 bracts, which are narrow-linear, free, 3–5mm long, and united at the base with the calyx; petals 2–4cm long, emarginate, pale pink; mericarps densely hairy on the back. Fl.6–10. Poor grassland, sunny meadows, waysides, hedgebanks; fairly common to rather rare. Britain, C. and S. Europe.

St. John's-wort Family, Hypericaceae

3. Perforate St. John's-wort, *Hypericum perforatum* L., plant 30–60cm; stem 2-angled, pithy; lvs ovate-oblong, densely dotted with translucent spots; flowers 5-merous; sepals lanceolate, acute, 4–5mm long, longer than the ovary; petals golden-yellow, 10–15mm long. Fl.7–8. Poor grassland, fallow fields, sunny wood margins; common or widespread. Europe. Similar species: **Imperforate St. John's-wort**, *H. maculatum* Crantz, with 4-angled, hollow stem and obtuse, elliptic sepals; lvs without, or with very few, translucent dots; corolla golden-yellow; siliceous grassland, damp wood margins, hedgebanks; local; Europe. **Square-stalked St. John's-wort**, *H. tetrapterum* Fries, with 4-winged, hollow stem, acute sepals and pale yellow corolla; river banks, ditches, damp meadows, marshes, level moorland; widespread; Britain, S. Scandinavia, C. and S. Europe. See also *H. humifusum*, Pl.22 and *H. pulchrum*, Pl.161.

Rock-rose Family, Cistaceae

4. Common Rock-rose, *Helianthemum nummularium* (L.)Mill.(*H. chamaecistus* Mill.), plant 10–30cm; lvs leathery, ovate, margins inrolled downwards, grey-felted below, with lanceolate stipules; flowers 15–25mm across, yellow. Fl.6–9. Calcareous grassland, dry pinewoods; fairly common. Britain, C. and S. Europe. **White Rock-rose**, *H. apenninum* (L.)Mill., has white flowers and lvs grey- or white-felted on both sides; stipules subulate; dry calcareous grassland; rare; Britain, S. Germany, France, S. Europe.

5. Common Fumana, *Fumana procumbens* (Dun.)Gr.et Godr., plant 5–20cm, prostrate, woody at base; lvs needle-shaped, 2–12mm long; flowers usually solitary in the axils of bracts, 15–20mm across; petals 5, yellow, unequally large. Fl.6–10. Dry calcareous grassland, stony slopes; likes warm places; rare. S. Germany, Thuringia, Moravia, very local in S. Sweden, France, valleys of C. and S. Alps, S. Europe. Protected!

6. Teesdale Violet, *Viola rupestris*, see p.176.

Plate 64   Carrot Family, Umbelliferae or Apiaceae

1. Honewort, *Trinia glauca* (L.)Dum., plant 20–80cm, much-branched at the base, dioecious, male plant smaller than the female, with a tuft of fibres at the base; lvs blue-green, 2–4 pinnate, with narrowly linear to filiform segments, the uppermost leaf with an inflated leaf sheath and fewer segments; flower umbels numerous; petals of the male flowers with narrow, green midribs, those of the female flowers with broad, red midribs. Fl.4–5. Dry sunny calcareous grassland, dunes, stony slopes; rare. Britain, C. and S. Europe. Protected!

2. Wild Carrot, *Daucus carota* L., plant 30–100cm; root thick, white, smelling of carrot; lvs 2–3 pinnate, hairy; umbels flat-topped or convex at flowering time, becoming hollow in the top, like a nest, when fruiting; bracts 3-lobed or pinnate, ciliate; bracteoles usually simple; marginal flowers white, the outermost enlarged, the central flower usually rudimentary and dark red. Fl.6–9. Poor grassland, paths, quarries, cliffs and dunes; common. Europe.

3. Spignel, *Meum athamanticum* Jacq., plant with a strong spicy smell, with a tuft of fibres at the base, 15–20cm, with a thick rhizome; lvs 2–4 pinnate, with filiform segments, 2–6mm long, which are finely pointed; stem ribbed and angled; umbel with 5–15 rays; bracts 0–6; bracteoles several, not secund; flowers white, pale yellow or pink; fruit 6–8mm long and 3–4mm wide, scarcely compressed, with strong, protruding ribs. Fl.5–6. Siliceous grassland and pasture, mountain meadows; calcifuge; scattered. Medicinal plant and vegetable. Principally in the mountainous parts of Europe.

For other species in this family, see also *Orlaya*, Pl.23, *Pimpinella*, Pl.82 and *Astrantia*, Pl.162.

Primrose Family, Primulaceae. See *Primula veris*, Pl.167.

Plantain Family, Plantaginaceae. See *Plantago coronopus*, Pl.199.

Violet Family, Violaceae

6. Teesdale Violet, *Viola rupestris* F.W. Schmidt (Pl.63), plant 3–8cm; lvs at base and on stem, roundish-cordate, obtuse, 15–20mm long, grey-green, weakly crenate to entire; lvs and stem usually pubescent; petals pale violet, white at base, more rarely all white, spur pale violet, 5mm long; sepals acute. Fl.5–6. Calcareous grassland, open dry pinewoods, sandy soils; rather rare. Britain, N. Europe, S. Germany, Alps and hinterlands, southwards to the Pyrenees, Cevennes, Caucasus. Protected! Fairly common in siliceous grassland, heaths and wood margins is **Heath Dog-violet**, *V. canina* L., plant 5–15cm, calcifuge; lvs ovate-lanceolate, cordate-based, 2–4cm long, long-stalked, rough, dark green, occurring on the stem only, basal lvs absent; corolla blue-violet, quadratic in outline, with a whitish or yellowish, usually straight spur, 5–7mm long. Europe. See also *V. tricolor*, Pl.22, *V. hirta*, Pl.160.

Plate 65   Gentian Family, Gentianaceae

1. Field Gentian, *Gentianella campestris* (L.)Börner (*Gentiana c.* L.), plant 5–30cm, often branched at the base; lvs spathulate; corolla and calyx 4-merous; corolla bearded in the throat, violet, rarely white; calyx lobes unequal, with 2 broad lanceolate outer lobes and 2 narrow lanceolate inner lobes. Fl.6–9. Protected! Siliceous grassland, waysides, dunes; rather rare. Britain, Scandinavia, C. European mountains, E. Alps, Pyrenees, Abruzzi.

2. Chiltern Gentian, *Gentianella germanica* (Willd.)Börner (*Gentiana germanica* Willd.), plant 5–40cm, similar to *G. aspera*; stem usually only branched above; stem lvs ovate-lanceolate, usually acute; flowers 5-merous, 2–4cm long; calyx teeth rough on the margins; calyx narrowly winged; the sinuses between the calyx lobes acute; corolla reddish-violet, bearded within, with lobes 3–5mm wide; ovary shortly stalked. Fl.6–10. Protected! Calcareous grassland, stony or loamy soils; scattered to rare. England, C. Europe, the Alps. Similar species: **Austrian Gentian**, **G. austriaca* (Kern.)Holub (*Gentiana a.* Kern.), but calyx unwinged, sinuses between the calyx lobes obtuse, lobes smooth on the margins and longer than the calyx tube; lower stem branches lengthy; poor grassland; rare; S. Germany, E. Alps; protected! **Autumn Gentian**, *G. amarella* (L.)Börner, has small, 4 (rarely 5-)-merous flowers, 10–18mm long, and stem with short branches; corolla lilac, lobes 5–7mm long; calcareous grassland and dunes; fairly common to rare; Britain, Scandinavia, France, mountains of C. Europe, Alps (very rare), Carpathians, Caucasus.

3. Rough Gentian, **Gentianella aspera* (Hegetschw.)Dostal (*Gentiana a.* Hegetschw.), plant 5–30cm; stem branched from the base; stem lvs triangular-ovate, obtuse, usually ciliate on the margin; flowers usually 5-merous, bearded within, blue-violet, 2–4cm long; corolla lobes 9–15mm long and 5–10mm wide; calyx narrowly winged, sinuses between the lobes acute; calyx lobes roughly ciliate on the midrib and margins; ovary long-stalked. Fl.5–10. Protected! Dryish calcareous grassland, scree, up to 2800m; widespread to rare; E. Alps, alpine hinterland, French Jura.

4. Fringed Gentian, **Gentianella ciliata* (L.)Borkh. (*Gentiana c.* L.), plant 10–25cm; stem 4-angled; lvs linear-lanceolate, 1-nerved; flowers 4-merous, blue; corolla lobes fringed on the margin. Fl.6–9. Protected! Calcareous grassland, open pinewoods, stony subalpine grassland; scattered. C. Europe and the mountains of S. Europe.

5. Spring Gentian, *Gentiana verna* L., plant 3–10cm, with a basal leaf rosette; lvs lanceolate, acute, 1–3 nerved, up to 3cm long; stem lvs smaller; stem erect, 1-flowered; corolla deep blue, with a cylindrical tube and spreading lobes, between each of which there is an appendage with 2 points. Fl.3–8. Protected! Calcareous grassland, dry to damp, stony or pure clay soils, also level moorland, scattered in the alpine zone and hinterland. N. England, Ireland; mountains of C. and S. Europe.

6. Bladder Gentian, **Gentiana utriculosa* L., plant 8–20cm; stem many-flowered, angled; lvs ovate; calyx with wings 2–4mm wide, finally inflated; corolla tubular, with 5 spreading, lanceolate lobes, dark blue, the outside often greenish. Fl.5–8. Protected! Calcareous moorland and flushes, poor grassland and stony, subalpine grassland; rare. Mainly the mountains of C. and S. Europe.

7. Cross Gentian, **Gentiana cruciata*, see p.180.

Plate 66   Gentian Family, Gentianaceae

1. **Common Centaury**, *Centaurium erythraea* Rafn (*C. minus* Moench, *C. umbellatum* Gilib.), plant 10–30cm, branched above, with basal rosette; rosette lvs ovate, over 5mm wide, upper lvs ovate-lanceolate to oblong-ovate, usually 5-veined; flowers in lax cymes; corolla 5-merous, pinkish-red, with a tube 9–15mm long and ovate, obtuse lobes 5–8mm long. Fl.7–9. Dryish grassland, dry wood margins, dunes; common or scattered. Britain, S. Scandinavia, C. and S. Europe. Similar species: **Lesser Centaury**, *C. pulchellum* (Sw.)Druce, but plant 3–15cm, without basal rosette; stem dichotomously branched from the base; corolla lobes 3–4mm long, lanceolate, acute; river banks, paths, gravel pits, meadows near the sea; scattered. **Seaside Centaury**, *C. littorale* (D.Turner)Gilmour (*C. vulgare* Rafn), plant with linear, usually 3-veined stem lvs and linear-spathulate rosette lvs, up to 5mm wide; coasts of W. Europe, dunes and sandy places.

2. **Yellow-wort**, *Blackstonia perfoliata* (L.)Huds., plant erect, 10–40cm, bluish pruinose; stem 4-angled above; lvs opposite, triangular-ovate, connate at the base in pairs; flowers in cymose inflorescences; flowers 6–8-merous; calyx divided almost to the base, shorter than the corolla; corolla yellow, with a short campanulate tube and 6–8 spreading lobes. Fl.6–8. Calcareous grassland, dwarf rush communities, seasonally damp loamy or clayey soils, dunes; likes warm places; fairly common to rare. C. and S. Europe, northwards to England and Ireland (western).

Bindweed Family, Convolvulaceae

3. **Dodder**, *Cuscuta epithymum* (L.)Nath., reddish parasitic plant, 10–30cm, with scale-shaped lvs; flowers 8–10 together in clusters 5–8mm across; corolla pink or white; corolla tube closed by connivent scales; style longer than the ovary, exceeding the flower. Fl.7–8. Parasitic on Wild Thyme, Broom, Heather and other plants; widespread. Europe. See also Greater Dodder, *C. europaea*, Pl.26.

Bedstraw Family, Rubiaceae

4. **Dyer's Woodruff**, **Asperula tinctoria* L., plant 30–50cm; stem 4-angled; lvs in whorls of 4–6, narrow-lanceolate, 1-veined, 2–6cm long; inflorescence few-flowered; corolla white, 3-merous, 3–4mm wide; fruit smooth. Fl.6–8. Open pine and oak woods, sunny scrub, dryish grassland; rather rare. C. Europe, northwards to S. Sweden. Protected!

5. **Squinancywort**, *Asperula cynanchica* L., plant ascending, 10–30cm; lvs usually in whorls of 4, narrowly lanceolate, 1-nerved, mucronate; inflorescence few-flowered; corolla funnel-shaped, 4-merous, pink, 2–3mm; fruit rough, tuberculate. Fl.6–8. Calcareous grassland, sunny wood and path margins, pinewoods, dunes; local. Britain, C. and S. Europe.

---

7. **Cross Gentian**, **Gentiana cruciata* L. (Pl.65), plant 15–50cm; lvs lanceolate, usually 3-veined, up to 10cm long, leaf pairs united below in sheaths; flowers 1–3 together in the axils of the uppermost lvs; corolla 4-merous, narrowly campanulate, blue, greenish outside. Fl.7–8. Protected! Poor calcareous grassland, open pine and oak woods, path and wood margins; rare. Principally C. Europe, in the south only in the mountains.

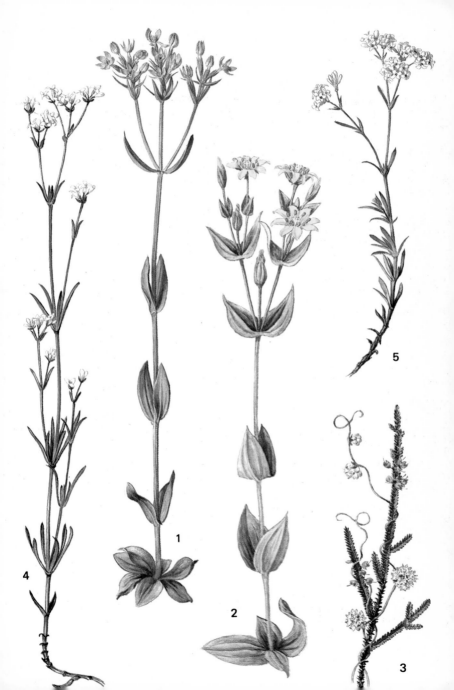

Plate 67   Bedstraw Family, Rubiaceae

1. Lady's Bedstraw, *Galium verum* L., plant 20–70cm; stem with 4 raised lines above; lvs 6–12 per whorl, linear, 1mm wide, 1-veined, revolute at the margins, softly hairy below; flowers in dense, many-flowered, panicle-like, scarcely interrupted inflorescences; corolla 4-merous, yellow, 2–4mm across, with 4 acute lobes. Fl.6–9. Calcareous grassland, waysides, pinewoods, also on moorland pastures, dunes; common. Europe. Similar is *G. verum* ssp. **wirtgenii* (F.Schultz)Oborny, but lvs up to 2mm wide, scarcely revolute, glabrous below, inflorescences lax, interrupted; dry pastures; S.W. Germany, valleys of C. and S. Alps.

2. Glaucous Bedstraw, **Galium glaucum* L. (*Asperula glauca* (L.)Bess.), plant 30–80cm, blue-green; stem branched, with 4 raised lines; lvs linear, 8–10 in a whorl, revolute at the margins, 1-veined; inflorescence lax, corymbose; corolla white, funnel-shaped, 4-merous, 4–5mm across; fruit smooth. Fl.5–7. Calcareous grassland, sunny, stony slopes; rather rare. C. and S. Europe. **Heath Bedstraw**, *G. saxatile* L., is also white-flowered, but plant 10–20cm, lvs obovate, in whorls of 6–8, corolla 2.5–4mm across, lobes acute, fruit rough; pastures, heaths and scrub; calcifuge, common. W. Europe. See also *G. aparine*, Pl.25, and *G. album*, Pl.84.

Jacob's-ladder Family, Polemoniaceae. See *Polemonium*, Pl.170.

Borage Family, Boraginaceae

3. Greater Golden Drop, **Onosma taurica* Willd., plant 20–50cm; lvs linear-lanceolate, with bristly hairs 1–4mm long, with swollen disc-shaped bases, which are also surrounded by short, stiff, radially-spreading hairs; flowers in the axils of bracts; calyx cylindrical, deeply 5-lobed; corolla 15–25mm long, yellow. Fl.5–7. (Genus with many similar species which are difficult to tell apart.) Dry grassland, stony steppes, open pinewoods; rare. S. Alps, Apennines, S.E. Europe. Similar species: **Golden Drop**, **O. arenarium* W.et K., but swollen bristle-bases glabrous, with the bristle-hair only; corolla 12–18mm long, pale yellow; dry grassland; rare. S. Germany, S. Europe. Protected!

For other species of this family, see Pl.26.

Mint Family, Labiatae or Lamiaceae

4. Basil Thyme, *Acinos arvensis* (Lamk.)Dandy (*Calamintha a.* (L.)Clairv.), plant smelling of peppermint, 10–30cm; stem hairy; lvs opposite, ovate-lanceolate, almost sessile, 10–20mm long, with 1–4 teeth on each side, veins very prominent beneath; flowers in whorls of about 6, rather one-sided, with short bracts; corolla 8–10mm long, blue-violet; calyx 2-lipped, 5–7mm long, densely spreading-hairy, calyx teeth broadened below and tube thus inflated. Fl.6–9. Walls, rocks, stony grassland, arable fields; likes warm places; scattered. C. and S. Europe, S. Scandinavia, Britain.

5. Blue Bugle, **Ajuga genevensis* L., plant 10–30cm, villous-hairy, without stolons; basal lvs long-stalked; upper lvs 3-lobed or deeply crenate, often blue-tinged; flowers dark blue; corolla 12–18mm long, upper lip very short, 2-lobed, lower lip much longer, 3-lobed; stamens 4. Fl.4–6. Calcareous grassland, sunny slopes, path margins, also in weed communities; scattered. Mainly C. and S. Europe.

6. Wall Germander, *Teucrium chamaedrys*, see p.184.

Plate 68   Mint Family, Labiatae or Lamiaceae

1. Mountain Germander, *Teucrium montanum* L., plant prostrate, 5–35cm, aromatically scented; lvs narrow lanceolate, 5–20mm long, entire, evergreen, leathery, densely white-felted below; flowers bunched in heads at the ends of the twigs, whitish or pale yellow; calyx almost regularly 5-toothed; corolla with only a 5-lobed lower lip. Fl.6–9. Sunny calcareous grassland, communities of rocks and boulders; scattered. C. and S. Europe. Protected!

2. Large Self-heal, *Prunella grandiflora* (L.)Scholl., plant 10–30cm; lvs oblong-ovate, 2–6cm long, the upper ones often pinnately lobed, sparsely hairy; flowers 20–25mm long, blue-violet, corolla tube curved. Fl.6–8. Calcareous grassland, sunny wood margins, open woods; scattered. Mainly C. Europe, northwards to S. Sweden, southwards to N.E. Spain, the Po valley, Bulgaria. See also *P. vulgaris*, Pl.83.

3. Yellow Woundwort, *Stachys recta* L., plant 20–60cm; lvs ovate-lanceolate, roughly hairy, the lower stalked, the upper sessile; flowers in a whorl of 6–10; calyx teeth triangular, with a glabrous, spiny tip. Fl.6–10. Calcareous grassland, sunny wood margins, open pine and oak woods; scattered. C. and S. Europe. Similar species: **Annual Woundwort**, *S. annua* L., but plant 10–30cm, all lvs stalked; flowers in a whorl of 4–6; calyx teeth lanceolate, with a hairy point; arable fields, vineyards; rare. Mainly C. and S. Europe. Protected! Both these species are rare casuals in Britain.

4. Large Thyme, *Thymus pulegioides* L., plant creeping or ascending, 5–30cm, aromatically scented; stem 4-angled, shortly hairy on the angles; lvs oval, glabrous, or only ciliate at the base; inflorescence cylindrical or globose; corolla 3–6mm long, pale to dark purple; upper calyx teeth usually ciliate. Fl.6–10. With many different forms and microspecies. Dry grassland and pasture, usually calcareous, waysides, slopes, gravel pits; widespread. Europe. Similar species: **Breckland Thyme**, *Thymus serpyllum* L., plant 5–15cm, creeping or curved-ascending; stem almost terete, usually hairy, hairs all round the stem; lvs narrowly elliptic, glabrous or hairy; flowers arranged in heads at the ends of the twigs; corolla 3–6mm long, pale to dark purple, the upper 3 calyx teeth broadly triangular, the lower 2 narrowly lanceolate; also with many forms and microspecies; dry sandy grassland, dry pinewoods; rare. Protected!

5. Meadow Clary, *Salvia pratensis* L., plant 30–60cm; stem few-flowered; lvs mainly basal, long-stalked; leaf blade oval, 6–12cm long, irregularly crenate; flowers in whorls of 4–8; corolla 20–25mm long, dark blue, with sickle-shaped upper lip; inflorescence glandular. Fl.4–8. Calcareous grassland, dryish grassland, sunny meadows, waysides; widespread. Mainly in C. and S. Europe, also introduced to N. Europe, a rare native in Britain.

---

6. Wall Germander, *Teucrium chamaedrys* L. (Pl.67), plant 15–30cm, with underground rhizomes; stem hairy; lvs crenate, with cuneate base, 1–2cm long; flowers usually in whorls of 6 in the axils of the upper lvs; calyx almost regularly 5-lobed, lobes shortly awned; corolla 10–15mm long, pink; upper lip apparently missing, deflexed, lower lip 5-lobed. Fl.7–8. Calcareous grassland, open pine and oak woods; scattered. Grown in gardens, and sometimes escaping. Britain (introduced), C. and S. Europe.

Plate 69   Mint Family, Labiatae or Lamiaceae

1. Sage, *Salvia officinalis* L., plant 20–80cm, woody at the base, with strong aromatic smell; lvs oblong-ovate, long-stalked, reticulately wrinkled, weakly crenate, felted below; flowers violet, in whorls of 4–8; corolla 17–25mm long, with a straight upper lip. Fl.6–7. Dry meadows, stony steppes; cultivated as a medicinal and pot herb, and naturalised. S. Europe, in C. Europe only as an escape.

2. Whorled Clary, *Salvia verticillata* L., plant 30–60cm; lvs cordate to triangular, with lobes on the petiole, 4–12cm long and 3–10cm wide, irregularly serrate, the lower long-stalked; whorls with 15–30 flowers, almost spherical; bracts short, scarious; corolla 10–15mm long, pale violet, with a straight upper lip; calyx narrowly campanulate, 4–7mm long, hairy; stamens fixed. Fl.6–9. Dryish grassland, banks, paths, weed communities; scattered. Originally from S.E. Europe but now spread far and wide. Similar species: **Wild Clary**, *Salvia verbenaca* L. (*S. horminoides* Pourr.), plant 30–80cm; lvs oblong to ovate, variable in toothing from crenate to pinnatifid; flowers violet-blue and up to 15mm, usually with white spots on lower lip, but often not opening. Dry places, pastures; local. Britain, France, S. Europe. **Wild Sage**, *S. nemorosa* L., has lanceolate lvs, cordate at base, 4–8cm long, which are wrinkled, finely and regularly serrate and grey-felted below; flowers in whorls of 2–4, set closely in the spike; corolla violet, 10–15mm long; dryish grassland, paths, waste places; rare; S. Germany, Austria, S.E. Europe.

3. Garden Lavender, *Lavandula angustifolia* Mill. (*L. officinalis* Chaix), plant 20–60cm, aromatically scented; lvs narrow-lanceolate, revolute at the margins, densely grey-felted below, 3–4cm long; flowers in interrupted, distantly leafy, terminal spikes; corolla blue-violet, calyx tubular, obscurely 5-toothed. Fl.7–8. Stony slopes, stony steppes; likes warmth. S.W. Europe, sometimes cultivated, rarely naturalised.

For other species in this family, see *Stachys, Clinopodium* and *Origanum*, Pl.172.

Figwort Family, Scrophulariaceae

4. Purple Mullein, *Verbascum phoeniceum* L., plant 30–80cm; stem usually unbranched, glandular-hairy above; basal lvs oval, stalked, sinuate-dentate, dark green; stem lvs sessile; flowers violet, 2–3.5cm across; calyx glandular-hairy, with lanceolate lobes; filaments with violet hairs. Fl.5–7. Dry grassland, sunny slopes, scrubby woodland; rare. C. Germany, plain of the upper Rhine, Alps (e.g. Aosta valley), Apennines, S.E. Europe.

5. Ivy-leaved Toadflax, *Cymbalaria muralis* G.M.Sch. (*Linaria cymbalaria* (L.)Mill.), plant creeping, 30–60cm; stem filiform; lvs long-stalked, alternate, roundish-cordate, coarsely 5–7 lobed or toothed; flowers long-stalked, solitary in the leaf axils; corolla pale violet, palate yellowish, about 1cm; spur blunt, about half as long as the rest of the corolla. Fl.6–9. Shady, humid walls, rocks; widespread. Originally from the S. Alps, Apennines, mountains of the Balkan peninsula; cultivated plant, nowadays naturalised in almost all of Europe. Also a garden escape is the **Snapdragon**, *Antirrhinum majus* L., an erect perennial; flowers in terminal racemes, 3–4cm long, variously red-purple, yellow or white, with a palate closing the mouth of the corolla. Old walls, widespread.

See also *Euphrasia*, Pl.83 and *Rhinanthus*, Pl.84.

Plate 70   Figwort Family, Scrophulariaceae

1. Large Speedwell, *Veronica austriaca* L., plant 10–50cm, erect; lvs lanceolate, distantly serrate to entire, narrowed into a short stalk, often revolute at the margins; stem hairy all the way round; flowers in opposite, many-flowered, long-stalked racemes; calyx 5-lobed; corolla dark blue, 11–15mm across; capsule acutely emarginate. Fl.5–6. Calcareous grassland, stony slopes, dunes; scattered to rare. C. and E. Europe. Protected! Similar are *Veronica austriaca* ssp. *teucrium* (L.)Webb, but lvs ovate to broadly lanceolate, rounded or cordate at base, sessile, coarsely toothed, softly hairy; capsule acutely emarginate; dryish grassland, wood margins and waysides; scattered. **Prostrate Speedwell**, *V. prostrata* L., plant with prostrate, evergreen shoots and ascending flowering stems; lvs lanceolate, finely serrate, densely short-hairy; corolla pale lilac, 4–11mm across; capsule weakly emarginate; dry grassland, stony steppes; rather rare; S. Germany, valleys of the C. Alps, E. Europe. Protected!

2. Spiked Speedwell, *Veronica spicata* L. (*Pseudolysimachia spicata* (L.)Opiz), plant 15–30cm; stem hairy; lvs opposite, almost sessile, ovate-oblong, usually bluntly serrate, entire at the tip; flowers shortly stalked, in long, dense, spike-like racemes at the end of the stem; corolla 4-merous, 8–12mm across, blue, funnel-shaped, weakly 2-lipped; calyx 4-merous, glandular-hairy. Fl.6–8. On rocks, sunny, dry grassland, dunes; rare. C. and S. Europe, northwards to England, S. Sweden. Protected! Similar species: **Long-leaved Speedwell**, *V. longifolia* L., but lvs 2–4 in whorls, linear-lanceolate, sharply serrate, acute; plant 40–100cm; damp grassland, marshy meadows. Garden escape in Britain. See also *V. arvensis* and *V. filiformis*, Pl.32, *V. serpyllifolia*, Pl.33, and see Pl.173.

Broomrape Family, Orobanchaceae

The genus *Orobanche* has many species of parasites which grow exclusively on certain host plants, and are difficult to distinguish. If the host plant is known, this makes identification easier.

3. Slender Broomrape, *Orobanche gracilis* Sm., parasitic on the pea family; without green lvs, 15–40cm, brownish; flowers in spikes; the corolla tube at the back regularly curved throughout; corolla 15–25mm long, yellow, red inside, scented like pinks; stigma yellow, red-bordered; filaments hairy below. Fl.5–8. On leguminous plants (*Lotus, Genista, Hippocrepis, Trifolium*). Dryish grassland, sunny pastures; scattered to rare. S. Germany, warm valleys of the Alps (Rhône, Rhein), S. Europe. Protected! Similar species: **Common Broomrape**, *O. minor* Sm., parasitic on pea or daisy family; corolla yellowish with purple veins, unscented; stigma purple; filaments practically glabrous. Often in clover crops. Britain, S. and W. Europe.

4. Germander Broomrape, *Orobanche teucrii* Hol., plant 15–40cm, parasitic on *Teucrium*; bracts almost as long as the flowers; corolla 20–35mm long, brownish-lilac, abruptly curved on the back; stigma purplish to brown; filaments hairy below; Fl.6–7. Calcareous grassland in warm spots, sunny slopes; scattered. Mountains of C. and S. Europe. Protected!

Plantain Family, Plantaginaceae

5. Hoary Plantain, *Plantago media*, see p.190.

Plate 71   Valerian Family, Valerianaceae

1. **Red Valerian**, *Centranthus ruber* (L.)DC., plant 30–80cm; lvs ovate-lanceolate, opposite, the upper with cordate base, sessile, entire, or with a few blunt teeth, glabrous, blue-green; inflorescence cymose; corolla red, with tube 7–9mm long, with a slender spur at the base; stamens 1. Fl.5–7. Rock outcrops, walls; cultivated plant, sometimes naturalised, widespread in the south. S. Alps, S. Europe; S.W. England and Ireland. Similar species: **Narrow-leaved Valerian**, **C. angustifolius* (Cav.)DC., but lvs narrowly lanceolate, sessile with a cuneate base, usually entire; S. Alps, Abruzzi, S.W. Europe.

Teasel Family, Dipsacaceae

2. **Devil's-bit Scabious**, *Succisa pratensis* Moench, plant 20–80cm; rootstock short, as if bitten off; basal lvs oval, stalked, usually entire; stem lvs sessile; flower heads hemispherical, later spherical, 15–25mm across; receptacle with receptacular scales; florets 4-merous, blue-violet; corolla tube 4–7mm long; inner calyx with 4–5 black bristles, 1mm long; epicalyx 4-angled, roughly hairy, with 4 triangular teeth. Fl.7–9. Poor grassland, moorland pastures, level moorland, marshes and fens; common. Almost all of Europe. Similar species: **Moor Scabious**, **Succisella inflexa* (Kluk) G.Beck (*Succisa inflexa*(Kluk)Jundz.), but rootstock creeping, branched; lvs lanceolate; corolla pale blue to whitish; inner calyx without bristles; epicalyx almost terete, 8-ribbed, glabrous; upland pasture, wet places; very rare. Protected!

3. **Small Scabious**, *Scabiosa columbaria* L., plant 20–60cm; stem usually branched, many-headed; stem lvs opposite, 1–2 pinnate, with fine, crisped hairs, dull, segments 1–3mm wide; flower heads 2–3cm wide, hemispherical; receptacles with scales; marginal flowers enlarged; corolla 5-lobed, blue-violet; calyx bristles setaceous, blackish. Fl.7–11. Calcareous grassland, moorland pasture, open pinewoods; widespread. Britain, S. Scandinavia, C. and S. Europe. Similar species: **Shining Scabious**, **S. lucida* Vill., but lvs glabrous, faintly shining; stem usually 1-headed; calyx bristles winged below; subalpine and alpine stony grasslands, rock rubble, in the Alps at about 1200–2500m; mountains of C. and S. Europe. **Yellow Scabious**, **S. ochroleuca* L., but corolla pale yellow; calyx bristles reddish-brown; dry meadows, banks, stony steppes; rare; S. Germany, E. Europe. **Southern Scabious**, **S. triandra* L., but lvs 2–3 pinnate, with segments 0.5–2mm wide; calyx bristles pale to dark brown, 1–3mm long; flowers lilac; dry, stony slopes, pinewoods; C. and S. Alps, Carpathians, S. Europe. See also *Knautia*, Pl.84.

---

5. **Hoary Plantain**, *Plantago media* L. (Pl.70), plant 20–40cm; lvs in basal rosettes, usually flat to the ground; lvs broad ovate, narrowed into the short, broad stalk, densely short-hairy when young; flower spike cylindrical, 2–6cm long, much shorter than the stem (in *P. major* longer than the stem); flowers scented; corolla 4mm long, with white corolla lobes; anthers lilac. Fl.5–7. Dryish grassland, poor pastures and meadows; widespread. Almost all of Europe, rare in the south.

1

2

3

3

Plate 72 Bellflower Family, Campanulaceae

1. Sheep's-bit, *Jasione montana* L., plant 20–50cm, without stolons; stem much-branched, leafy in the lower part; lvs narrowly ovate, undulate on the margin, glabrous or stiffly hairy; flower heads spherical, 15–25mm across; florets 6–15mm long, blue. Fl.6–10. Dry sandy grassland, dunes, banks, on rocks, fallow land, dry pinewoods; calcifuge; scattered. Britain, S. Scandinavia, C. and S. Europe (western). Similar species: **Perennial Sheep's-bit**, **J. laevis* Lam. (*J. perennis* L.), but plant with stolons; stem usually simple; lvs entire or bluntly serrate, not undulate; siliceous grassland, waysides, open woods; rather rare; S. Germany, S.W. Europe. See also *Globularia*, Pl.192.

2. Round-headed Rampion, **Phyteuma orbiculare* L., plant 10–30cm; basal lvs lanceolate or cordate-ovate, long-stalked, crenate-serrate; stem lvs ovate-lanceolate, upper sessile; all lvs with an inconspicuous network of veins beneath; flower heads globose, blue; corolla tube 10–15mm long; corolla lobes 5, strap-shaped, at first united both at the tip and at the base, acute, later remaining joined only at the base; styles usually 3; bracts ovate, acute about as long as the flower. Fl.5–7. (Species with many microspecies.) Sunny, calcareous grassland, also moorland pasture, in the Alps to over 2500m. C. and S. Europe. Similar species: **English Round-headed Rampion**, *P. tenerum* R.Schulz, but lvs with a prominent network of veins beneath; bracts much shorter than the flower; styles usually 2; dry grassland; local to rare; Britain, S.W. Germany, S.W. Europe. See also *Phyteuma*, Pl.177.

3. Clustered Bellflower, *Campanula glomerata* L., plant 20–60cm, shortly hairy; stem simple; lvs ovate-lanceolate, bluntly serrate, lower lvs with cordate or rounded base, petioled, upper lvs sessile; flowers clustered at the end of the stem and in the axils of the upper lvs; corolla funnel- to bell-shaped, blue, 15–30mm long; calyx teeth narrow-lanceolate, acute. Fl.6–9. Calcareous grassland, wood and path margins; widespread. C. Europe, northwards to S. Sweden, Britain, and southwards to C. Italy. Similar species: **Bristly Bellflower**, **C. cervicaria* L., but plant with stiff prickly hairs; lower lvs gradually narrowed into the stalks; calyx lobes obtuse; style longer than the corolla; open oak and pine woods, moorland pasture; likes warm places; rare; S. Scandinavia, C. Europe, southwards to C. Italy. Protected!

4. Rampion Bellflower, *Campanula rapunculus* L., plant 50–80cm, with a turnip-shaped root; lvs ovate-lanceolate, narrowed close to the base; flowers in narrow racemes or panicles; corolla narrowly funnel-shaped, 15–25mm long, pale blue-violet, divided almost to the middle; calyx lobes narrowly lanceolate, more than half the length of the corolla. Fl.6–8. Dryish grassland, dry pastures, scrub, wood margins; scattered. Almost all of Europe.

5. Harebell, *Campanula rotundifolia* L., plant 10–40cm; basal lvs roundish, cordate or reniform, usually withered by flowering time; stem lvs narrowly linear; flowers in lax racemes or solitary, 10–20mm long, pale blue or violet; calyx teeth narrowly lanceolate. Fl.6–9. Poor grassland, heaths, wood and path margins, open oakwoods, dunes; widespread. Europe, in the south only in the mountains.

See also *Campanula*, Pl.192.

Plate 73 Daisy Family, Compositae or Asteraceae

1. Goldilocks Aster, *Aster linosyris* (L.)Bernh., plant 15–45cm; lvs linear, 1–2mm wide, 1-nerved; flower heads 8–10mm wide, in an umbellate raceme; ray florets absent, disc florets golden-yellow; pappus yellowish; fruit 3mm long. Fl.8–9. Dry sandy grassland, dry and semi-dry grasslands, sunny slopes, wood margins, cliffs; rare. Britain, C. and S. Europe.

2. Blue Fleabane, *Erigeron acer* L., plant 10–40cm; lvs lanceolate, the lower ones stalked, usually entire, undulate; flower heads 6–13mm across, in few-flowered racemes or panicles with few branches; involucre densely hairy; ray florets erect, pale violet, scarcely longer than the yellowish disc florets, which turn a dirty red; fruit 2–3cm long; pappus white, yellowish or reddish. Fl.6–9. Calcareous grassland, dry sandy grassland, waysides, gravel pits, dunes, walls; local or scattered. Almost all of Europe. See also *E. annuus*, Pl.35.

3. Mountain Everlasting or **Cat's-foot**, *Antennaria dioica* (L.)Gaertn., plant 5–20cm, with stolons and a basal leaf rosette; lower lvs oval, upper spathulate, 1-nerved, white-felted beneath; flower heads 5–8mm wide, 3–12 together; bracts of the male heads white, those of the female heads pink. Fl.5–6. Siliceous grassland, heaths, pinewoods, dry mountain slopes; fairly common. Europe, in the south only in the mountains. Similar species: **Carpathian Cat's-foot**, **A. carpatica* (Wahlb.)Bluff et Fingerh., but plant without stolons; bracts brownish; lvs grey-felted on both sides, weakly 3-nerved; wind-exposed ridges and stony grasslands of the Alps. See also *Omalotheca*, Pl.178.

4. Yellow Milfoil, **Achillea tomentosa* L., plant 5–20cm, densely villous, with spreading non-flowering shoots and erect, unbranched flowering stems; lvs lanceolate in outline, several times pinnatisect up to the midrib, finely divided, densely woolly-hairy; flower heads tiny, in dense umbellate panicles; ray florets 4–6, golden-yellow; bracts obtuse, with pale brown borders, white-woolly. Fl.5–7. Dry grassland, rocky steppes, open pinewoods. Alps (e.g. Wallis, Aosta valley), Pyrenees, S. Europe. Naturalised in Scotland and Ireland. See also *A. millefolium*, Pl.85.

5. Sand Everlasting, **Helichrysum arenarium* (L.)Moench, plant 10–30cm, white-woolly; rosette lvs spathulate, stalked; stem lvs linear, obtuse, sessile, all lvs woolly-hairy; flower heads 6–7mm across, globose, in dense terminal corymbs; bracts shining yellow or orange; ray florets absent, disc florets yellowish. Fl.7–8. Dry sandy grassland, open pinewoods, waysides, dunes; scattered to rare. C. and S. Europe. Protected! Similar species: **Goldy-locks** or **Stinking Everlasting**, **H. stoechas* (L.)DC., but lvs linear, with revolute margins, with a smell of curry when rubbed; involucre hemispherical, inner bracts twice as long as the outer; stony steppes; S.W. Europe. **Italian Everlasting**, **H. italicum* (Roth)Guss., with a cylindrical to campanulate involucre; inner bracts 3–4 times as long as outer; lf margins revolute; S.W. Europe.

6. Irish Fleabane, *Inula salicina* L., plant 25–80cm; stem glabrous; lvs oblong-lanceolate, amplexicaul with a cordate base, glabrous on the blade but ciliate, reticulate-veined; flower heads 25–35mm across; bracts lanceolate, ciliate on the margins, tips reflexed; outer florets ligulate, inner tubular, yellow. Fl.6–10. Dryish grassland, moorland pasture, wood margins; scattered. C. and E. Europe, southwards to Italy, northwards to Ireland.

For other similar plants, see *Filaginella*, Pl.36, *Inula*, Pl.178, *Solidago*, Pl.179.

Plate 74   Daisy Family, Compositae or Asteraceae

1. Yellow Ox-eye, *Buphthalmum salicifolium* L., plant 20–60cm; lvs ovate-lanceolate, silky-hairy, entire or finely serrate; flower heads 3–6cm wide, solitary, terminal; lanceolate receptacular bracts present; marginal florets ligulate, 2–3mm wide, yellow, the inner tubular, numerous; fruit from the edge of the disc winged and 3-angled. Fl.6–9. Calcareous grassland, wood margins, open pine and oak woods; widespread. Mountains of C. and S. Europe.

2. Arnica, *Arnica montana* L., plant 20–50cm, with a basal rosette, usually of 4 lvs; stem lvs opposite, ovate, acute; flower heads 1 (rarely up to 5), 5–8cm wide; marginal florets ligulate, 2–3cm long and 4–6mm wide. Fl.6–7. Protected! Siliceous grassland, upland pasture; in the Alps up to about 2500m; scattered. Mainly in the mountains of C. Europe, northwards to S. Sweden, southwards to N. Spain, Pyrenees.

3. Hoary Ragwort, *Senecio erucifolius* L., plant 30–120cm; lvs ovate in outline, deeply 1–2 pinnate, with linear segments; flower heads about 1.5cm across, in a corymbose panicle; involucres usually campanulate, with a row of 4–6 external bracts; marginal florets ligulate, 6–8mm long, yellow; fruit 1–2mm long, 6–8 ribbed, pappus 6mm long. Fl.8–10. Calcareous grassland, dryish grassland, dry upland pasture, waysides; common. Europe, northwards to S. Scandinavia, Britain. Very similar to it is **Common Ragwort,** *Senecio jacobaea* L., but lvs with broader segments and larger terminal lobes; external bracts 1–3; waysides, waste land, poor pastures, dunes, wood margins and scrub; common. Almost all of Europe.

4. Eastern Marsh Ragwort, *Senecio aquaticus* Hill ssp. *barbaraefolius* (Wimm. et Grab.) (*S. erraticus* Bert.), plant 30–100cm, similar to the last 2 species, but exterior bracts very short, only 1–2; lvs dark green, unequally divided, with large terminal lobes, lateral lobes spreading almost at right angles; stem branches squarrose. Fl.7–8. Weed communities, damp woods and wood margins; scattered. C. and S. Europe (see also Marsh Ragwort, *S. aquaticus* ssp. *aquaticus*, Pl.118).

Plate 75 Daisy Family, Compositae or Asteraceae

1. Yellow Chamomile, *Anthemis tinctoria* L., plant 20–50cm, erect or ascending; lvs pinnate, with segments toothed like a comb, hairy; flower heads 2–5cm wide; receptacles of flower heads hemispherical, with lanceolate, acute scales; ray florets ligulate, 8–15mm long, yellow, disc florets tubular; bracts felted, overlapping; involucre hemispherical. Fl.6–9. Dry grassland, waysides, rocky steppes; scattered; the flowers were once used for dyeing. Europe, naturalised garden escape in Britain.

2. Stemless Carline Thistle, **Carlina acaulis* L., plant with a thick taproot and very short stem, 10–30cm, usually with just 1 flower head; lvs in a rosette, pinnatisect as far as, or nearly up to, the midrib, spiny (in ssp. *simplex* (W.et Kit.)Arc., lvs crisped, closely packed, with subulate segments, 2–6mm wide; in ssp. *acaulis*, lvs flat, not crowded, with ovate segments 6–14mm broad); flower heads 4–7cm across; inner bracts silvery-white, 3–4cm long; only disc florets present, which are whitish or pink; pappus 10–15mm long. Fl.7–9. Protected! Sunny, poor grassland, open woods. Mountains of C. and S. Europe. Similar species: **Acanthus-leaved Carline Thistle**, **C. acanthifolia* All., but lvs undulate, only pinnatisect to the middle or little more, with 5–7 broad, sinuate and spiny-toothed segments; flower heads 10–15cm across; inner bracts yellowish; pappus 20–25mm long; dry grasslands, open woods; S. Alps, E. Pyrenees, Apennines, mountains of the Balkan peninsula.

3. Carline Thistle, *Carlina vulgaris* L., plant 15–40cm; stem with many heads; lvs oblong-lanceolate, spiny sinuate-serrate, undulate; flower heads 2–3cm broad; inner bracts straw-yellow. Fl.7–9. Sunny, calcareous grassland, path and wood margins, open oak and pine woods; widespread to common. Europe.

4. Greater Knapweed, *Centaurea scabiosa* L., plant 30–100cm; lvs 1–2 pinnate, segments oblong-lanceolate, dark green; flower heads 2–3cm across; bracts green, with black, triangular appendages with margins fringed like combs; florets tubular, the marginal ones the largest, purple. Fl.6–9. Calcareous grassland, wood margins and waysides, cliffs; widespread. Almost all of Europe. **Panicled Knapweed**, **C. rhenana* Bor., likewise has pinnate lvs, but the segments linear, grey to white-felted; flower heads 1cm across, arranged in a panicle; bracts brown above or black-spotted; dry grassland, stony steppes; likes warm places; rare; S. Germany, dry Alpine valleys, Pyrenees, southwards to C. Italy, Bulgaria. See also *Centaurea*, Pl.85.

5. Cat's-ear, *Hypochaeris radicata* L., plant 20–60cm, with a basal leaf rosette; lvs deeply toothed, glabrous or roughly hairy; stem blue-green, simple or branched, leafless or with scale-like lvs; stem a little thickened under the flower head; flower heads 2–4cm across, yellow; receptacular scales present; fruit long-beaked; pappus yellowish-white, 2-rowed, the inner pappus hairs feathery, interwoven with one another. Fl.5–8. Poor grassland, dry grassland, poor pastures. Almost all of Europe. Similar species: **Spotted Cat's-ear**, *H. maculata* L., with red-brown-spotted lvs, flower heads 4–5cm wide and a 1-rowed pappus; stem roughly hairy at the base; dry and poor grassland, cliffs; scattered to rare. Protected!

For other thistle-like plants, see *Carduus nutans*, Pl.37 and *Cirsium tuberosum*, Pl.119.

Plate 76   Daisy Family, Compositae or Asteraceae

1. Purple Viper's-grass, *Scorzonera purpurea* L., plant 25–50cm, with abundant blackened remains of the previous year's lvs at base; stem 1–3 headed; basal lvs grass-like; flower heads 3–5cm wide, with ligulate florets only, pale violet; bracts over-lapping. Fl.5–6. Dry or dryish grassland; rare. Mountains of C. and S. Europe, Germany, in warm and dry localities. Protected!

2. Austrian Viper's-grass, *Scorzonera austriaca* Willd., plant 5–30cm, with brown-coloured remains of last year's lvs at base; stem with 1 head; lvs radical, linear-lanceolate, bluish-green, glabrous or woolly-hairy; heads with ligulate florets only, 3–4cm long, yellow. Fl.4–5 (like *S. humilis*, Pl.120). Rocky steppes, dry pinewoods, calcareous soils in warm places; rare. S. Germany, S. Europe, dry valleys of the Alps (Rhône, Valais, Aosta).

3. Mouse-ear Hawkweed, *Hieracium pilosella* L., plant 5–25cm, with stolons having distant and small lvs; stem leafless or only with 1–2 scale-like lvs, 1-headed; basal lvs narrowly ovate, entire, grey-felted below; bracts 1–2mm wide, linear, grey-felted; heads with ligulate florets only, which are pale yellow, often red-striped outside; fruit 1–2.5mm long with white pappus. Fl.5–10. Poor grassland, heaths, open pine woods. Almost all of Europe. Similar species: **Eared Mouse-ear Hawkweed**, *H. lactucella* Wallr., but stem 2–5 headed, leafless or with 1 leaf only; plant 10–25cm, with stolons, whose lvs become larger towards the tip of the stolon; basal lvs spathulate, blue-green, hairy on the margin and at the base; poor grassland, upland pastures; widespread. Europe, only casual in Britain. See also *H. aurantiacum* Pl.194.

4. Cymose Mouse-ear Hawkweed, *Hieracium cymosum* L., plant 30–60cm, without stolons; stem 1–3 leaved, 10–50 headed, with short, rough hairs; basal lvs lanceolate, gradually narrowed to the base, stellate-hairy on both sides; involucre many-rowed, 5–10mm long; flowers golden-yellow to brownish-red; fruit 1.5–2.5mm long. Fl.5–7. Dryish grassland, calcareous, dry, stony soils; rare. C. and E. Europe. Similar species: **Meadow Mouse-ear Hawkweed**, *H. caespitosum* Dum., but plant with stolons; flower heads arranged in a panicle; flowers yellow, marginal florets reddish-striped; stem hollow, with spreading hairs, 3–5mm long; lvs with the underside stellate-hairy, at the most; upland pasture, dryish grassland, waysides; scattered to rare; protected! **Floren-tine Mouse-ear Hawkweed**, *H. piloselloides* Vill., plant 20–80cm, without stolons; lvs blue-green, linear-lanceolate, rough, sparsely hairy with stiff bristles; inflorescence many-headed, lax, flowers yellow; poor and dry grasslands, quarries, gravel heaps; widespread. **Bugloss Mouse-ear Hawkweed**, *H. echioides* Lumn., stem with 5–15 ovate-lanceolate, rough, bristly- and stellate-hairy lvs; basal lvs usually withered by flowering time; flower heads 10–30; sandy and siliceous grassland; rare; E. and C. Germany, E. Europe. See also *Hieracium*, Pl.181.

5. Blue Lettuce, *Lactuca perennis* L., plant 20-60cm; stem glabrous, hollow, branched above; lvs glabrous, blue-green, pinnately lobed, with lanceolate segments; flower heads in a lax panicle; bracts with a narrow white border; flower blue to lilac, 2cm across; fruit black, with a beak 10–15mm long. Fl.5–6. Sunny, dry grassland, rock outcrops, walls; rare. S. Germany, S. Europe. Likewise blue-flowered is the **Russian Blue Lettuce**, *L. tatarica* (L.)C. A. Mey., with red-spotted bracts; fruit greenish-brown, spotted; dunes of North Sea coast, and naturalised in Britain.

For other species in this family, see *Cichorium*, Pl.38, *Picris*, Pl.39, *Crepis*, Pl.39, *Tragopogon*, Pl.40, *Taraxacum*, Pl.86 and *Mycelis*, Pl.180.

Plate 77   Grass Family, Gramineae or Poaceae

1. Smooth Brome, *Bromus racemosus* L., plant 30–80cm, pale green; lvs 3–4mm wide; panicle 3–10cm long, erect, raceme-like; spikelets ovate, glabrous, 10–15mm long, 5–8 flowered, lower glume 3–6 nerved, upper 5–9 nerved, lemma parchment-like, ovate, awn 6–7mm long. Fl.5–6. Damp meadows; local to fairly common. Europe.

2. Meadow Fescue, *Festuca pratensis* Huds., plant 30–80cm, dark green; lvs 3–5mm wide, 10–20cm long, with glabrous auricles; panicle slender, 10–20cm long, secund, contracted; lowest panicle branch with 4–5 spikelets and short branch at base with 1(–3) spikelets; spikelets 9–11mm long, usually 7–8 flowered, yellow-green or violet spotted, awnless. Fl.6–7. Rich meadows, dryish grassland; common. Europe. Similar species: **Tall Fescue**, *F. arundinacea* Schreb., but larger; auricles minutely hairy; lowest panicle branches each with spikelets. See *Lolium*, Pl.1.

3. Red Fescue, *Festuca rubra* L., plant 30–80cm, forming a loose turf, grey-green; basal lvs bristle-like, 0.5–1mm wide; stem lvs flat, 1–3mm wide, not auricled at base; ligule short, blunt; panicle erect, 6–15cm long; spikelets 7–10mm long, 4–6 flowered, reddish-violet or brownish; awn of the lemma 1–2mm long, lowest panicle branch about half as long as the panicle. Fl.6–7. Meadows, pastures, dryish grassland, waysides, dunes, saltmarshes and mountains; common. Europe.

4. Smooth Meadow-grass, *Poa pratensis* L., plant 10–80cm, dark green; stem usually smooth; lvs 3–5mm wide, flat or folded (not rolled), tip suddenly contracted, like the prow of a boat, with double central grooves (tramline); ligules 0.5–2mm long, blunt; panicle 5–10cm long, many-flowered; lower panicle branches 3–5 together, rough; spikelets flat, 4–6mm long; glumes almost equal, hairy like the lemma. Fl.5–6. Meadows, dunes; a good fodder grass; common. Europe. See also *P. annua*, Pl.1.

5. Rough Meadow-grass, *Poa trivialis* L., very similar to Smooth Meadow-grass, but ligule 3–7mm long, pointed; glumes very unequal in length; stem rough above. Fl.5–6. Damp meadows, riverside woods, arable fields, waste places; common. Europe.

6. Cock's-foot, *Dactylis glomerata* L., plant 30–120cm, tufted, grey-green; lvs 4–10mm wide; spikelets 3–4 flowered, green or violet tinged, clustered on the axis of the panicle; panicle triangular in outline, erect, with the lowest branch spreading; glumes rough, not translucent. Fl.5–7. Rich meadows, dryish grassland, waste places; common. Europe. Similar species: **Wood Cock's-foot**, *D. glomerata* ssp. *aschersoniana* (Graebn.)Thell (*D. polygama* Horvatovszky), but pale green; lvs 3–6mm broad; panicle slender, usually drooping; spikelets 5–6 flowered; glumes whitish, translucent. Mixed deciduous woodland; C. Europe, introduced to Britain.

7. Crested Dog's-tail, *Cynosurus cristatus* L., plant 20–60cm; lvs 2–3mm wide, finely ribbed, shining above, dull green below; ligules 1mm long; spike linear, 4–10cm long, secund; spikelets arranged in 2 rows, green, 3–4mm long; each fertile spikelet with a sterile, comb-like spikelet. Fl.6–7 and 9–10. Rich meadows and pastures, both lowland and upland; common. Almost all of Europe.

8. Yorkshire-fog, *Holcus lanatus* L., plant 30–100cm, grey-green, in dense tufts; stem nodes, leaf sheaths and leaf blades softly hairy; ligule 2mm long, ciliate; panicle 6–12cm long, reddish tinged; spikelets 4–5mm long, glumes ciliate on the keel and margins, whitish, reddish above; lemma shining white, that of the male florets with a shorter glume, which scarcely exceeds spikelet. Fl.6–8. Damp meadows, level moorland, waste land, open woodland; common. Europe. See *Deschampsia*, Pl.124.

Plate 78   Grass Family, Gramineae or Poaceae

1. False Oat-grass, *Arrhenatherum elatius* (L.) J.et C.Presl., plant 50–150cm; lvs flat, 4–8mm wide, upper surface shortly hairy, leaf sheaths glabrous; ligule short, obtuse, toothed; panicle 10–20cm long; spikelets 8–12mm long, 2-flowered, with only 1 geniculate awn; outer glume 1-nerved, inner 3-nerved. Fl.6–7. Rich meadows, waysides; productive fodder grass; common. Almost all of Europe.

2. Yellow Oat-grass, *Trisetum flavescens* (L.)P.B., plant 30–80cm, forming a loose turf; stem hairy at and below the nodes; leaf sheaths pubescent; leaf blades flat, ciliate on the margin; ligules 1–2mm long; panicle 10–20cm long, branches very slender, with 3–12 spikelets; spikelets roundish, 5–8mm long, 2–3 flowered, golden-yellow, with 2–3 awns; outer glume 1-nerved, inner 3-nerved, lemma 5-nerved, with 2 tiny bristle points, and with geniculate awn, 5–7mm long. Fl.5–6 and 8–9. Rich meadows, mountain pastures, especially on dry calcareous soils; good fodder grass; common. Almost all of Europe.

3. Hairy Oat-grass, *Avenula pubescens* (Huds.)Dum.(*Helictotrichon p.* (Huds.)Pilger), plant 30–100cm, in loose tufts; lvs flat, 5–10mm wide; leaf sheaths and lower lvs softly hairy; ligules triangular, acute, 4–6mm long; panicle 10–20cm long, lower branches 3–5 together; spikelets with 2–3 awns, greenish, violet and golden-yellow mottled; outer glume 1-nerved, about 12mm long, inner 3-nerved, about 16mm long; awn 10–20mm long, thick below, twisted. Fl.5–6. Poor pasture, calcareous grassland; common. Almost all of Europe.

4. Timothy, *Phleum pratense* L., plant 20–100cm, forming tufts; lvs light green, rough, 3–8mm wide; ligules 3–5mm long; panicle spike-like, dense, cylindrical, 5–20(30)cm long, 5–10mm wide, usually green, with very short branches, not appearing lobed when bent over; spikelets lyre-shaped; glumes not united below, long-ciliate; awns on the glumes $\frac{1}{4}–\frac{1}{2}$ their length; anthers violet. Fl.6–8. Rich meadows and pastures; often sown; rather common. Europe.

5. Meadow Foxtail, *Alopecurus pratensis* L., plant 30–100cm, mid-green; lvs 6–10mm wide, rough above; leaf sheaths smooth, the upper ones somewhat inflated; ligules 3–5mm long; panicle spike-like, dense, cylindrical, axis hidden, up to 10cm long and 1cm wide; spikelets ovate to elliptic, shortly stalked, 4–6 together per branch; glumes united to $\frac{1}{3}$, 5mm long, ciliate on the keel; awn of the lemma up to 9mm long, usually inserted near to the base, weakly geniculate. Fl.4–6. Damp meadows, river banks, grassy places; common. Almost all of Europe. See also *A. geniculatus*, Pl.91.

6. Sweet Vernal-grass, *Anthoxanthum odoratum* L., plant 10–50cm, turf-forming; lvs 3–6mm wide, blue-green, bitter-tasting and sweet-smelling, auricled and with a ring of hairs at the base; ligules 1–2mm long, obtuse and toothed; spikelets 1-flowered, finally yellow-brown, apparently with 4 glumes, in a short, narrow, ovoid panicle, 2–4cm long. Fl.4–6. Poor meadows and pastures, heaths and moors, open deciduous woodland, waysides; common. Europe.

Sedge Family, Cyperaceae. See *Carex ovalis*, Pl.43, *C. hirta*, Pl.96, *C. panicea* and *C. flava*, Pl.97.

Rush Family, Juncaceae. See Pl.98.

Plate 79   Lily Family, Liliaceae

1. Meadow Saffron, *Colchicum autumnale* L., plant 5–30cm, with a bulb; leafless when it flowers in autumn; lvs broad-lanceolate, basal, rather fleshy, 12–20cm long and 2–5cm long, appearing in the spring along with the fruiting capsule; perianth segments 6, lilac, united into a tube up to 20cm long below, the free part 4–6cm long. Fl.8–11. Poisonous! Damp meadows and woods; widespread. Mainly C. Europe; Britain. Similar species: **Alpine Saffron**, *C. alpinum* Lam.et DC., but all plant parts smaller, free part of the perianth segments 2–3cm long; lvs linear-lanceolate, 7–12mm wide. Fl.7–9. Mountain meadows of the W. Alps and the mountains of S. Europe. In **Spring Bulbocodium**, *Bulbocodium vernum* L., the lvs and the pinkish-violet flowers appear at the same time; plant like a crocus, but stamens 6. Fl.3–5. Mountain meadows of the S.W. Alps and the mountains of S. Europe.

Also in this family, see *Fritillaria*, Pl.100, *Tulipa*, Pl.127 and *Leucojum*, Pl.130.

Daffodil Family, Amaryllidaceae

2. Pheasant's-eye Narcissus, *Narcissus poeticus* L., plant 20–40cm high, 1-flowered; lvs linear, fleshy, blue-green, 5–10cm wide; flowers 3–6cm wide, white, with a short, dish-shaped corona with a crisped, red margin; perianth segments 6, united into a tube below. Fl.4–5. Mountain pastures, gardens and naturalised. Mainly S.W. Europe.

Iris Family, Iridaceae

3. Purple Crocus, *Crocus vernus* (L.)Hill ssp. *albiflorus* (Kit.) Asch.et Graebn. (*C. albiflorus* Kit.), plant 8–15cm, with a corm, without a stem above ground; lvs radical, grass-like, narrowly-linear, with white midrib; flowers white, violet or striped; perianth segments united below into a tube; stamens 3. Fl.3–4. Mountain meadows and pastures; in the Alps up to 2800m; widespread. Mountains of C. and S. Europe, naturalised in Britain.

Orchid Family, Orchidaceae. See *Dactylorhiza*, Pl.101; *Listera ovata*, Pl.132.

Dock Family, Polygonaceae

4. Common Sorrel, *Rumex acetosa* L., plant 30–100cm; basal lvs long-stalked, hastate, ovate-oblong, sharp-tasting, rough, upper lvs sessile; leaf sheaths split or toothed; inflorescence branched, lax, interrupted, erect; flowers unisexual, inconspicuous, small; outer 3 perianth segments reflexed in fruit, inner 3 perianth segments 3–4mm long, roundish or broader than long, not toothed, with a green or red tubercle at the base. Fl.5–7. Meadows, pastures, waysides, open woodland; common. Europe. See also *R. crispus* and *R. obtusifolius*, Pl.4.

5. Common Bistort, *Polygonum bistorta* L., plant 30–80cm, with a snake-like, contorted rhizome; lvs ovate-oblong, rather undulate, 10–20cm long, the lower with winged petioles, the upper sessile; leaf sheaths long, acute; inflorescence spike-like, 1–2cm wide; perianth segments 5, reddish-white, 4–5mm long. Fl.5–7. Damp meadows, grassy roadsides, shrub communities, wet woods; local to common. Mainly C. Europe, southwards to the mountains of N. Spain, Apennines, mountains of the Balkan peninsula, northwards to Britain.

Purslane Family, Portulacaceae. See *Montia fontana*, Pl.102.

Plate 80 Pink Family, Caryophyllaceae

1. Red Campion, *Silene dioica* (L.) Clairv. (*Melandrium rubrum* (Weigel) Garcke), plant 30–80cm, softly hairy; lvs ovate, acute, sessile; flowers 2–3cm long, in lax inflorescences, unisexual; petals 5, red, deeply bifid; calyx 10-nerved, densely hairy, more or less inflated; styles 5; teeth of the open capsule rolled back outwards. Fl.4–6. Damp pastures and wood margins, hedgerows, cliff ledges; widespread. Most of Europe.

2. Ragged-robin, *Lychnis flos-cuculi* L., plant 30–80cm; basal lvs spathulate, often ciliate, stalked, upper lvs linear-lanceolate; inflorescence lax, dichotomously branched; flowers 3–4cm across; petals 5, deeply divided into 4 segments, pinkish-red; calyx 10-veined, often reddish. Fl.5–7. Damp meadows, upland pasture, moorland, marshes and fens; common to widespread. Europe.

3. Common Mouse-ear, *Cerastium fontanum* Baumg. (*C. holosteoides* Fries, *C. vulgatum* auct.), plant 10–40cm, densely spreading-hairy; lvs oblong-ovate, dark grey-green, 1–3cm long; petals white, 2-lobed, about as long as the sepals, which are 4–6mm long, and with scarious margins. Fl.4–10. Meadows, pastures, waysides, arable fields, dunes, waste places; common. Europe.

Buttercup Family, Ranunculaceae

4. Meadow Buttercup, *Ranunculus acris* L., plant 30–100cm, appressed-hairy or glabrous; basal lvs 3–5 lobed, the lobes further divided into linear-lanceolate segments; peduncles terete (not furrowed); flowers 5-merous, 2–3cm across; sepals appressed to petals; achene with short beak. Fl.5–9. Meadows, waysides, rock ledges; common. Europe. See *R. repens*, Pl.10; *R. bulbosus*, Pl.51; *R. ficaria, R. polyanthemos*, Pl.144.

Cabbage Family, Cruciferae or Brassicaceae

5. Cuckooflower, *Cardamine pratensis* L., plant 10–40cm, stem hollow, round; lvs imparipinnate; basal lvs with roundish-ovate lobes, the stem lvs with linear ones; sepals 4, 3–5mm long; petals 4, 8–14mm long, white, pink or pale lilac; anthers yellow; siliquae 2–4cm long. Fl.4–6. Rich and wet meadows, moorland, river banks, wet woods; widespread. Almost all of Europe.

Rose Family, Rosaceae

6. Great Burnet, *Sanguisorba officinalis* L., plant 30–100cm; basal lvs in a rosette; lvs imparipinnate, with 7–15 pairs of leaflets; leaflets ovate, stalked, 2–5cm long, blue-green below, with about 12 teeth on each side; flowers small, dark red, in cylindrical-ovoid heads, 1–3cm long; stamens 4, short, as long as the red-brown calyx. Fl.7–9. Water meadows and upland pasture, alpine meadows; widespread. Almost all of Europe, northwards to S. Norway, Britain, southwards to C. Spain, Calabria.

7. Lady's-mantle, *Alchemilla vulgaris* L., plant 10–30cm; lvs roundish, 2–12cm wide, divided to $\frac{1}{3}$ to $\frac{2}{3}$ into 7–11 toothed, semi-circular, quadrilateral or triangular lobes; lvs glabrous, or sparsely to densely appressed or spreading hairy; the hairiness of the stem and the much-branched inflorescence also just as variable; flowers 2–4mm wide, yellow-green, in lax clusters; petals absent, calyx 4-merous with an epicalyx with 4 lobes also; stamens 4. Fl.5–8. (Exceptionally variable aggregate species with numerous microspecies.) Rich meadows, water meadows, flush and shrub communities, wood margins, rock ledges; widespread. Most of Europe. See also *Potentilla*, Pl.17.

Plate 81   Pea Family, Papilionaceae or Fabaceae

1. White Clover, *Trifolium repens* L., plant 15–45cm, creeping and rooting at the nodes; lvs trifoliate, leaflets ovate, finely serrate, glabrous; stipules scarious; flower heads globose, solitary; individual flowers 2–5mm long, stalked, drooping after flowering; calyx 10-veined; corolla 7–12mm long, white, later brownish. Fl.5–9. Rich meadows and pastures, park grassland, waysides, arable fields; common. Europe.

2. Red Clover, *Trifolium pratense* L., plant 10–30cm; lvs trifoliate, leaflets ovate, entire, usually spotted; stipules ovate, with a ciliate and awned, pointed tip; flower heads globose, usually in 2s, with 2 bracts; flowers red; calyx 10-veined, calyx teeth hairy. Fl.6–9. Meadows, pastures, waysides; common. Cultivated for hay. Europe. See also *T. medium*, Pl.152.

3. Alsike Clover, *Trifolium hybridum* L., plant 20–40cm, ascending, not rooting at the nodes; lvs trifoliate, leaflets finely serrate, stipules herbaceous; flower heads globose or ovoid; pedicels of the individual flowers 2–3 times as long as the 5-veined calyx tube; corolla white to reddish. Fl.5–9. Rich pasture, water meadows, waysides; cultivated; widespread. Almost all of Europe. See also *T. fragiferum*, Pl.198.

4. Lesser Trefoil, *Trifolium dubium* Sibth. (*T. minus* Sm.), plant 5–15cm, prostrate; lvs trifoliate, bluish-green, glabrous, central leaflet with a longer stalk than the 2 lateral ones, without a mucron (compare *Medicago lupulina*, Pl.57); stipules ovate-lanceolate; flower heads 10–20 flowered, 6–8mm wide; corolla yellow, 3–4mm long. Fl.5–9. Meadows, pastures, waysides; widespread. Almost all of Europe.

5. Meadow Vetchling, *Lathyrus pratensis* L., plant 30–100cm; lvs with 2 oblong-lanceolate leaflets, with parallel veins, and a terminal tendril; stipules almost as large as the leaflets; stem angled; flowers 10–15mm long, yellow, in 3–10 flowered racemes. Fl.6–8. Rich meadows, upland pasture, river banks, wood margins, hedges; widespread. Almost all of Europe. See also *Lotus corniculatus*, Pl.58.

6. Tufted Vetch, *Vicia cracca* L., plant 30–100cm, prostrate, ascending or climbing; lvs with 12–20 lanceolate leaflets, 2–6mm wide, and a branched tendril; flowers 10–30 together in a long-stalked raceme, the peduncle about as long as the subtending leaf; corolla 8–12mm long, blue-violet, blade of the standard about as long as its claw; pod 10–25mm long, 4–6mm wide. Fl.6–8. Meadows, pastures, wood margins, river banks, bushy places; widespread to common. Europe. See also *V. sepium*, Pl.153.

Crane's-bill Family, Geraniaceae

7. Meadow Crane's-bill, *Geranium pratense* L., plant 20–60cm, somewhat glandular-hairy above; lvs usually 7-lobed, with narrow, pinnately divided lobes; flowers usually in 2s; petals 12–20mm long, blue-violet; filaments with broadened triangular bases; flower stalk bent downwards after flowering, and often becoming erect again later. Fl.6–8. Rich pasture, waysides; scattered. Almost all of Europe. Similar species: **Wood Crane's-bill**, *G. sylvaticum* L., but leaf lobes broadly rhomboid, deeply cut, serrate; flowers red-violet, filaments lanceolate; flower stalks remaining erect after flowering; meadows in the mountains, hedgebanks, damp woods, rock ledges, mountain scrub communities. Europe.

Plate 82 Carrot Family, Umbelliferae or Apiaceae

1. Cow Parsley, *Anthriscus sylvestris* (L.)Hoffm., plant 60–150cm; biennial; lvs 2–3 pinnate, segments lanceolate and pointed; flower umbel with 8–15 rays; bracts absent; bracteoles 4–8, broadly lanceolate, suddenly contracted to a long point, ciliate on the margin, 2–5mm long; flowers white, equal-sized; fruit 6–10mm long, rather shorter than its stalk. Fl.4–6. Rich meadows, waysides, hedges, edges of woods, waste places; common. Europe. Similar species: **Garden Chervil,** *A. cerefolium* (L.)Hoffm., but plant smaller, aromatically scented, annual; lvs soft, pale green, 2–4 pinnatisect, with ovate-lanceolate segments, which are crenate-serrate or further divided; branches of the umbel densely pubescent; fruit linear, smooth, 6–10mm long, longer than its stalk. Fl.5–8. Hedges, gardens, vineyards, wood margins, waste places; S.E. Europe, occasionally introduced and naturalised; a pot herb.

2. Greater Burnet-saxifrage, *Pimpinella major* (L.)Huds., plant 40–100cm; lvs once pinnate, shining, with 2–4, ovate, coarsely toothed leaflets on each side, 1–4cm long, and a weakly 3-lobed, serrate terminal leaflet; stem angled and furrowed; inflorescence 10–15 rayed; bracts and bracteoles absent; corolla 2–3mm wide, white or pink; style after flowering longer than the ripening fruit. Fl.6–9. Meadows, wood margins, hedgebanks, shrub communities; local to common. Europe, northwards to S. Scandinavia, Britain, and southwards to N. Spain, Calabria.

3. Burnet-saxifrage, *Pimpinella saxifraga* L., plant like *P. major*, but 15–50cm; stem finely ribbed, downy, almost leafless above; segments of basal lvs ovate, obtuse, 10–15mm long, dull; segments of stem lvs linear; petals white or pink; style after flowering shorter than the fruit. Fl.7–9. Siliceous and calcareous grassland, dry pinewoods, heaths; widespread; former medicinal plant. Almost all of Europe. Similar species: **Pignut,** *Conopodium majus* (Gouan) Loret, with tuberous, edible root; stem glabrous; segments of upper lvs filiform. Open woods and grassland; widespread. Britain, N. and W. Europe.

4. Pepper-saxifrage, *Silaum silaus* (L.)Schinz.et Thell., plant 30–100cm; lvs 3-pinnate; segments narrow-lanceolate, 1–3mm wide, with fine points; umbel with 5–10 rays; bracts 0–3; bracteoles many, white-hyaline bordered; petals yellowish-green, flowers 2mm across; fruit ovoid, 4mm long, with sharp-angled, wing-like ribs. Fl.6–9. Rich and water meadows, upland pasture, grassy banks; scattered. Mainly C. Europe; Britain.

5. Masterwort, *Peucedanum ostruthium* (L.)Koch (*Imperatoria o.* L.), plant 50–100cm, with a stout rhizome; stem terete, hollow; lvs 2-ternate, 10–30cm long; leaf segments broad ovate, 2–7cm wide, serrate or divided, pale green beneath; leaf sheaths inflated; bracts 0–1; umbel 20–50 rayed; bracteoles tiny, thread-like; petals white to pink; fruit roundish, 4–6mm, with broad lateral ribs. Fl.7–8. Subalpine shrub communities, mountain meadows, scrub of Green Alder, river banks; once cultivated as a pot herb and medicinal plant and naturalised; widespread. Mainly the Alps and the mountains of C. Europe, S. Europe; Britain.

Plate 83   Carrot Family, Umbelliferae or Apiaceae

1. Caraway, *Carum carvi* L., plant 30–80cm; lvs 2–3 pinnate, with narrowly linear, acute segments, 1mm wide, the basal segments reflexed and looking like stipules; leaf sheaths with scarious margins; flower umbel with 8–16 rays; bracts 0, bracteoles 0–2; flowers white or pink, 2–3mm wide; fruit ovoid, 3–4mm long. Fl.4–6. Meadows and pastures, especially in mountains, waysides, waste places; seeds used for flavouring; local. Almost all of Europe.

2. Hogweed, *Heracleum sphondylium* L., plant 80–150cm; stem 5–20mm wide, with stiff, bristly hairs, angled and furrowed; basal lvs roundish or ovate in outline, 20–50cm long, pinnatisect or pinnate, with large, broadly ovate to lanceolate, deeply lobed or coarsely toothed segments; umbel 15–30 rayed; bracts 0–3; flowers white or greenish-yellow, outermost flowers with very unequal petals; fruit 7–11mm long, oval, flattened, winged on the margins. Fl.6–9. Rich meadows, river banks, ditches, wet woods, alpine shrub communities; common. Europe.

Mint Family, Labiatae or Lamiaceae

3. Bugle, *Ajuga reptans* L., plant 15–30cm, with stolons; basal lvs stalked, spathulate, entire or bluntly serrate; stem lvs gradually becoming smaller, often tinged with red-violet; inflorescence spike-like; flowers 2–6 together in the axils of the upper stem lvs; corolla blue or reddish, 10–15mm long. Fl.5–8. Meadows, waysides, deciduous woods; common. Almost all of Europe.

4. Self-heal, *Prunella vulgaris* L., plant 10–25cm, with stolons; lvs oblong-lanceolate, sparsely hairy, 2–4cm long; inflorescence spike-like or in a head; calyx 2-lipped, upper lip with 3 short, unequal, aristate teeth, lower lip with 2 lanceolate, awned teeth; corolla blue-violet, 8–15mm long, about twice as long as the calyx (compare Large Self-heal, Pl.68). Fl.6–9. Meadows and pastures, upland pasture, paths in woods, river banks; common. Europe.

Figwort Family, Scrophulariaceae

5. Germander Speedwell, *Veronica chamaedrys* L., plant 10–30cm; stem with 2 rows of hairs; lvs opposite, shortly stalked or sessile, ovate, crenate; flowers in long-stalked, lax racemes; calyx 4-lobed; corolla sky-blue, with dark veins; fruiting capsule triangular or cordate, hairy, 4–5mm wide. Fl.5–6. Meadows, wood margins and waysides, open oakwoods, hedges; common or widespread. Europe.

6. Eyebright, *Euphrasia officinalis* L., plant 5–25cm, branched, sometimes glandular-hairy above; lvs ovate, opposite, with 3–6 acute teeth on each side; flowers solitary in the axils of the upper lvs; calyx 5–6mm long; corolla 8–14mm long, white, or with a violet upper lip, or all violet, usually with a yellow spot on the lower lip. Fl.5–10. (Aggregate species with many microspecies.) Poor meadows and pastures, moorland, calcareous grassland, mountain meadows, dunes, rock ledges, sea cliffs; common. Almost all of Europe.

Plate 84   Figwort Family, Scrophulariaceae

1. Greater Yellow Rattle, *Rhinanthus alectorolophus* (Scop.) Poll., plant 10–60cm; stem simple or branched, villous-hairy above; stem lvs oval to ovate-lanceolate, sharply serrate, shortly hairy, opposite; flowers solitary in the axils of pale green, villous-hairy bracts, which have sharp teeth all of the same size (**Aristate Yellow Rattle**, *R. aristatus* Čelak, has suddenly, sharply upwards-curved corolla tube and glabrous calyx, bracts with awn-like sharp-pointed teeth at the base); calyx villous-hairy; corolla 18–23mm long, with an upwards-curved tube, yellow, tooth on the upper lip longer than wide, blue-violet, 1–2mm long. Fl.5–7. Rich meadows, dryish grassland, cornfields; widespread. Mainly C. Europe. Similar species: **Narrow-leaved Yellow Rattle**, *R. angustifolius* Gmel. ssp. *angustifolius* (*R. serotinus* (Schönh.)Oborny), with likewise a slightly curved corolla tube, but stem almost glabrous; calyx and bracts glabrous, the latter long-pointed and unequally toothed. Fl.5–9. Damp meadows; widespread, but rare in Britain.

2. Yellow Rattle, *Rhinanthus minor* L., plant 15–40cm, scarcely hairy; lvs lanceolate, glabrous; bracts green, triangular, acute; calyx glabrous; corolla 13–15mm long, pale yellow, with a straight tube, tooth of the upper lip wider than long, 0.5–1mm long, rounded, whitish or pale lilac. Fl.5–8. Poor pasture, level moorland; fairly common. Almost all of Europe.

Also in this family, see *Pedicularis palustris*, Pl.116.

Plantain Family, Plantaginaceae

3. Ribwort Plantain, *Plantago lanceolata* L., plant 10–40cm; lvs in basal rosettes, lanceolate, entire, with 5–7 parallel veins; flowers 4-merous, inconspicuous, in ovoid-oblong spikes on a long, 5-angled, erect stalk; calyx lobes 4, unequal; corolla 2–4mm long, glabrous, with brownish segments; filaments very long, anthers yellowish. Fl.4–8. Meadows, pastures, waysides, arable fields; common. Almost all of Europe.

Primrose Family, Primulaceae. See *Primula veris* and *Lysimachia nummularia*, Pl.167.

Bedstraw Family, Rubiaceae

4. Hedge Bedstraw, *Galium album* Mill. (*G. mollugo* L.), plant 25–80cm, prostrate, ascending or clambering; stem 4-angled; lvs linear-lanceolate, 2–8mm wide, with a mucronate tip, 6–9 to a whorl; inflorescence pyramidal, paniculate; corolla 4-merous, rotate, 2–5mm across; corolla lobes with awn-like points; pedicels rather longer than the flowers. Fl.5–9. Rich meadows, waysides, scrub, wood margins, hedgerows; common. Europe.

Teasel Family, Dipsacaceae

5. Field Scabious, *Knautia arvensis* (L.) Coult., plant 30–80cm; stem simple or branched, shortly hairy and deflexed bristly-villous; basal lvs ovate-lanceolate, stalked, entire or toothed; stem lvs sessile, pinnatisect, opposite, grey-green, dull; flowers in flat heads, 2–4cm across; marginal flowers larger; calyx with 8–10 bristles; corolla 4-merous, blue to red-violet. Fl.7–8. Rich meadows, wood margins and waysides, arable fields; common. Almost all of Europe. See also *Succisa*, Pl.71.

1

2

3

4

5

Plate 85   Bellflower Family, Campanulaceae

1. Spreading Bellflower, *Campanula patula* L., plant 20–50cm; basal lvs oblong-lanceolate to spathulate, shortly stalked; stem lvs lanceolate; flowers in lax, erect panicles; bracts shorter than the pedicels, lateral flowers with 2 bracteoles; corolla 15–25mm long, pale blue or blue-violet, 5-lobed to halfway; calyx teeth narrowly lanceolate. Fl.5–7. Meadows, waysides, margins of scrub, shady woods; local to common. Almost all of Europe.

Daisy Family, Compositae or Asteraceae

2. Daisy, *Bellis perennis* L., plant 5–15cm; stem appressed-hairy, leafless, 1-headed; lvs in a basal rosette, spathulate, bluntly toothed; flower head 1–3cm across; marginal florets ligulate, usually in 1 row, female, white or pink, inner florets tubular, hermaphrodite, yellow. Fl.3–10. Meadows and pastures, lawns; common. Almost all of Europe.

3. Yarrow, *Achillea millefolium* L., plant 15–50cm, aromatically scented; lvs oblong in outline, 2–3 pinnate, cut up to the midrib, 2–3cm wide; flower heads 3–6mm across in corymbose inflorescences; marginal florets ligulate, 3–5, shorter than the involucre, white or reddish; bracts with brown borders. Fl.6–10. Meadows, pastures, dryish grassland, arable fields, hedgerows, waysides; common. Europe. See also *A. ptarmica*, Pl.118.

4. Ox-eye Daisy, *Leucanthemum vulgare* Lamk. (*Chrysanthemum leucanthemum* L.), plant 20–50cm; lvs oblong-lanceolate, coarsely serrate, the lower stalked and the upper sessile; stem lvs smaller; stem simple or branched; flower heads 3–6cm across; receptacular scales absent; ray florets white, disc florets yellow; bracts with brown or black borders. Fl.6–10. (Many forms and numerous microspecies.) Pastures, meadows, waysides, arable fields; common. Almost all of Europe.

5. Brown Knapweed, *Centaurea jacea* L., plant 20–80cm; stem erect, angled, rough; lvs ovate-lanceolate to lanceolate, the lower narrowed into the stalk, sometimes pinnatifid, the upper sessile; flower heads solitary, 3–6cm wide; marginal flowers enlarged, red-violet; tips of the bracts with a definite appendage, distinct from the rest of the bract, appendage hyaline, roundish, brownish and usually fringed. Fl.6–10. Meadows, pastures, poor grassland, upland pasture; widespread. Almost all of Europe, but rare in Britain. Similar species: **Common Knapweed**, *C. nigra* L., but the marginal florets not larger than the rest; appendage of the bract with a long comb-like fringe, usually black and covering the green part of the rest of the bract. Fl.7–9. Siliceous grassland, waysides, heaths, mountain pasture; calcifuge: common or scattered. Mainly W. Europe.

6. Wig Knapweed, *Centaurea phrygia* L. ssp. *pseudophrygia* C.A.Mey., plant 30–80cm; stem few-headed; lvs ovate, finely serrate, the lower shortly stalked, the upper amplexicaul; bracts of the flowering head 15–20mm long and the same width; marginal florets spreading radially; appendage of the bracts brownish, long-pinnate fringed, ending in a feathery awn 1cm long. Fl.7–9. Mountain meadows, scrub; scattered. Mainly in the mountains of C. Europe, northwards to S. Scandinavia.

1

5

4

3

6

2

Plate 86 Daisy Family, Compositae or Asteraceae

1. Autumn Hawkbit, *Leontodon autumnalis* L., plant 10–40cm; stem branched, many-headed, thickened under the heads, with many scale lvs above; basal lvs deeply pinnate-lobed, with narrowly lanceolate segments, glabrous; heads erect before flowering; involucre glabrous to densely hairy; flowers yellow; fruit with yellowish-white pappus. Fl.7–9. Rich meadows, park lawns, waysides; common or widespread. Europe.

2. Rough Hawkbit, *Leontodon hispidus* L., plant 10–40cm; stem 1-headed, often thickened under the head, with 0–2 scale lvs; basal lvs sinuate-dentate, glabrous or with forked hairs; flower heads 2–4cm across, drooping before flowering, with ligulate florets only, yellow; involucre bristly or glabrous; all fruits with dirty-white or brownish pappus and feathery pappus hairs. Fl.6–10. (Many forms.) Rich meadows and pastures, upland pasture, water meadows, screes in the high mountains; widespread. Almost all of Europe.

3. Goat's-beard, *Tragopogon pratensis* L., plant 30–70cm; lvs narrowly lanceolate, long-pointed, entire; peduncles not, or scarcely, thickened under the flower heads; bracts usually 8, 2–3cm long, constricted above the base; flower heads 3–5cm wide, with ligulate florets only, yellow; fruit beaked, pappus hairs feathery, intertwined with one another. Fl.5–7. (With many forms and numerous microspecies, see also *T. dubius*, Pl.40.) Rich meadows, waysides, dryish grassland, dunes, waste places; widespread. Almost all of Europe.

4. Dandelion, *Taraxacum officinale* Web., plant 10–40cm, with a long taproot; lvs in a basal rosette, oblong, deeply lobed, pinnatifid or serrate, grass-green; stem 1-headed, hollow, leafless; flower heads 3–6cm across; exterior bracts reflexed to erect, linear to narrow ovate; flowers yellow; fruits long-beaked. Fl.4–7. (With many forms and numerous microspecies.) Rich meadows and pastures, calcareous and sandy grassland, upland pasture, gardens, waste places, paths, arable fields; common. Europe.

5. Rough Hawk's-beard, *Crepis biennis* L., plant 30–100cm, biennial; stem branched, many-headed; lvs oblong, sinuate-dentate to pinnatifid, lower narrowed into the stalk, the upper sessile or amplexicaul; flower heads 20–35mm wide, yellow; involucre campanulate, hairy, often with yellow or black glandular hairs; fruit beakless, 4–6mm long, tapered above, pappus white. Fl.5–6. Rich meadows, waysides, waste places; widespread. Almost all of Europe. Similar species: **Smooth Hawk's-beard**, *C. capillaris* (L.)Wallr., but stem lvs clearly sagittate-amplexicaul at base; flower heads 10–15mm across; fruit 10-ribbed, 2mm long, abruptly narrowed above. Fl.6–9. Meadows, pastures, heaths, weed communities; common or widespread. In damp meadows and pastures, especially in the mountains, is also the **Northern Hawk's-beard**, *C. mollis* (Jacq.)Aschers., plant 30–75cm; basal lvs ovate-oblong, usually entire, narrowed into the winged stalk; stem lvs sessile, semi-amplexicaul; involucre with black glands; fruit 20-ribbed, 3–4mm long. Fl.6–8. Rare to scattered. C. Europe, Britain.

6. Golden Hawk's-beard, *Crepis aurea* (L.) Cass., plant 5–20cm; lvs in basal rosettes, deeply sinuate-dentate, like dandelion lvs, glabrous; stem unbranched, leafless, with black hairs above; flower heads 2–4cm wide, only with ligulate florets, orange-yellow to red; involucre spreading black-hairy; fruit 6mm long, tapered above, 15–20 ribbed. Fl.7–9. Mountain meadows, alpine pastures, about 900–2900m; widespread. Mountains of C. and S. Europe (eastern). See also *Hieracium aurantiacum*, Pl.194.

For other species in this family, see also *Petasites*, Pl.118, *Inula*, Pl.120.

Plate 87   Pondweed Family, Potamogetonaceae

1. Opposite-leaved Pondweed, *Groenlandia densa* (L.) Fourr. (*Potamogeton densus* L.), plant 20–40cm; lvs submerged, in pairs, opposite, close together, 1–2cm long, lanceolate, serrate; flower spikes few-flowered. Fl.5–8. Slow-flowing, cool, little-polluted waters; scattered. Almost all of Europe. Protected!

2. Fennel Pondweed, *Potamogeton pectinatus* L., plant up to 2m; stem filiform, much dichotomously branched; lvs grass-like, linear, 1–4mm wide, 3-veined and cross-veined, with basal sheaths 5cm long; flower spikes lax, 3–5cm long; fruit almost semi-circular, keeled on the back, with a short beak, yellow-brown, about 4mm long. Fl.6–8. Nutrient-rich ponds, lakes, reservoirs, ditches, down to 4m; rather common. Europe. Similar species: **Slender-leaved Pondweed**, *P. filiformis* Pers., but stem branched only at the base; lvs filiform, 1-veined; fruit obliquely ellipsoid, rounded on the back, greenish, 2mm long; clear lakes and streams; protected!

3. Curled Pondweed, *Potamogeton crispus* L., plant 30–120cm, usually branched; lvs submerged, oblong, up to 10cm long, crisped-undulate, sessile; flower spikes short, few-flowered; fruits united at the base, 5–6mm long, with a curved, falcate beak. Fl.5–9. In standing or slow-flowing, nutrient-rich, often polluted waters down to a depth of 3m; rather common. Almost all of Europe.

4. Perfoliate Pondweed, *Potamogeton perfoliatus* L., plant 1–6m, much-branched; lvs submerged, roundish-oval, amplexicaul, 3–7cm long; flower spikes up to 3cm long, on a stalk up to 10cm, which is not thicker than the stem. Fl.6–8. Nutrient-rich waters down to about 6m deep; scattered. Protected! Almost all of Europe. **Long-stalked Pondweed**, *P. praelongus* Wulf., has similarly amplexicaul lvs, but which are oblong-lanceolate, with a hooded apex, 5–15cm long and 2–3cm wide; stem flexuous; flower spike up to 20cm long; clear, unpolluted waters; rare; protected! **Shining Pondweed**, *P. lucens* L., has oval, acute, shining, submerged lvs, 10–25cm long and 3–5cm wide, which are undulate on the margins and shortly stalked; plant 1–4m long, branched; flower stalks up to 6cm long on a stalk up to 30cm; stalk thickened above. Fl.6–9. Nutrient-rich, often polluted waters; fairly common; Europe.

5. Broad-leaved Pondweed, *Potamogeton natans* L., plant 50–150cm; floating lvs leathery, dark green or brownish, oval, up to 12cm long, usually cordate at base; submerged lvs linear, rush-like, rotted by flowering time; flower spikes up to 8cm long, peduncle up to 10cm, of equal thickness throughout; fruits 4–5mm long. Fl.6–8. In slow-flowing, nutrient-poor, base-rich waters; rather common. Europe.

6. Bog Pondweed, *Potamogeton polygonifolius* Pourr. (*P. oblongus* Viv.), plant 30–60cm, similar to *P. natans*, but floating lvs 2–5cm long, elliptic, narrowed at the base, 1–2 times as long as wide; submerged lvs lanceolate; petioles flattened above; flower spike not thickened upwards. Fl.6–8. Non-calcareous, nutrient-poor waters, bogs, pools on moorland; local to common. C. and S. Europe (western), Britain. Similar species: **Loddon Pondweed**, *P. nodosus* Poir. (*P. fluitans* Roth.), but the floating lvs oblong-lanceolate, 2–4 times as long as wide; flower stalk thickened above; reservoirs, slow-flowing streams; scattered. Both species protected!

7. Red Pondweed, *Potamogeton alpinus* Balb., plant 30–200cm; floating lvs rare, ovate; submerged lvs lanceolate, 10–20cm long, 2–2.5cm wide, narrowed to base, upper ones often reddish; stalk of flower spike thickened above; stipules up to 6cm long. Fl.6–8. Cool, unpolluted, still waters; scattered. C. and N. Europe, Alps, Pyrenees.

Plate 88   Pondweed Family, Potamogetonaceae

1. Lesser Pondweed, *Potamogeton pusillus* L., plant 30–80cm, submerged; stem slender, thread-like, branched; stem sections 2–4cm long; lvs narrowly linear, 15–30mm long, blunt, 3-veined; peduncle 2–3 times as long as the spike; fruit obliquely elliptic, up to 1.5mm long. Fl.6–9. Clear, moderately nutritious lakes or slow-flowing water; widespread, but rarer than in the past. Europe. Similar species: **Hairlike Pondweed**, *P. trichoides* Cham.et Schlecht., but lvs 1-veined, scarcely 0.5mm wide, finely pointed. Protected! **Flat-stalked Pondweed**, *P. friesii* Rupr., lvs 3–5 veined, up to 2.5mm wide, blunt; peduncle of flower spike thickened above. See also *Ruppia maritima*, Pl.196.

Horned Pondweed Family, Zannichelliaceae

2. Horned Pondweed, *Zannichellia palustris* L., plant up to 50cm, much-branched, submerged; lvs opposite or whorled, 1–10cm long, with large ligules and 2 small stipules; lvs up to 1.5mm wide, finely pointed; 1 male flower and 1–6 female flowers per leaf axil; fruits 1–4 together, 2–4mm long. Fl.5–9. Standing or slowly-flowing, nutrient-rich water over a muddy bottom, in 0.5–2.5m depth. Europe.

Naiad Family, Najadaceae

3. Holly-leaved Naiad, *Najas marina* L., plant submerged, rigid, brittle, 10–100cm; lvs 1–4cm long and 1–3mm wide, serrate; stem and leaf margins usually spiny; leaf sheaths entire, rarely with 1–4 small teeth; flowers inconspicuous, sessile in leaf axils. Fl.6–8. Nutritious, calm waters, usually over muddy bottom; rare. Europe. Protected!

Rannoch-rush Family, Scheuchzeriaceae

4. Rannoch-rush, *Scheuchzeria palustris* L., plant 10–20cm; stem leafy; lvs grass-like, yellowish-green, long-sheathing at base; flowers in 3–10 flowered racemes, with large bracts; perianth segments 6, yellow-green, inconspicuous; fruits usually 3, obliquely ovoid, inflated. Fl.5–6. Moorland at moderate to high levels, quaking bogs; rare. Britain, N. and C. Europe, southwards to the Pyrenees, N. Italy and the northern part of the Balkan peninsula. Protected!

Water-plantain Family, Alismataceae

5. Arrowhead, *Sagittaria sagittifolia* L., plant 30–100cm; lower lvs submerged, strap-shaped; floating lvs with oval to sagittate blade, upper lvs erect, deeply sagittate; flowers whorled; inner 3 perianth segments white, twice as long as the 3 green outer ones. Fl.6–8. Ponds, river edges, ditches, reed communities, nutrient-rich waters; local to rather rare. Almost all of Europe.

6. Water-plantain, *Alisma plantago-aquatica* L., plant 20–90cm, emerging from the water; lvs ovate, rounded or cordate at base, leaf blade 5–25cm long; flowers in whorls, the inner perianth segments white, 2–3 times as long as the 3 green outer ones. Fl.6–8. Ponds, banks of lakes and slow-flowing, nutrient-rich waters, reed and larger sedge communities; scattered. Europe. The less frequent **Narrow-leaved Water-plantain**, *A. lanceolatum* With., has lanceolate lvs, which are narrowed into the stalk, and acute perianth segments, and the rare **Ribbon-leaved Water-plantain**, *A. gramineum* Lej., has strap-shaped submerged lvs and rounded perianth segments.

Plate 89 Arrow-grass Family, Juncaginaceae

1. Marsh Arrow-grass, *Triglochin palustre* L., plant 10–50cm; lvs all basal, grass-like, with 2 furrows above; flowers in lax, bractless racemes; perianth absent; styles 3; stamens 6; fruit linear, with 3 mericarps. Fl.6–8. Marshy meadows, ditches, moors and flushes; local to rather rare. Europe. Protected!

Flowering-rush Family, Butomaceae

2. Flowering-rush, *Butomus umbellatus* L., plant 50–150cm; lvs basal, linear, 3-angled, sheathing at the base, 1cm wide and up to 100cm long, emergent from the water; flower stem leafless, terete, longer than the lvs; inflorescence umbellate; perianth segments 6, reddish-white, with darker veins. Fl.6–8. River margins, ditches, reed beds, still and slowly flowing nutrient-rich water; local to rare. Almost all of Europe.

Frogbit Family, Hydrocharitaceae

3. Water-soldier, *Stratiotes aloides* L., water plant, 15–40cm, free-floating, with stolons; lvs sword-shaped, 3-angled, prickly toothed, arranged in a funnel-shaped rosette; flowers 3–4cm wide; male flowers stalked, female flowers sessile; outer 3 perianth segments oval, green, inner 3 roundish, white, 2cm wide. Fl.5–7. Standing or slow-flowing, nutrient-rich water, reservoirs, ditches; rather rare, but very abundant in some places. Usually only female plants in Britain. Almost all of Europe. Protected!

4. Frogbit, *Hydrocharis morsus-ranae* L., water plant, 15–30cm, with stolons, which produce new rosettes; lvs stalked, roundish kidney-shaped, deeply cordate at the base, 2–6cm wide, floating; flowers with stalks 5cm long, outer 3 perianth segments narrowly elliptic, pink or green, the inner 3 roundish, white, with yellow bases. Fl.6–8. Reservoirs, ponds, bays in lakes, nutrient-rich water in places protected from the wind; scattered to rare. Europe. Protected!

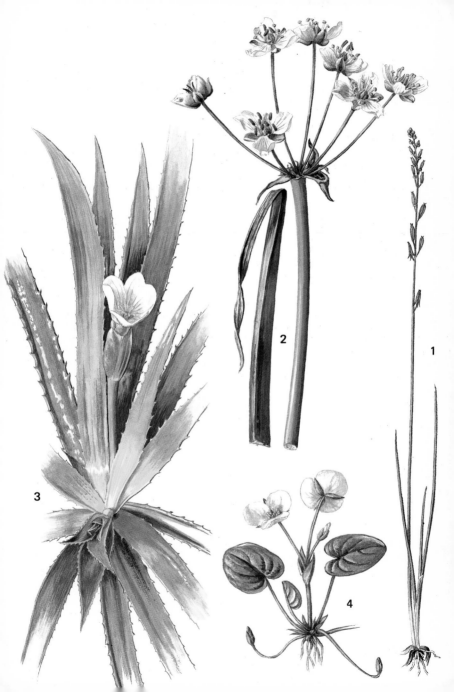

Plate 90   Frogbit Family, Hydrocharitaceae

1. Canadian Waterweed, *Elodea canadensis* Michx., water plant, 30–60cm, creeping or floating, often much-branched, with short stem internodes; lvs usually in whorls of 3, oval, 1cm long, 2–5mm wide, finely serrate; in Europe plants nearly all female; female flower 6-merous, white, 4–5mm wide; reproduces vegetatively. Fl.5–8. Still and slow-flowing, nutrient-rich water; widespread, often occurring in great abundance. Origin N. America, in Europe since 1840. Similar species: **Large-flowered Water-thyme**, *Egeria densa* Planch. (*Elodea densa* (Planch.)Casp.), but plant more robust, 30–300cm; lvs oblong-lanceolate, suddenly pointed, 2cm long, in whorls of 4; warm, nutrient-rich waters; aquarium plant, occasionally naturalised; origin S. America.

Reed-mace Family, Typhaceae

2. Bulrush, *Typha latifolia* L., plant 100–250cm; lvs in 2 rows, 10–20mm wide, flat on both sides, blue-green; flowers in spikes 10–20cm long, 2–3cm wide, the female flowers below in the blackish-brown, wider part, and the males above in the yellow-brown, narrower part; the male and female sections contiguous and each about the same length. Fl.7–8. Still or slowly flowing, nutrient-rich waters, reed beds on river banks, on muddy soil, in water 20–150cm deep; rather common, but becoming rarer because of drainage. Almost all of Europe, northwards to S. Norway, Britain. Similar species: **Lesser Bulrush**, *T. angustifolia* L., but lvs 3–10mm wide, convex on the back; male and female parts of the spike about 3–5cm apart; river banks, ditches; local to rather rare; Europe. **Dwarf Bulrush**, *T. minima* Hoppe, plant 30–70cm; lvs grass-like, 1–2mm wide, radical; stem lvs prolonged below into leaf sheaths; spikes short, ovoid, 2–4cm long, the male spike somewhat narrower; very rare. Protected!

Bur-reed Family, Sparganiaceae

3. Branched Bur-reed, *Sparganium erectum* L. (*S. ramosum* Huds.), plant 30–50cm; stem branched; lvs in 2 rows, sword-shaped, 5–15mm wide, trigonous below; flowers in unisexual, globose heads, the lower female and larger, the upper male and smaller; heads arranged in a panicle. Fl.7–9. River bank reed beds, standing nutrient-rich waters, on muddy soils; widespread. Europe.

4. Unbranched Bur-reed, *Sparganium simplex* Huds. (*S. emersum* Rehm.), plant erect, rarely floating, 20–60cm, unbranched; lvs 3–10mm wide, keeled, trigonous at base; male flower heads 2–8, female 1–5, in spikes or racemes; styles filiform; fruit with a long beak. Fl.6–7. Standing or sluggish waters, muddy soils; rather rare. Europe.

Grass Family, Gramineae or Poaceae

5. Floating Sweet-grass, *Glyceria fluitans* (L.)R.Br., plant 40–100cm, with long stolons; lvs grey to mid-green, 5–10mm wide, rather rough; ligules up to 10mm long, torn; inflorescence a one-sided, often contracted panicle; panicle branches with 1–4 spikelets, which are 10–25mm long; lemmas 6–7mm long, tapered at tip, entire. Fl.5–8. Streams, ditches, reed beds, wet woods; common. Europe. Similar species: **Small Sweet-grass**, *G. declinata* Brebiss., but lemmas 3–5mm long, with 3–5 acute teeth at the tip; panicle branches with 1–4 spikelets; lvs blue-green, suddenly contracted into short-pointed tips; ditches, wet tracks in woods; scattered. **Plicate Sweet-grass**, *G. plicata* Fr., lemmas 3–4mm long, crenate or bluntly 3-lobed at tip; lvs mid-green, long-pointed; panicle branches with 5–15 spikelets; ditches, streams, reed beds; scattered.

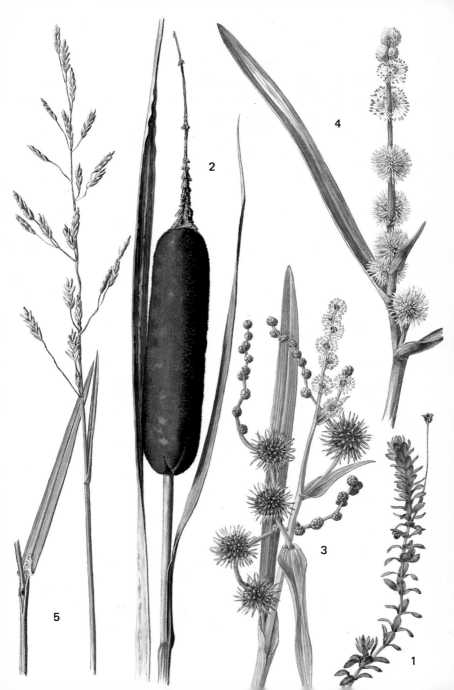

Plate 91   Grass Family, Gramineae or Poaceae

1. Reed Sweet-grass, *Glyceria maxima* (Hartm.)Holmbg., plant 80–200cm; stem thick, erect, reed-like; lvs pale green, 10–15mm wide; ligules 1–3mm long, truncate, leaf sheaths keeled, rough; panicle dense, many-flowered, 20–40cm long; spikelets 5–8mm long. Fl.7–8. Reed of river banks, in still or slow-flowing, nutrient-rich waters; common. Almost all of Europe.

2. Brown Bent, *Agrostis canina* L., plant 20–60cm; with either stolons above ground, or rhizomes beneath, stolons with dense tufts of lvs, and often rooting at the nodes; lvs soft, grey-green, the basal bristle-like, the stem lvs flat, 1–3mm wide, with ligules 2–4mm long and pointed; panicle contracted after flowering; spikelets 2–3mm long, 1-flowered; glumes 1-veined, acute; lemmas 4–5 veined, blunt, awned on the back from below the middle; palea absent or very small. Fl.6–8. In acid grassland, moors, bogs, river banks; common. Europe. Similar species: **Common Bent**, *A. capillaris* L., but panicle spreading after flowering; palea half as long as the lemma; ligules 1–1.5mm long, obtuse; see Pl.42. **Creeping Bent**, *A. stolonifera* L., is always stoloniferous, but stolons without the dense leaf tufts; ligules up to 6mm; panicle contracted after flowering. Grassy and waste places, common. Most of Europe.

3. Purple Moor-grass, *Molinia caerulea* (L.)Moench, plant 50–100cm, stems with bulbous-thickened bases; lower stem internodes very short, uppermost very long, forming the whole of the aerial stem, the stem therefore without nodes; lvs blue-green, flat, rough, weakly ciliate on the margins, with a ring of hairs in place of the ligule; panicle narrow, up to 30cm, usually slate-blue, more rarely violet or green; spikelets 6–8mm long; lemmas 3–4mm long, unawned. Fl.6–9. Moorland pasture, damp sandy and peaty soils, heaths, in the Alps up to over 2000m; common. Almost all of Europe. See also *Deschampsia*, Pl.124.

4. Reed Canary-grass, *Phalaris arundinacea* L., plant reed-like, 50–200cm; lvs 6–12mm wide, flat; ligules 3–6mm long, often split; panicle narrowly oblong and lobed in outline, 10–12cm long, greenish-white or reddish; spikelets 5–7mm long, clustered in spreading branches, 1-flowered, apparently with 4 glumes (because of sterile florets). Fl.6–7. Reed beds on river banks, alder swamps, wet woods, often forming pure stands; common. Europe. Often planted to stabilise banks beside flowing water. See also *Calamagrostis*, Pl.124.

5. Marsh Foxtail, *Alopecurus geniculatus* L., plant 15–40cm; stem prostrate, often floating in water, rooting at the nodes; lvs grey-green, 2–8mm wide; leaf sheaths somewhat inflated; spikes 2–5cm long and 4–5mm wide; awn inserted below the middle of the lemma; anthers pale yellow, then a rusty brown. Fl.5–9. River banks, wet depressions, wet paths, ditches, meadows; widespread. Europe. Similar species: **Orange Foxtail**, *A. aequalis* Sobol., but awn inserted above the middle of the lemma; anthers whitish, then brick-red; leaf sheaths often blue-violet; ditches, pond margins; rarer than the last; Europe.

1

2

3

4

5

Plate 92   Grass Family, Gramineae or Poaceae

1. Common Reed, *Phragmites australis* (Cav.)Trin. (*P. communis* L.), plant 100–400cm, spreading by underground rhizomes by up to 4m; lvs grey-green, 2–3cm wide; a ring of hairs takes the place of the ligule; panicle 15–40cm long, weakly one-sided, brownish or reddish, finally drooping; spikelets 6–10mm long, 3–8 flowered, with silky hairs of 1cm length at the base of the floret; lemmas glabrous. Fl.7–9. River bank reed beds, in up to about 2m of water, upland pastures, damp arable fields, wet woods; often forming extensive pure stands; common. Europe. Often planted for the stabilisation of banks of rivers, streams and lakes. Similar species: **Giant Reed**, *Arundo donax L., plant 2–5m, with a thick stem; lvs 3–4cm wide; ligules very short; panicle 40–70cm long, whitish-green or violet, finally silvery; spikelets 12mm long; lemma with 3 points, covered with dense silky hairs up to 1cm long; Fl.9–12; marshy places, river banks, pond margins; S. Europe.

Sedge Family, Cyperaceae

2. Hare's-tail Cotton-grass, *Eriophorum vaginatum* L., plant forming tussocks, 30–60cm, forming tufts with its bristle-like lvs; rhizomes present; stem lvs rough on the margins, with inflated sheaths; stem round below, trigonous above, with a single terminal flower head; heads oblong-ovoid, up to 2cm long while flowering, forming a white-woolly fruiting head with white bristles 2–4cm long. Fl.4–5. High-level moorland, pine and birch moorland, acid, nutrient-poor, peaty soils; scattered, but locally abundant. C. and N. Europe, Britain, Pyrenees and Spanish mountains.

3. Scheuchzer's Cotton-grass, **Eriophorum scheuchzeri* Hoppe, plant 10–40cm, forming a turf; stem round, with a single terminal flower head which is round, 1cm long at flowering; stem lvs smooth, without an inflated sheath. Fl.6–9. Flush communities, alpine moors; scattered. Alps, Pyrenees, Tatra mountains, Carpathians, Scandinavia. Similar to the last species.

4. Broad-leaved Cotton-grass, *Eriophorum latifolium* Hoppe, plant 30–60cm, tufted, with rhizomes; stem bluntly trigonous; leaf sheaths of the lower lvs blackish-brown, becoming net-veined; stem lvs narrowly lanceolate, 3–8mm wide; spikes 4–10 together, with rough peduncles; perianth consisting of soft bristles and hairs, forming a white-woolly head at fruiting time. Fl.4–6. Level moorland and fens; scattered to rare. Europe. Protected! Similar species: **Slender Cotton-grass**, *E. gracile* Koch, has channelled lvs 1–2mm wide, rough peduncles and a bluntly trigonous stem, which is often bent over; spikes 3–4 together, small; wet acid bogs; rare. Protected!

5. Common Cotton-grass, *Eriophorum angustifolium* Honck., plant 30–60cm, with underground rhizomes; plant tinged pinkish-red at base; stem terete; stem lvs linear, channelled, 3–6mm wide, narrowed into a triquetrous point; upper stem lvs with inflated sheaths; peduncles smooth. Fl.4–5. Level moorland, flushes, bogs; widespread. Almost all of Europe.

Plate 93   Sedge Family, Cyperaceae

1. Deergrass, *Scirpus cespitosus* L. (*Trichophorum cespitosum* (L.)Hartm.), plant 5–30cm, forming thick, dense tufts; stem round, furrowed; spikes 4–6mm long, 3–6 flowered, without woolly hairs. Fl.5–6. Blanket bogs, heaths, on acid peaty soils; scattered. Europe.

2. Alpine Deergrass, *Scirpus hudsonianus* (Michx.)Fernald (*Trichophorum alpinum* (L.)Pers.), plant 10–30cm; stem 3-angled, rough; spikes terminal, 5–7mm long, 8–12 flowered, with wavy, white, woolly hairs 2cm long. Fl.4–5. High and intermediate level moorland; rare. Mainly the mountains of C. and S. Europe, N. Germany, Scandinavia. Protected!

3. Bristle Club-rush, *Scirpus setaceus* L. (*Isolepis setacea* (L.)R.Br.), plant 2–15cm; stem round, leafy at the base; lvs bristle-like; spikes 2–4mm long, 1–4 together, lateral; glumes brownish, with a green keel; bracts far exceeding inflorescence. Fl.7–10. Wet paths in woods, river banks, ditches on moorland, bare sandy or gravelly places; scattered. C. and S. Europe, north to Britain and Scandinavia.

4. Round-headed Club-rush, *Scirpus holoschoenus* L. (*Holoschoenus vulgaris* Link), plant 30–100cm; stem terete, leafless, ribbed; inflorescence of 3–10 spherical heads, 5–15mm across, of which 1 is sessile, the others stalked. Fl.6–8. Marshy meadows, clayey river banks, damp sandy ground near the sea; very rare. Britain, C. and S. Europe. Protected!

5. Common Club-rush, *Scirpus lacustris* L. (*Schoenoplectus lacustris* (L.)Pall.), plant 100–400cm; stem round, leafless, erect; lvs linear, floating at the base of the stem; spikes red-brown, 5–10mm long, in an apparently lateral, head-like panicle; styles 3; fruit 3-angled. Fl.6–7. Reed beds at the water's edge, in still or slow-flowing waters; widespread. Europe. Similar to the **Grey Club-rush**, *S. lacustris* ssp. *tabernaemontani* (Gmel.)Syme (*Schoenoplectus tabernaemontani* (Gmel.)Pall.), but stem grey-green, styles 2, fruit flat; habitat like the last, but often near the sea.

6. Common Spike-rush, *Eleocharis palustris* (L.)R.et Sch., plant 10–80cm; stem leafless, terete, 1–4mm wide; spike terminal, 5–20mm long, 20–30 flowered, with two flowerless glumes at the base, each of which half encircles the stem; styles 2. Fl.5–8. (Variable, with several subspecies.) Reed beds, sedge marshes, wet meadows, ditches, pond margins; fairly common. Europe. Similar species: **Ovate Spike-rush**, *E. ovata* (Roth.)R.et Sch., spike 20–30 flowered, but 3–6mm long, globose-ovoid; styles 2; stem thin, finely striate; lower leaf sheaths purple; plant 5–15cm, densely tufted, without rhizomes; pond margins, muddy soils; rather rare; protected! **Many-stalked Spike-rush**, *E. multicaulis* Sm., spikes about 20-flowered, ovoid, 10–13mm long, styles 3, plant densely tufted, 10–40cm; stem often prostrate and rooting at the tip; pond margins, acid bogs, wet sandy heaths, on wet peaty soils; local to rare; protected!

7. Needle Spike-rush, *Eleocharis acicularis* (L.)R.et Sch. plant 2–10cm, forming a sward; stem 3–4 angled, under 0.5mm wide; lower leaf sheaths often purple; spikes terminal, 2–4mm long, 3–7 flowered; glumes brown with white borders; styles 3. Fl.6–10. Seasonally dry pond bottoms, reservoirs; local. Europe. The **Few-flowered Spike-rush**, *E. quinqueflora* (F.X.Hartm.)O.Schwarz (*E. pauciflora* (Lightf.)Link), has stem 0.5–1mm wide and spikes brownish-red, 3–7 flowered and 5–8mm long; plant 5–20cm; damp places on moors and fens; local to rare. Protected!

8. Flat-sedge, *Blysmus compressus*, see p.236.

Plate 94 Sedge Family, Cyperaceae

1. Wood Club-rush, *Scirpus sylvaticus* L., plant 30–100cm, with sterile leafy shoots; stem triquetrous, leafy; lvs 8–20mm wide, shining green; inflorescence much-branched, up to 20cm wide; spikelets 3–4mm long, blackish-green, arranged in heads of 3–5. Fl.5–7. Wet meadows and woods, marshes, beside streams; local to common. Almost all of Europe. Similar species: **Creeping Club-rush**, **S. radicans* Schk., but spikelets usually solitary, long-stalked; inflorescence very lax; plant often sterile and with long, arching stolons which curve earthwards and tip-root; river banks, muddy soils; rare.

2. Black Bog-rush, *Schoenus nigricans* L., plant densely tufted, 15–50cm; lvs bristle-like, basal sheaths blackish-brown, shining; inflorescence in a head, 10–15mm long, consisting of 5–10 blackish-brown spikelets, exceeded by a bract of 2–5cm long. Fl.5–7. Damp, base-rich peaty soils, fens, moors; locally abundant to rare. Europe. Protected!

3. Brown Bog-rush, *Schoenus ferrugineus* L., plant 10–30cm; basal leaf sheaths dark red-brown; spikelets 2–3, also dark red-brown, scarcely exceeded by the short bract. Fl.5–6. Non-calcareous bogs and moors, flushes; rare. Almost all of Europe.

4. Great Fen-sedge, *Cladium mariscus* (L.)Pohl, plant 80–200cm; stem trigonous; lvs 7–15mm wide, grey-green, sharply serrate-toothed on the margin and the midrib beneath; inflorescence much-branched, composed of many red-brown heads with 3–10 spikelets each; spikelets 1–3 flowered, 3–5mm long. Fl.6–7. River bank red beds, ditches, pools in blanket bogs, fens; local to rare. Almost all of Europe. Protected!

5. White Beak-sedge, *Rhynchospora alba* (L.)Vahl., plant 15–40cm, with or without short rhizomes; stem trigonous; lvs bristle-like or channelled, 1–2mm wide; spikelets 4–5mm long, densely clustered, 2-flowered, white, later reddish; bracts as long as the spikelet cluster. Fl.6–8. Wet peaty places on acid soils, bogs, moors; local to rather rare. Almost all of Europe. Protected!

---

8. Flat-sedge, *Blysmus compressus* (L.)Panzer (Pl.93), plant mid-green, 10–40cm; stem somewhat compressed; lvs flat, keeled, shining, 1–3mm wide, rough on the margins; spikelets red-brown, in 2 rows of 5–12 in a compressed spike, 2–3cm long. Fl.6–8. Marshy places, moors, wet meadows; local or rather rare. Almost all of Europe. Protected! Similar species: **Saltmarsh Flat-sedge**, *B. rufus* (Huds.)Link, but plant grey-green; stem round; lvs smooth, not keeled; spikelets 3–6; saltmarshes, coasts.

Plate 95   Sedge Family, Cyperaceae

1. Brown Beak-sedge, *Rhynchospora fusca* (L.)Ait.fil., plant with long underground rhizomes, 10–30cm; lvs 1mm wide; spikelets densely clustered, red-brown, upper bracts 2–4 times as long as the spikelet clusters. Fl.5–7. Wet peaty places on heaths, edges of bogs: rare. C. and N. Europe, Britain, Pyrenees. Protected!

Single-spiked sedges: plants with one terminal spike only.

2. Davall's Sedge, **Carex davalliana* L., plant 0–40cm; dioecious, male and female spikes on different plants, in dense tufts; lvs bristle-like; lvs and stem triquetrous, rough; fruiting utricle of the female spike 3–5mm long, spreading, long-beaked, brown, somewhat curved. Fl.5–6. Base-rich moors and fens; scattered. C. Europe. Protected! Similar species: **Dioecious Sedge**, *C. dioica* L., but plant with rhizomes; stem and lvs smooth; fruiting utricle shortly beaked, straight; fens, base-rich marshes and moors; local to rare. Europe. Protected!

3. Few-flowered Sedge, *Carex pauciflora* Lightf., plant 5–10cm, monoecious; spikes female at base, male above; stem bluntly 3-angled; spikes few-flowered, with 3–5 spreading or reflexed, yellowish, long-beaked fruiting utricles, 7mm long; styles 3. Fl.5–6. Moors and bogs, at higher levels; rather rare. N. Europe, Britain, W., C. and S.E. European mountains. Similar species: **Flea Sedge**, *C. pulicaris* L., with 5–10 brown, shining, short-beaked fruiting utricles, 4–5mm long; styles 2; fens, base-rich moors; locally common. Europe. Protected! **Bristle Sedge**, *C. microglochin* Wahlenb., but a stout bristle 1–2mm long projects from the tip of the ripe utricle; bogs in mountains; rare. N. Europe, Britain, Alps, Pyrenees. Protected!

Similarly spiked sedges: plants with several spikes, all of the same appearance, each spike with both male and female flowers.

4. Bohemian Sedge. **Carex bohemica* Schreb. (*C. cyperoides* L.), plant 5–30cm, pale green; lvs 1–2mm wide, flat; spikes green, male at the base, female at the tip, head-like, with a bract 2–15cm long; fruiting utricle long-beaked, up to 1cm long; styles 2. Fl.6–9. Pond margins, reservoirs, muddy soils; rare. Mainly C. Europe. Protected!

5. White Sedge, *Carex curta* Good. (*C. canescens* L.), plant tufted, 20–40cm; lvs grey-green, 2–3mm wide; stem sharply angled; spikes 4–6, grey-green, ovoid, 5–9mm long, distant from each other below; utricles pale green, with short beaks. Fl.5–6. Bogs and marshes, on acid soils, river banks; rather common. Europe. Similar species: **Brownish Sedge**, **C. brunnescens* (Pers.)Poir., but spikelets brownish, spherical, 3–5mm long, utricles brown, beak with a long slit down the back; level moorland, damp poor grassland; rare; N. Europe, Alps (up to 2500m), Carpathians.

6. Star Sedge, *Carex echinata* Murray (*C. stellulata* Good.), plant 10–30cm; lvs stiff, grey-green, 1–2mm wide, shorter than the sharply 3-angled stem; spikes 3–5, spherical, 4–6mm long; utricles convex on the outside, flattened on the inside, spreading like a star. Fl.5–7. Moorland, flushes, ditches, damp meadows on acid soils; common. Europe.

Plate 96   Sedge Family, Cyperaceae

Similarly spiked sedges: plants with several spikes, all of the same appearance, each spike with both male and female flowers.

1. String Sedge, *Carex chordorrhiza* Ehrh., plant 10–30cm, with creeping overground stolons up to 1m long, with an ascending, non-flowering shoot at each node and an erect flowering stem at the end; stem leafy mainly at the base; lvs 1–2mm wide, rough on the margins; inflorescence 1cm long, consisting of 2–5 few-flowered, head-like spikes; spikes female below, male above; fruiting utricle brown, smooth on the margin, shortly beaked; styles 2. Fl.5–6. In quaking bogs, on muddy, peaty soils; very rare. Mainly C. and N. Europe, Scotland.

2. Greater Tussock-sedge, *Carex paniculata* L., plant 40–100cm, often building large tussocks; lvs 3–7mm wide; stem very rough, the basal leaf sheaths shining brown, not fibrous; spikes in lax panicles 2–10cm long, with elongated, spreading branches; utricles ovoid, pale brown, shining. Fl.5–6. Ditches, springs, alder carr, sedge fens, on base-rich soils; scattered. Almost all of Europe. Similar species: **Fibrous Tussock-sedge**, *C. appropinquata* Schum., but lvs 2–3mm wide, yellow-green, leaf sheaths black, fibrous; panicle with erect branches; alder carr, moorland, peat diggings; rather rare.

Differently spiked sedges: plants with two distinct kinds of spike.

3. Common Sedge, *Carex nigra* (L.)Rich. (*C. fusca* auct.), plant 5–40cm; lvs 2–5mm wide, grey-green; basal leaf sheaths shining brown; stem rough above; male spikes 1–2, female several, almost sessile, 1–3cm long; styles 2; utricles 2–3mm long, scarcely beaked; lowest bract almost as long as the inflorescence. Fl.4–6. Wet meadows and upland pasture, moorland; common. Europe.

4. Tufted Sedge, *Carex elata* All. (*C. stricta* Good.), plant 50–120cm, forming large tufts or tussocks; basal leaf sheaths yellow-brown, with a network of fibres; lvs 4–5mm wide, stiff, grey-green; spikes 2–6cm long, usually sessile, the lowest spike entirely female, the uppermost entirely male; styles 2; utricle 3–4mm long, soon falling away. Fl.4–5. River banks, bogs, alder carr, beside lakes; locally common. Almost all of Europe.

5. Slender Tussock-sedge, *Carex acuta* L. (*C. gracilis* Curt.), plant 40–150cm; basal leaf sheaths shining brown; stem sharply 3-angled, rough; lvs 5–10mm wide, up to 150cm long; female spikes 3–6, 2–10cm long, male spikes 1–4; utricles 3mm long, pointed, indistinctly beaked; styles 2; lowest bract far exceeding the inflorescence. Fl.5–6. River banks, wet meadows, flooded hollows; rather common. Europe. Similar species: **Lesser Pond-sedge**, *C. acutiformis* Ehrh., plant 30–120cm; lvs grey-green, 5–10mm wide; lower leaf sheaths strongly net-veined and fibrous; spikes 2–8cm long; utricles 4–5mm long, beaked; styles 3; wet meadows, river banks, ponds on moorland; fairly common. Europe.

6. Hairy Sedge, *Carex hirta* L., plant 10–60cm, with brownish to purple-red, weakly net-veined fibrous basal sheaths; lvs grey-green, hairy like the leaf sheaths; inflorescence with 2–3 male spikes and 2–4 long-stalked female spikes; fruiting utricle 5–7mm long, yellow-green, densely hairy; styles 3. Fl.4–6. Paths, river banks, slopes, chalk pits, open spots in meadows; common. Almost all of Europe.

Plate 97   Sedge Family, Cyperaceae

Differently spiked sedges: plants with two distinct kinds of spike.

1. Bog Sedge, *Carex limosa* L., plant 20–40cm, far-creeping, with rhizomes or stolons; lvs 1–1.5mm wide, grey-green, channelled and folded; female spikes shortly cylindrical, 15–20mm long, on slender pedicels, nodding; utricles strongly ribbed; glumes pale brown; lowest bract shorter than the inflorescence as a whole. Fl.5–6. Quaking bogs, very wet bogs and moorland; local to rare. Britain, N. and C. Europe, Alps, Pyrenees, Caucasus. Protected! Similar species: **Tall Bog-sedge**, *C. magellanica* Lam. (*C. paupercula* Michx.), but plant with short rhizomes; lvs flat, 2–4mm wide, green; lowest bract as long or longer than the inflorescence; utricles scarcely ribbed, soon falling away; moorland flushes, moors in the Alps; rare. Britain, N. Europe, Bohemia, Carpathians. Protected!

2. Carnation Sedge, *Carex panicea* L., plant 20–50cm, grey- to blue-green; basal leaf sheaths brown; lvs 2–4mm wide, flat; female spikes 2–3cm long, cylindrical, lax-fruited; styles 3; utricles ovoid-globose, grey- or yellow-green, with a short, thick beak; glumes blackish. Fl.5–6. Fens, level moorland, flushes, wet heaths and meadows; common. Almost all Europe. Often grows with *C. flacca*, Pl.125.

3. Tawny Sedge, *C. hostiana* DC., plant loosely tufted, pale green, 20–50cm; lvs 2–3mm wide, flat, about half as long as the smooth stem; female spikes short, cylindrical, about 1cm long, distant from each other; utricles yellow-green, gradually tapered into the bifid beak; styles 3; glumes red-brown, with white hyaline margin and green midribs. Fl.5–7. Wet moorland pasture and mountain flushes, fens; local. Britain, mainly in the European mountains. Protected! Similar species: **Green-ribbed Sedge**, *C. binervis* Sm., but glumes purple-brown and without white margins; utricles with 2 prominent marginal green nerves. Heaths, moors, upland pasture; common. Britain, W. Europe.

4. Yellow Sedge, *Carex flava* L., plant 5–50cm, yellow-green; basal leaf sheaths yellow-brown; lvs 1–5mm wide, channelled or flat; female spikes 1–6, shortly cylindrical to globose, male spike 1, terminal; utricles 2–7mm long, yellow-green, gradually or suddenly narrowed into the long beak; bracts exceeding the inflorescence. Fl.6–8. (Aggregate species with many microspecies.) Moors, flushes, ditches, wet meadows; scattered. Europe.

5. Cyperus Sedge, *Carex pseudocyperus* L., plant 40–90cm, yellow-green; stem sharply 3-angled, leafy to the top; lvs 6–12mm wide, flat; female spikes 3–6, lax-fruited, drooping on slender pedicels, the upper ones close together; utricles 5–6mm long, shining yellow, deflexed when ripe; teeth of the beak 1mm long, spreading. Fl.6–7. Banks of ponds, reservoirs, ditches, alder carr, wet woods; local to rare. Almost all of Europe.

Plate 98   Sedge Family, Cyperaceae

Differently spiked sedges: plants with two distinct kinds of spike.

1. Bottle Sedge, *Carex rostrata* Stokes (*C. inflata* auct.), plant 30–80cm, basal leaf sheaths brownish-red, with a network of black veins; stem bluntly triquetrous; lvs 3–5mm wide, channelled, grey-green; female spikes 2–8cm long; utricles spreading nearly horizontally, ovoid-globose, suddenly narrowed into the bifid beak, brownish when ripe. Fl.5–6. River banks, ditches, hollows on moors, communities of large sedges; rather common. Almost all of Europe.

2. Bladder Sedge, *Carex vesicaria* L., plant 30–80cm, similar to *C. rostrata*, but stem sharply angled, rough; basal leaf sheaths red, strongly net-veined; lvs pale green, 4–7mm wide, flat; utricles obliquely erect, gradually narrowed into the beak, straw-yellow when ripe. Fl.5–6. River banks, ditches on moors, communities colonising open water, alder carrs; local to common. Almost all of Europe.

Rush Family, Juncaceae

3. Sharp-flowered Rush, *Juncus acutiflorus* Ehrh., plant 30–100cm, pale green; lvs bristle-like, not furrowed; stem flattened, especially at the base; inflorescence terminal, dense, much-branched; perianth segments 6, brown, the outer shorter than the inner, tapering to awn-like points, the outer recurved. Fl.7–9. Wet meadows, springs, ditches, moorland, wet woods; common to local. Almost all of Europe. Similar species: **Jointed Rush**, *J. articulatus* L., but all perianth segments equally long, the inner with a broad, colourless margin; lvs with prominent cross-walls inside; inflorescence with long, little-branched branches; plant 20–50cm; common. **Blunt-flowered Rush**, *J. subnodulosus* Schrank (*J. obtusiflorus* Ehrh.), but perianth segments all obtuse, equally long; inflorescence spreads quadrately; non-flowering shoots like the stems, with bladeless sheaths at the base; plant forms patches, 30–100cm.

4. Thread Rush, *Juncus filiformis* L., plant 10–40cm, creeping, loosely tufted; stem slender, finely striate, leafless, with brown leaf sheaths at the base; inflorescence few-flowered, its bract about as long as the stem, so that the inflorescence appears to be in the middle of the stem; perianth segments whitish, finely pointed. Fl.6–8. Near water, calcifuge; widespread to local. Europe, but not the Mediterranean.

5. Soft Rush, *Juncus effusus* L., plant 30–120cm; stem shining, smooth, grass-green; inflorescence lax, 4–10cm long, apparently lateral. Fl.6–8. Wet meadows, upland pasture, flushes, bogs, wet woods; common. Europe. Similar species: **Compact Rush**, *J. conglomeratus* L., but stem grey-green, finely striate, dull; inflorescence in contracted clusters, 2–4cm long. **Hard Rush**, *J. inflexus* L., see Pl.3. **Heath Rush**, *Juncus squarrosus* L., forms dense tufts; lvs all radical; inflorescence lax, terminal, flowers in clusters of 2–3. Moors, bogs, on acid soils; common. Britain, most of Europe.

Duckweed Family, Lemnaceae

6. Common Duckweed, *Lemna minor* L., plant not differentiated into stem and leaf, but consisting of free-floating, roundish plates (thalli); thallus 2–3mm long, flat below, with only one root; flowers very rarely, without perianth. Fl.4–5. Still, nutrient-rich waters, often covering the whole of the water surface; widespread. Europe.

7. Ivy-leaved Duckweed, *Lemna trisulca*, and **8. Greater Duckweed**, *Spirodela polyrhiza*, see p.246.

Plate 99 Arum Family, Araceae

1. Sweet-flag, *Acorus calamus* L., plant 60–120cm, with a strong sweet smell when crushed; stem trigonous, with 2 rows of lvs; lvs 5–20mm wide, with undulate margins; flower spike (spadix) apparently lateral, 4–10cm long, with a green, stem-like flower sheath (spathe) at its base, which appears to continue the stem above the spadix; flowers pale green, minute; perianth segments 6, under 1mm long. Fl.5–7. River banks, reservoirs, ditches, nutrient-rich waters; scattered. Almost all of Europe.

2. Bog Arum, *Calla palustris* L., plant 15–30cm, creeping; lvs cordate, long-stalked; spathe ovate, 3–7cm long, greenish outside, white within; spadix 2–4cm long, with flowers up to the tip; berries red. Fl.5–9. Poisonous! Protected! Moorland hollows, alder carr, sedge communities; rare. Britain (introduced), N. and C. Europe.

Lily Family, Liliaceae

3. Bog Asphodel, *Narthecium ossifragum* (L.)Huds., plant 10–30cm; lvs in 2 rows, sword-shaped; flowers yellow within, greenish outside, long-stalked, in lax racemes 6–7cm long; anthers orange-red, densely yellow-hairy; styles 3-lobed; bracts lanceolate. Fl.7–8. Heather moorland, bogs, wet acid places on mountains; common in Britain but rarer in C. Europe. Mainly W. Europe. Protected!

4. German Asphodel, *Tofieldia calyculata* (L.)Wahlb., plant 10–30cm; basal lvs in 2 rows, grass-like, 2–4mm wide, 5–10 veined; flowers in the axils of bracts, yellowish, 3–5mm wide, with a 3-lobed, calyx-like epicalyx, in a raceme 2–8cm long; perianth segments 6, lanceolate. Fl.5–7. Base-rich marshes and fens; scattered. S. Scandinavia, mountains of C. Europe, Alps, Pyrenees, Balkan peninsula. Protected! Similar species: **Scottish Asphodel**, *T. pusilla* (Michx.)Pers. (*T. palustris* Huds.), but flowers whitish, without a calyx-like epicalyx, in heads 5–10mm across; lvs 3–5 veined; plant 5–12cm; moorland flushes to 2600m, by streams; local to rare; N. Europe, Britain.

5. Chives, *Allium schoenoprasum* L., plant 10–40cm, with oblong bulbs; lvs radical, tubular, hollow; inflorescence dense-flowered, globose; perianth segments pink. Fl.6–8. Ssp. *schoenoprasum*, the garden herb, which is sometimes naturalised on river banks, has stem 1–2mm wide and perianth up to 10mm long; ssp. *sibiricum* (L.)Hartm., on damp, stony slopes, places where snow lies late, river shingle and level mountain moorland, has a stem 3–5mm wide and perianth segments up to 15mm. Mainly in the European mountains, Britain (rare).

6. False Helleborine, *Veratrum album* L., plant 50–150cm; lvs alternate (in contrast to gentian species), broadly ovate, deeply folded longitudinally; panicle 30–60cm long; flowers star-shaped, 8–15mm across, white or yellowish, greenish outside, the lower hermaphrodite, the upper usually only male. Fl.6–8. Upland pasture, alpine meadows, subalpine scrub communities; widespread. Mountains of C. and S. Europe.

---

7. Ivy-leafed Duckweed, *Lemna trisulca* L. (Pl.98), thalli usually submerged, lanceolate, attached to each other crosswise, 3-nerved, occasionally with rounded floating lvs, 5–10mm long. Fl.5–6. Ponds, lakes, ditches; scattered. Europe.

8. Greater Duckweed, *Spirodela polyrhiza* (L.)Schleid. (*Lemna p.* L.) (Pl.98), each thallus with a tuft of roots, floating at the upper surface of the water, roundish, 3–10mm across, 2–10 attached together, flat on both sides, underside usually red. Fl.5–6. Lakes, ponds, reservoirs; scattered. Europe.

Plate 100 Lily Family, Liliaceae

1. Fritillary, *Fritillaria meleagris* L., plant 15–30cm, with an almost spherical bulb; lvs alternate, usually 4–5, linear, channelled, grey-green, distributed throughout the stem; flowers usually solitary, more rarely 2–3 together, inflated and bell-shaped, drooping, 3–4cm long and about 2cm across, purple-brown, mottled in a chessboard pattern; perianth segments obtuse, curved over at the tip, with a nectar gland at the base; style 3-lobed. Fl.4–5. Poisonous! Protected! Level moorland pasture, wet and frequently flooded marshy water meadows; rare. S. Scandinavia, Britain, C. and S. Europe; cultivated and often naturalised. Similar species: **Tyrolean Fritillary**, **F. tubiformis* Gren.et Godr. ssp. *burnatii* (Planch.)Rix, but the lvs disposed in the upper part of the stem; style 3-lobed only right at the tip; stony grassland of the S. Alps.

Iris Family, Iridaceae

2. Yellow Iris, *Iris pseudacorus* L., plant 50–100cm, with a thick rhizome; lvs sword-shaped, 1–3cm wide, shorter than the round, many-flowered stem; flowers yellow; the outer 3 perianth segments 4–8cm long, ovate, the inner 3 linear, much shorter, not exceeding the styles; capsule bluntly 3-angled, 4–5cm long. Fl.5–6. Protected! Marshes, river banks, ditches, reservoirs, in communities colonising bare mud; widespread. Almost all of Europe.

3. Siberian Iris, * *Iris sibirica* L., plant 30–80cm; lvs less than 1cm wide, shorter than the 1–3 flowered stem; flowers blue to violet, yellow at the base; outer 3 perianth segments 4–5cm long, the inner shorter, but much exceeding the styles; capsule 3–4cm long, transversely wrinkled. Fl.5–6. Protected! Moorland pasture, ditches; has become rarer because of collecting. Most of Europe, northwards to S. Scandinavia, southwards to C. Italy.

4. Marsh Gladiolus, **Gladiolus palustris* Gaud., plant 30–60cm, with a corm; corm with a fibrous, net-veined tunic; lvs 4–9mm wide; flowers about 3cm across, 2–6 per stem, purple; capsule about 1.5cm long, oblong to obovoid; seeds broadly winged. Fl.6–7. Protected! Moorland pasture; rare. C. and S. Europe.

Orchid Family, Orchidaceae

5. Marsh Helleborine, *Epipactis palustris* (Mill.)Crantz, plant 20–50cm; lvs oblong-lanceolate, 1–2cm wide; outer 3 perianth segments 10–12mm long, brownish, the 2 inner shorter, white; labellum (lower lip) 10–12mm long, the terminal part of which (epichile) is white with pink veins, roundish, and with an undulate margin; flowers in a raceme of 8–15 flowers. Fl.6–8. Protected! Fens, dune slacks, level moorland and upland pasture; local to rather rare. Almost all of Europe, northwards to S. Scandinavia, Britain.

Plate 101 Orchid Family, Orchidaceae

1. **Summer Lady's-tresses**, *Spiranthes aestivalis* (Poir.)Rich., plant 10–30cm; stem with narrow-lanceolate, yellowish-green, net-veined lvs at the base, which are 6–10cm long; stem lvs scale-like; flower spike lax; flowers white with greenish veins, arranged in a spiral, pleasantly scented; bracts lanceolate; perianth segments 4–6mm long, lip oblong-ovate, the tip broadened and rounded, margin crenate. Fl.7. Protected! Base-rich moorland; rare. C. and S. Europe, possibly extinct in Britain.

2. **Single-leaved Bog Orchid**, *Microstylis monophyllos* (L.)Lindb. (*Malaxis monophyllos* (L.)Sw.), plant 8–30cm, usually with a single basal, oblong-lanceolate leaf, 4–6cm long, and a many-flowered lax raceme; flowers small, greenish-yellow; bracts lanceolate, acute; the 3 outer perianth segments lanceolate, both the lateral ones linear; lip ovate, abruptly sharp-pointed, placed at the top. Fl.6–7. Protected! Wooded streams and woods by rivers, shady moorland meadows; rare. N. and C. Europe, above all in the Alps and hinterlands, Black Forest. Similar species: **Bog Orchid**, *Hammarbya paludosa* (L.) O.Kuntze (*Malaxis paludosa* (L.)Sw.), plant 5–20cm; lvs 2–3, oval, the upper the largest, up to 3cm long; flowers small, greenish-yellow; lip lanceolate, hollowed out like a spoon, placed at the top. Fl.7–8. Protected! Sphagnum bogs, moorland at moderate altitudes; rare; Britain, N. and C. Europe.

3. **Fen Orchid**, *Liparis loeselii* (L.)Rich., plant 6–20cm, yellow-green, with 1–3 shining, greasy, lanceolate, basal lvs and 3–7 small yellow-green flowers; lip 4–5mm long, elliptic, folded and channelled, faintly crenate on the margin. Fl.6–7. Protected! Base-rich moorland at low and moderate altitudes, wet fens, dune slacks; rare. Britain, N. and C. Europe, southwards to the Pyrenees, Apennines and the northern Balkans.

See also *Listera cordata*, Pl.132.

4. **Loose-flowered Orchid**, *Orchis laxiflora* Lam. ssp. *palustris* (Jacq.)Bonnier et Layens (*O. palustris* Jacq.), plant 30–50cm; lvs linear-lanceolate, channelled, up to 1cm wide, unspotted; inflorescence 5–10cm long, lax-flowered; bracts scarious, 3–5 veined; flowers dark red, lateral perianth segments ascending; lip 3-lobed, 12–15mm long. Fl.4–5. Level moorland, moorland pasture; rare. C. and S. Europe. Protected! Very similar is ssp. *laxiflora*, but central lobe of lip shorter than lateral lobes.

5. **Early Marsh-orchid**, *Dactylorhiza incarnata* (L.)Soó (*Orchis incarnata* L.), plant 20–60cm; stem hollow, angled, with 3–6 lvs; lvs yellow-green, stiffly erect, unspotted, broadest at the base, hooded at the tip; bracts scarious, large; flowers flesh-coloured, rarely yellowish; lip entire or weakly 3-lobed; spur conical, shorter than the ovary. Fl.5–7. Protected! Marshy meadows, level moorland, fen peat; scattered. Mainly N. and C. Europe, Britain.

6. **Irish Marsh-orchid**, *Dactylorhiza majalis* (Rchb.)Hunt.et Summer. (*Orchis latifolia* auct., *O. majalis* Rchb.), plant 15–60cm; lvs broadly lanceolate, 5–10cm long, widest at about the middle, usually spotted; inflorescence dense; bracts green to red, longer than the flowers; flowers red with darker markings; outer 2 perianth segments spreading; lip 3-lobed, the lateral lobes deflexed. Fl.5–6. Protected! Wet meadows, marshes, fens, ditches, level moorland, dune slacks; widespread. Mainly N. Europe, northwards to S. Scandinavia, Ireland and N. Scotland, southwards to N. Italy, N. Spain.

See also *Dactylorhiza*, Pl.134, and *Gymnadenia*, Pl.46.

Plate 102   Bog Myrtle Family, Myricaceae. See *Myrica*, Pl.135.

Willow Family, Salicaceae

1. Creeping Willow, *Salix repens* L., dwarf shrub, 20–100cm, with creeping underground stock and prostrate, arched-ascending branches; lvs narrowly elliptic or lanceolate, 1–5cm long, entire; catkins 10–15mm long, appearing shortly before the lvs; ovaries stalked, silky-felted. Fl.4–5. (Variable and often hybridising.) Moorland pasture, dry grassland, heaths, dunes; scattered. Most of Europe. See *S. aurita*, Pl.136.

Birch Family, Betulaceae

2. Dwarf Birch, *Betula nana* L., dwarf shrub, 20–70cm; lvs almost circular, broader than long, bluntly crenate, about 1cm wide; catkins erect, up to 1cm long. Fl.4–6. High-level moorland, acid peaty soils, in pinewoods on moorland; rare; ice-age relict: Britain, N. Europe, N. Germany, Jura, Alps, Bohemia, Carpathians. Protected!

3. Lesser Birch, **Betula humilis* Schrank, dwarf shrub, 50–200cm; lvs ovate, unequally serrate, 2–3cm long, stalked; young twigs glandular-hairy. Fl.4–5. Birchwoods on moors, moors, peaty soils; rare; ice-age relict. Germany, Alps, Carpathians. Protected!

Dock Family, Polygonaceae

4. Amphibious Bistort, *Polygonum amphibium* L., plant 30–100cm, the water form up to 300cm; lvs of water form oblong-ovate, glabrous, with cordate base, stalks up to 10cm long, lvs of land form oblong-lanceolate, hairy, rounded at the base, shortly stalked; flower stalk arising above the middle of the leaf sheath; inflorescence dense, compact, cylindrical; flowers pink; stamens 5. Fl.6–9. Ponds, ditches, wet meadows, damp arable fields, reed and pondweed communities; common or scattered. Europe.

Purslane Family, Portulaceae

5. Blinks, *Montia fontana* L., plant 10–30cm, prostrate to ascending or floating in water; lvs opposite, spathulate, 5–25mm long; flowers inconspicuous, white, 2–5 together; sepals 2; petals 3–5, 1–2mm long; stamens 3. Fl.5–10. (Aggregate species with many subspecies, distinguished by the size and surface patterning of the seeds.) River banks, ditches, flushes, damp acid arable fields; common or local. Europe.

Pink Family, Caryophyllaceae

6. Water Chickweed, *Myosoton aquaticum* (L.)Moench (*Malachium aquaticum* (L.)Fries), plant 20–60cm, prostrate or clambering; stem 4-angled; lvs opposite, ovate-lanceolate, acute, sessile with a cordate base, 3–8cm long; petals 5, white, bifid almost to the base, 7–10mm long; styles 5 (*Stellaria* has 3 styles); capsule springing open by 5 teeth. Fl.6–9. River banks, ditches, damp loamy and muddy soils, wet woods, marshes and fens; common. Almost all of Europe.

7. Lesser Stitchwort, *Stellaria graminea* L., plant 10–50cm; stem weak; lvs linear-lanceolate, grass-green, 2–4cm long; petals 3–5mm long, white, bifid almost to the base, as long as the 3-veined sepals; styles 5; capsule opening by 6 teeth. Fl.4–6. Moorland pasture, dry grassland, heaths, waysides; common or local. Europe.

8. Bog Stitchwort, *Stellaria alsine*, see p.254.

Plate 103   Water-lily Family, Nymphaeaceae

1. White Water-lily, *Nymphaea alba* L., water plant with a thick rhizome; lvs floating, roundish, 10–30cm long; lateral leaf veins joined together at the leaf margin; petiole up to 3m long, according to water depth; flowers 10–12cm wide; sepals 4, green; petals 15–25, white, as long or longer than the sepals; filaments of the inner stamens filiform; stigmatic disc flat, 12–24 rayed, usually yellow. Fl.6–8. Protected! Nutrient-rich, still or slowly flowing waters, lakes and ponds; scattered. Almost all of Europe. Similar species: **Shining Water-lily**, **N. candida* Presl., but flowers 6–8cm wide, usually half-closed; petals shorter than the sepals; filaments of the inner stamens lanceolate; stigmatic disc concave, 6–14 rayed, reddish; rare. Protected!

2. Yellow Water-lily, *Nuphar lutea* (L.)Sm., water plant with thick, branched rhizome; lvs floating, broadly ovate, 10–30cm long, lateral nerves not joined together at the margin; petioles up to 3m long; flowers 3–5cm wide; perianth segments 5, yellow, as well as the 7–24 spathulate nectaries; stigmatic disc 15–20 rayed, with a funnel-shaped hollow in the centre; fruit pear-shaped, smelling of alcohol. Fl.6–8. Protected! Nutrient-rich, still or slowly flowing waters up to 4m in depth; rather common. Almost all of Europe. Similar species: **Least Water-lily**, *N. pumila* (Timm)DC., but flowers 2–3cm wide; lvs 5–12cm long; stigmatic disc 8–10 rayed, margin emarginate between the rays in a star-shape, flat; nutrient-poor moor and mountain lakes; rare; ice-age relict. Britain, N. and C. Europe. Protected!

Hornwort Family, Ceratophyllaceae

3. Rigid Hornwort, *Ceratophyllum demersum* L., plant 50–100cm, submerged; lvs whorled, forked dichotomously 1–2 times, segments linear, stiff, with many prickly teeth, dark green; flowers small, sessile and solitary in the leaf axils, unisexual; perianth segments 6–12, green; fruit with 2 spines. Fl.6–8. Still or slowly flowing waters, reservoirs, ponds, ditches; fairly common. Pollination takes place under water. Europe.

4. Soft Hornwort, *Ceratophyllum submersum* L., plant 50–100cm, submerged; lvs pale green, dichotomously forked 3–4 times, up to 3cm long, segments soft, linear, filiform, scarcely toothed; fruit spineless. Fl.6–8. Ponds, reservoirs, pondweed communities; local or rare. Britain, C. and S. Europe. Protected!

Buttercup Family, Ranunculaceae

5. Marsh-marigold, *Caltha palustris* L., plant 15–40cm; stem hollow, many-flowered; lvs cordate to kidney-shaped, crenate, dark green, shining, up to 15cm wide; perianth segments 5, glistening, shining yellow inside. Fl.4–6. Marshy meadows, river banks, ditches, wet woods in the Alps to over 2400m; common to widespread, but threatened in some places because of drainage. Europe.

---

8. Bog Stitchwort, *Stellaria alsine* Grimm (*S. uliginosa* Murr.) (Pl.102), plant prostrate, loosely tufted, 10–30cm; lvs blue-green, oblong-ovate, acute, almost sessile, 1–2cm long, ciliate at the base; petals 1–3mm long, white, bifid almost to the base, shorter than the sepals. Fl.6–7. Flush communities, streamsides, ditches, wet tracks and woodland rides; widespread. Europe, in the south only in the mountains.

Plate 104   Buttercup Family, Ranunculaceae

1. River Water-crowfoot, *Ranunculus fluitans* Lam., plant floating in water, up to 6m long; floating lvs absent; submerged lvs finely divided, 5–20cm long, with stretched-out, parallel, filiform segments, which stick together when out of the water like the bristles of a paint brush; flowers 1–2cm across, white; fruiting receptacle (almost) glabrous. Fl.6–8 Streams, rivers, usually in fast-flowing oxygen-rich water; rather common. Britain, S. Sweden, C. Europe, southwards to S. France, N. Italy.

2. Common Water-crowfoot, *Ranunculus aquatilis* L., plant 10–20cm; floating lvs roundish, usually 3–5 lobed to about $\frac{2}{3}$, lobes again divided; submerged lvs up to 5cm long, several times tripartite, finally dichotomously branched; segments spread out in the water, sticking together like a paint brush when out of the water; flowers 1–2cm across, petals contiguous or overlapping, white. Fl.5–8. Nutrient-rich, still or slowly flowing waters, ponds, ditches and streams; scattered. Europe. Very similar species: **Pond Water-crowfoot**, *R. peltatus* Schrank, but floating lvs tripartite only to about the middle, segments coarsely crenate; flowers 2–3cm across; scattered. **Ivy-leaved Crowfoot**, *R. hederaceus* L., floating lvs roundish to kidney-shaped, shining, with 3–5 shallow and obtuse lobes; submerged lvs absent; stem creeping and rooting at the nodes; springs and flushes, wet sandy or muddy soils; local to rare; protected! **Brackish Water-crowfoot**, *R. baudotii* Lloyd, floating lvs with 3 deeply cuneate segments; petals 6–10mm long, yellow at the base; submerged lvs absent or present; salty and brackish water. Coasts of most of Europe.

3. Fan-leaved Water-crowfoot, *Ranunculus circinatus* Sibth., plant 50–100cm; segments of the submerged lvs stiff, not sticking together when taken out of the water; segments in one plane, tripartite at base, then forked dichotomously several times; floating lvs absent; petals white, overlapping one another, 6–10mm long. Fl.5–8. Nutrient-rich waters, water-lily stands; scattered. Almost all of Europe. Similar species: **Thread-leaved Water-crowfoot**, *R. trichophyllus* Chaix (*R. flaccidus* Pers.), but segments of the submerged lvs not all in one plane, usually sticking together like a brush when out of the water; floating lvs absent; petals 3–6mm long, not overlapping; plant usually brittle; nutrient-rich waters; widespread. Europe.

4. Celery-leaved Buttercup, *Ranunculus sceleratus* L., plant 20–60cm, glabrous; stem hollow, striate; lvs rather thick, shining, 3–5 lobed; flowers 5–10mm across, pale yellow; sepals reflexed; achenes glabrous, 1mm long, in cylindrical heads, 6–10mm long. Fl.6–10. Poisonous! Ditches, pond margins, muddy soils, by slow streams; common to local. Europe.

5. Creeping Spearwort, *Ranunculus reptans* L., plant 5–30cm; stem filiform, creeping, rooting at every node; lvs linear-lanceolate to spathulate, stalked, 1–5cm long; flowers solitary, 5–10mm across. Fl.6–8. River banks which are flooded at times, wet depressions, lake margins; rare. Britain (perhaps only as a hybrid with *R. flammula*, Pl.105), N. and C. Europe. Protected!

6. Greater Spearwort, *Ranunculus lingua* L., plant 50–150cm; all lvs undivided, entire or weakly toothed, lanceolate, up to 25cm long; stem thick, hollow, erect; flowers 3–4cm across, yellow. Fl.6–8. Poisonous! River banks, ditches, marshes and fens, reed beds; local to rare. Almost all of Europe. Protected!

Plate 105 Buttercup Family, Ranunculaceae

1. Lesser Spearwort, *Ranunculus flammula* L., plant 10–50cm; lower lvs elliptic, the upper lanceolate, sessile, up to 10cm long; stem prostrate to ascending, rooting only at the lower nodes (in Creeping Spearwort (see Pl.104), stem rooting at every node); flowers 8–15mm across, yellow. Fl.6–10. Poisonous! Marshy meadows, ditches, wet places generally, on muddy soils; colonist; common or widespread. Almost all of Europe.

2. Aconite-leaved Buttercup, **Ranunculus aconitifolius* L., plant 20–60cm; basal lvs 3–7 lobed, long-stalked, lobes broadly ovate, unequally serrate, the middle lobes shortly stalked; stem lvs sessile; flowers white, 1–2cm across; flower stalk hairy; stem with spreading branches. Fl.5–7. Shrub-rich mountain woods, by streams, springs, scrub communities, in the Alps to 2600m; scattered. Mountains of C. and S. Europe. Similar species: **Large White Buttercup**, **R. platanifolius* L., but leaf lobes narrower, central lobes not stalked; flower stalks glabrous; stem with erect branches. Fl.5–7. Mountain scrub communities.

3. Globeflower, *Trollius europaeus* L., plant 30–60cm; lvs palmately divided, the lobes again tripartite, with unequal segments; flowers 1–3, globose, 3–5cm across; sepals 10–15, yellow. Fl.5–7. Protected! Marshy meadows, level moorland, by streams, mountain meadows; in the Alps up to 2400m; scattered. N. Britain, N. and C. Europe, the mountains of S. Europe.

See also *Aquilegia vulgaris*, Pl.143.

Cabbage Family, Cruciferae or Brassicaceae

4. Water-cress, *Nasturtium officinale* R. Br. (*Rorippa nasturtium-aquaticum* (L.)Hayek), plant 30–80cm; stem creeping or ascending, hollow, branched, glabrous; lvs pinnate, with a large terminal lobe, sharp-tasting; flowers white; petals 4, 4–5mm long; anthers yellow; siliquae 13–18mm long, on stalks 8–12mm long. Fl.5–9. Streams, ditches, springs, in fast-flowing, cool waters; common or scattered. Europe. Similar species: **Large Bitter-cress**, *Cardamine amara* L., plant 10–60cm; stem pithy, angled; lvs pinnate, with 4–10 oval leaflets on each side and a larger terminal lobe; petals white (rarely reddish), 4–10mm long; stamens purple; siliquae 2–4cm long. Fl.4–6. By streams, ditches, in flush communities and alder carrs, fens, usually on peaty soil; widespread; almost all of Europe, but in the south only in the mountains.

5. Marsh Yellow-cress, *Rorippa islandica* (Oeder)Borb. (including *R. palustris* (L.)Bess.), plant 15–50cm; lvs pinnate, with 3–7 narrowly ovate, irregularly toothed leaflets on each side and a larger terminal leaflet; petals pale yellow, about as long as the sepals; siliquae 5–12mm long, bloated, as long as the stalks. Fl.6–9. River banks, ditches, damp arable fields and paths, reservoirs; colonist; rather common. Almost all of Europe. Similar species: **Great Yellow-cress**, *R. amphibia* (L.)Bess., but petals golden-yellow, longer than the sepals; basal lvs pinnatisect; stem lvs usually undivided, elliptic to lanceolate, entire or toothed; siliquae ovoid, 3–6mm long, their stalks spreading horizontally, 2–3 times longer; reservoirs, ditches, river banks; scattered.

Plate 106   Sundew Family, Droseraceae

1. Round-leaved Sundew, *Drosera rotundifolia* L., plant 5–15cm; lvs in rosettes, round, long-stalked, with long-stalked, sensitive, sticky glandular hairs (tentacles) and short digestive glands, for catching and digesting insects; inflorescence few-flowered; petals 5, 4–6mm long, white. Fl.7–8. Protected! Moorland at high or moderate levels, acid peaty soils, on bare peat, or in patches of Sphagnum (bog moss); scattered; becoming rarer because of the cultivation of moorland. Britain, N. and C. Europe, southwards to C. Spain, Italy, Corsica, the Balkan peninsula.

2. Great Sundew, *Drosera anglica* Huds., plant 8–20cm; lvs oblong-cuneate, 1–4cm long, inflorescence arising from the centre of the basal leaf rosette, erect, 2–3 times longer than the lvs. Fl.6–8. Hollows in moorland, bogs, often with Sphagnum moss; scattered or rare. Britain, N. and C. Europe, Alps, Pyrenees, Carpathians. Protected! Similar species: **Oblong-leaved Sundew**, *D. intermedia* Hayne, but plant 3–10cm; lvs 7–10mm long; flowering stem arising laterally from below the rosette and curved-ascending, scarcely longer than the lvs; damp peaty soil on moors and heaths, often in drier places than the 2 previous species; local or rare. Also like *D. intermedia* is *D. × obovata* Mert. et Koch, the hybrid between *D. rotundifolia* and *D. anglica*, which has a vertically ascending and straight (not curved) flower stem to distinguish it from *D. intermedia*; capsules without seeds.

Saxifrage Family, Saxifragaceae

3. Yellow Saxifrage, *Saxifraga aizoides* L., plant 3–15cm; lvs fleshy, linear, mucronate, shortly ciliate on the margin; flowers 5–12, lemon-yellow to dark orange, often with orange-red spots. Fl.6–8. Flush communities, by streams, rocks and screes where water trickles; in the Alps to over 3000m; widespread. Britain, N. Europe, Alps, Pyrenees, Carpathians. The **Marsh Saxifrage**, *S. hirculus* L., is likewise yellow-flowered, but flowers only 1–3; petals 10–15mm long, yellow, 2–3 times as long as the sepals, which are soon reflexed; lvs ovate-oblong, 1–3cm long, gradually narrowed into the stalk, flat on both sides, entire, with brownish-red hairs at the base; plant 10–40cm; moors, wet peaty soils; ice-age relict; very rare, has died out in many places. Britain, C. and N. Europe. Protected!

4. Starry Saxifrage, *Saxifraga stellaris* L., plant 4–15cm; lvs in rosettes, obovate, cuneate at base, coarsely toothed at tip, fleshy, shining; inflorescence 3–15 flowered; petals 5 (rarely 6), 2–3mm long, white, with 2 yellow spots. Fl.6–8. Protected! Wet rocks and flushes, damp stream banks; in the Alps to 3000m; widespread. Britain, European mountains, the Arctic. See also *Chrysosplenium*, Pl.146.

5. Grass-of-Parnassus, *Parnassia palustris* L., plant 10–30cm; basal lvs cordate, long-stalked; stem leaf 1, amplexicaul; flowers solitary, 1–3cm across; petals 5, oval, white. Fl.7–9. Marshes, wet moors, flushes, on wet rocks in the Alps to over 2500m; local. Europe, in the south only in the mountains. Protected!

Rose Family, Rosaceae

6. Water Avens, *Geum rivale*, and **7. Marsh Cinquefoil**, *Potentilla palustris*, see p.262.

Plate 107 Rose Family, Rosaceae

1. Tormentil, *Potentilla erecta* (L.) Räusch., plant creeping, prostrate to ascending, never rooting, 15–30cm, with a thickened rhizome with reddish flesh; lvs ternate with 3 cuneate, coarsely toothed leaflets, and a pair of large, pinnately lobed stipules, hence lvs appearing 5-lobed; stem lvs sessile or with a stalk of only up to 5mm; flowers long-stalked, usually with 4 yellow, ovate-cordate petals, 4–5mm long. Fl.6–8. Moorland pasture, heaths, poor grassland, open woods, especially on light acid soils; common. Medicinal plant. Europe. Similar species: **Trailing Tormentil**, *P. anglica* Laich., see Pl.150.

2. Meadowsweet, *Filipendula ulmaria* (L.)Maxim., plant 50–150cm, with large, pinnate lvs; leaf lobes ovate, 4–6cm long, doubly serrate, dark green above and pale green or white-felted beneath; terminal lobes usually ternate; flowers in dense, many-flowered inflorescences; petals 5–6, roundish to ovate, 2–5mm long, yellowish-white, sweet-scented. Fl.6–7. Wet meadows, upland pasture, ditches, river banks, wet woods, fens; common. Europe.

3. Cloudberry, *Rubus chamaemorus* L., plant dioecious, 5–20cm, with creeping underground rhizome; lvs cordate to kidney-shaped, with 5–7 rounded, serrate lobes and shallow sinuses; stipules ovate, papery; stem without prickles; flowers solitary, terminal, white, 2cm wide; petals 5 or more, much longer than the shortly glandular-hairy sepals; fruit pale red to orange, edible but rather tasteless. Fl.5–6. Mountain moors and bogs; locally abundant to rare. Britain, N.W. Germany, mainly N. Europe.

4. Arctic Bramble, *Rubus arcticus* L., plant 10–20cm, with a creeping underground rhizome; stem without prickles; lvs ternate, with ovate, doubly serrate leaflets; flowers 1–3 together, pinkish-red or red, 1.5 to 2.5cm wide; petals 5–7, often toothed; sepals not hairy; fruit dark red. Fl.6–7. Peaty moorland of the N. European mountains, the Arctic.

---

6. Water Avens, *Geum rivale* L. (Pl.106), plant 20–100cm; lvs roundish, divided $\frac{2}{3}$ of the way, or as far as base, in 3 lobes, lobes coarsely toothed; stem laxly branched, many-flowered, spreading-hairy; flowers 5–6-merous, nodding; sepals red-brown, lanceolate, the outer shorter; petals reddish outside, yellow inside; styles jointed above the middle and with hooked tips. Fl.5–6. Wet meadows, river banks, flushes, wet woods, in the Alps to over 2000m; common. Europe, in the south mainly in the mountains.

7. Marsh Cinquefoil, *Potentilla palustris* (L.)Scop. (*Comarum palustre* L.) (Pl.106), plant creeping, rhizomes up to 1m long; stems curved-ascending, 15–30cm high; lvs pinnate, with 3–7 serrate lobes, dark green above, grey- to blue-green below; flowers dark purple; petals remaining until fruiting time, 3–8mm long, finely pointed, half as long as the dull purple sepals; epicalyx green. Fl.6–7. Non-calcareous moors, hollows in heaths, fens, marshes and bogs; in the Alps to about 2000m; local. Europe.

Plate 108   Pea Family, Fabaceae or Papilionaceae

1. Marsh Pea, *Lathyrus palustris* L., plant 30–100cm; stem prostrate or climbing, narrowly winged; lvs with 4 or 6 lanceolate, 5-veined leaflets, 3–6cm long and a terminal tendril; flower stalk scarcely winged; flowers 3–6 together; corolla 15–20mm long, dirty blue or blue-violet. Fl.7–8. Moorland pastures, ditches, large sedge communities, fens, damp grassy places; rare. Mainly N. and C. Europe. Protected!

Crane's-bill Family, Geraniaceae

2. Marsh Crane's-bill, *Geranium palustre* L., plant 30–80cm, dichotomously branched, hairy, but not glandular; lvs 5–7 palmately lobed, lobes irregularly toothed or cut; inflorescence usually 2-flowered; peduncle and calyx appressed-hairy, without glandular hairs (glandular-hairy in *G. pratense* and *G. sylvaticum*, Pl.81); corolla 3–4cm across; petals emarginate, red-violet, Fl.6–9. Streams, ditches, upland pastures, wet scrub; scattered. S. Scandinavia, C. and E. Europe, south to the Po valley.

Mallow Family, Malvaceae

3. Marsh-mallow, *Althaea officinalis* L., plant grey-felted, 50–120cm; lvs weakly 3–5 lobed, grey-felted; flowers clustered in the leaf axils, 3–5cm broad, pale lilac; epicalyx lobes 7–9, lanceolate, shorter than the ovate, acute sepals. Fl.7–9. Ditches, wet, salty soils, coastal; locally common to rather rare; medicinal. Britain, C. and S.E. Europe.

Water-starwort Family, Callitrichaceae

4. Water-starwort, *Callitriche palustris* L., water plant, 5–40cm, with a filiform stem and rosette-like floating lvs; submerged lvs opposite, entire, linear to ovate; flowers small, in the leaf axils, without perianth segments, only with 2 sickle-shaped bracts, 1 stamen and an ovary. Fl.5–10. (Aggregate species with several microspecies which are hard to tell apart.) Still and flowing waters, widespread. Europe. Similar species: **Autumnal Water-starwort**, *C. hermaphroditica* L. (*C. autumnalis* L.), but flowers without bracts; plant always submerged and without floating lvs. Fl.6–9. Still, non-calcareous waters; local to rare; Britain, N. and C. Europe.

Violet Family, Violaceae

5. Marsh Violet, *Viola palustris* L., plant 3–10cm; lvs all radical, cordate to kidney-shaped, shining; flowers in the axils of the basal lvs, scented, lilac; spur straight; lateral petals directed downwards; sepals obtuse. Fl.5–6. Acid, nutrient-poor moorland, bogs, marshes and wet heaths, ditches; common to widespread. Almost all of Europe.

Loosestrife Family, Lythraceae

6. Purple Loosestrife, *Lythrum salicaria* L., plant 50–120cm, erect, woody at the base; lvs lanceolate, up to 10cm long, opposite or in whorls of 3; flowers purple-red, whorled, in long spikes; petals 6, 8–12mm long; stamens 12, of different lengths. Fl.6–9. Wet meadows, ditches, river banks, reed beds, wet scrub communities, fens and marshes; widespread. Most of Europe. Similar species: **Grass-poly**, *L. hyssopifolia* L., but lvs linear-lanceolate, usually alternate, gradually becoming larger from the tip of stem downwards; flowers pale red, 1 or 2 together in the axils, distributed along stem; stem 5–30cm, with curved-ascending branches; muddy river banks, damp arable fields and paths, dwarf rush communities; rare; mainly C. and S. Europe; protected!

Plate 109 Water Chestnut Family, Trapaceae

1. Water Chestnut, *Trapa natans* L., floating plant with a rosette of floating lvs, lvs 5–30, 2–6cm long, rhomboid, leathery, toothed at the tip, green or red above and brown below, with an inflated stalk; flowers 4-merous, white; fruit a nut with 2–4 thorns. Fl.7–8. Non-calcareous, still waters which are warm in summer; rare. Protected! C. and S. Europe, but has disappeared in many places recently because of water pollution.

Willow-herb Family, Onagraceae

2. Square-stemmed Willow-herb, *Epilobium tetragonum* L. (*E. adnatum* Griseb.), plant 30–100cm; stem 2 or 4-angled, more or less glabrous; lvs opposite, oblong-lanceolate, pale green, with many small teeth; petals 5–7mm long, red; stigma club-shaped (not 4-lobed). Fl.7–8. Ditches, river banks, wet scrub communities, woodland clearings, damp stony places, and cultivated land; scattered. C. and S. Europe, northwards to S. Sweden, Britain, rare in the Alps.

3. Great Willow-herb, *Epilobium hirsutum* L., plant 60–120cm; stem branched, spreading-hairy; lvs 6–12cm long, sharply toothed, semi-amplexicaul, softly hairy; corolla 15–22mm wide, red-violet; stigma 4-lobed. Fl.6–9. Ditches, springs, river banks, scrub communities on wet soil, fens and marshes; common to widespread. Almost all of Europe. Similar species: **Hoary Willow-herb**, *E. parviflorum* Schreb., but corolla 6–10mm, lvs weakly toothed, not amplexicaul. Fl.7–8.

Water-milfoil Family, Haloragaceae

4. Spiked Water-milfoil, *Myriophyllum spicatum* L., plant submerged or floating, 20–200cm; leaf whorls usually with 4 lvs; lvs with 13–35 opposite, bristle-like segments; flowers about 3mm across, pink, hermaphrodite and male and female all in the same erect, elongated spike; bracts shorter than the flowers, undivided in the upper part of the spike. Fl.6–8. Still or slow-flowing, nutrient-rich waters up to 6m deep; widespread. Europe. Similar species: **Whorled Water-milfoil**, *Myriophyllum verticillatum* L., but lvs 5–6 in a whorl; bracts all pinnately lobed, longer than the flowers; widespread. The **Alternate Water-milfoil**, *M. alterniflorum* DC., has 8–18 hair-like segments in each leaf; lvs usually 4 to a whorl; flower spike usually drooping at first, with alternate sessile flowers above; petals yellow; nutrient-poor and non-calcareous waters; rare. Protected!

Mare's-tail Family, Hippuridaceae

5. Mare's-tail, *Hippuris vulgaris* L., water plant, 20–100cm; stem usually erect, unbranched, often floating; lvs whorled, linear, submerged lvs flaccid, the aerial lvs stiff; flowers small, green, axillary, without petals, stamen 1, ovary 1, inferior. Fl.6–8. Slow-flowing or almost still, clear waters, especially when base-rich; scattered. Europe.

Carrot Family, Umbelliferae or Apiaceae

6. Marsh Pennywort, *Hydrocotyle vulgaris* L., plant 10–40cm; stem creeping, filiform, rooting at the nodes; lvs circular, peltate, crenate, 2–4cm across; flowers whitish, about 1mm across, 3–5 together in a head-like umbel, which is shortly stalked. Fl.7–8. Upland pastures, level moorland, ditches, river banks, fens and marshes; fairly common to rather rare. Almost all of Europe.

Plate 110 Carrot Family, Umbelliferae or Apiaceae

1. Cowbane, *Cicuta virosa* L., plant 60–120cm, with a stout rootstock which is chambered on account of cross-walls; lvs 2–3 pinnate, segments narrowly lanceolate, sharply serrate, up to 8cm long; umbel 15–25 rayed; bracts 0–2; bracteoles numerous, finally reflexed. Fl.7–9. Poisonous! River banks, ditches, reservoirs, marshes; scattered. Britain, N. and C. Europe, southwards to the Po valley.

2. Lesser Water-parsnip, *Berula erecta* (Huds.)Cov. (*Sium erectum* Huds.), plant 30–80cm, with stolons; stem round, hollow, striate; lvs once pinnate, segments 7–10 pairs, unequal, the lower ovate, the upper lanceolate, coarsely toothed, the terminal segments often ternate; flower umbels opposite to lvs; bracts several; fruit roundish, 2mm long. Fl.7–8. Streams, ditches, ponds, canals and marshes; widespread. Almost all of Europe. Similar species: **Fool's Water-cress**, *Apium nodiflorum* (L.)Lag., but leaf segments 4–6 pairs and bracts usually absent; common. Britain, C. and S. Europe.

3. Greater Water-parsnip, *Sium latifolium* L., plant without stolons, 60–120cm; stem angled and furrowed; stem lvs 1-pinnate, segments ovate-lanceolate, finely and sharply serrate, 3–6cm long, submerged lvs 2-pinnatisect, with linear segments; flower umbels terminal, 15–25 rayed; fruit oblong-ovoid, 4mm long. Fl.7–8. Ditches, river banks, nutrient-rich waters, fens, rather rare. Britain, S. Scandinavia, C. and S. Europe. See also *Chaerophyllum*, Pl.164.

4. Fine-leaved Water-dropwort, *Oenanthe aquatica* (L.)Poir., plant 30–120cm, with a thick, spongy rhizome and slender, tufted root fibres; lvs 2–3 pinnate, segments ovate-lanceolate, 4–6mm long; submerged lvs rare, with filiform segments; umbels shortly stalked, leaf-opposed, 8–15 rayed; bracts usually absent; bracteoles numerous; fruit about 4mm long. Fl.6–8. Ditches, ponds, reservoirs, marshes; scattered. Almost all of Europe. Similar species: **River Water-dropwort**, *O. fluviatilis* (Bab.)Coleman, but plant 100–200cm, the greater part floating in water; leaf segments 15–25mm long, narrowly rhomboid or cuneate, further divided; umbels leaf-opposed; fruit 5–6mm long; non-calcareous, slowly flowing, clear waters; local to rare. Britain, Denmark, Germany, Vosges, Alsace, Netherlands. Protected!

5. Tubular Water-dropwort, *Oenanthe fistulosa* L., plant 30–60cm, with stolons; stem and petioles tubular; lower lvs 2-pinnate, upper 1-pinnate, with linear, often 3-lobed segments; leaf blade much shorter than the petiole; umbels terminal, 2–5 rayed; bracts 0–2; flowers on the edge of the umbel long-stalked and somewhat larger, central flowers almost sessile; fruit bluntly angled, 3–4mm long. Fl.6–7. River banks, ditches, marshes; scattered. Britain, S. Scandinavia, C. and S. Europe (western). Protected! Similar species with terminal umbels: **Parsley Water-dropwort**, *O. lachenalii* C.C.Gmel., plant 40–60cm, without stolons; stem pithy, angled and furrowed; umbels 5–12 rayed; bracts 4–6; bracteoles as long as the pedicels; petals bifid to halfway; upper stem lvs 1-pinnate, segments ovate-lanceolate, 2–4cm long and up to 2mm wide; lower lvs with ovate or cuneate segments; saltmarshes, large sedge communities, marshes and fens; scattered; Britain, Germany, S. Sweden, W. Europe. Protected! **Milk-parsley Water-dropwort**, *O. peucedanifolia* Poll., plant 30–60cm; root turnip-shaped; bracts 0–1; bracteoles shorter than the pedicels; petals bifid to $\frac{1}{3}$; all lvs with linear segments; non-calcareous moorland pasture; rare; Germany, Netherlands, S. W. Europe. Protected!

Plate 111   Carrot Family, Umbelliferae or Apiaceae

1. Milk-parsley, *Peucedanum palustre* (L.)Moench, plant 80–150cm; stem tubular, angled and furrowed; lvs 2–3 pinnate, segments linear-lanceolate, 1–2mm wide, with white points; umbel 15–30 rayed; bracts and bracteoles broadly hyaline-margined; fruit ovoid, 3–5mm long. Fl.7–8. River banks, moors, marshes, fens, alder carr; scattered. All parts of the plant give a milky juice when young. Mainly C. Europe, southwards to N. Italy, S. France, and northwards to Scandinavia, Britain. See also *Angelica*, Pl.163.

Crowberry Family, Empetraceae

2. Crowberry, *Empetrum nigrum* L. ssp. *nigrum*, dwarf shrub, 15–50cm, forming extensive patches; plant dioecious; lvs evergreen, needle-shaped, 3–4 times as long as wide, with parallel margins, with a narrow white centre below; young shoots reddish, later red-brown; flowers solitary, unisexual, on short shoots, inconspicuous, 3-merous; pink or purple; fruit a black, shining, spherical berry; edible, but bitter. Fl.4–5. Peaty moorland, dunes, heaths, mountains, coniferous forests; calcifuge; common or local. N. Europe. N. Germany, Britain, Pyrenees, but absent from the Alps. Protected! Very similar species: **Mountain Crowberry**, *E. nigrum* ssp. *hermaphroditum* (Hagerup) Bocher, but plant with hermaphrodite flowers; the shrivelled stamens often visible at the base of the fruit; lvs 2–3 times as long as wide, oval, with a broad white central stripe below; young shoots green, the older brown; in the Alps to 3000m; N. and C. Europe, Britain, Alps, Pyrenees, Apennines, mountains of the Balkan peninsula.

Dogwood Family, Cornaceae. See *Cornus suecica*, Pl.165.

Heath Family, Ericaceae

3. Labrador Tea, *Ledum palustre* L., shrub 50–150cm, strong-smelling; lvs evergreen, leathery, linear, about 3cm long and 3mm wide, the underside, like the young shoots, rusty-felted; flowers white, in umbels; petals 5, free; ovary inferior. Fl.5–7. Poisonous! Medicinal plant. Protected! Non-calcareous pinewoods on moorland, moorland at high or moderate altitudes; rare. Mainly N. Europe. Scandinavia, Scotland, N. Germany.

4. Bog Rosemary, *Andromeda polifolia* L., plant 10–30cm; twigs curved-ascending; lvs evergreen, narrowly lanceolate, margin inrolled, upper side dark green, underside grey to blue-green; flowers 2–8 together, pink, nodding; corolla globose; stamens 10, enclosed within the corolla. Fl.5–8. Poisonous! Bogs, moorland at higher levels, wet heaths; scattered. N. and C. Europe, Britain, the Alps, Carpathians, Pyrenees. Protected!

5. Cranberry, *Vaccinium oxycoccus* L. (*Oxycoccus palustris* Pers.), plant with filiform, creeping stems up to 80cm long; lvs evergreen, ovate, 4–8mm long, broadest at the middle to $\frac{1}{3}$ the way up; pedicels red, finely hairy; corolla pink, lobes reflexed; filaments glabrous outside; berries globose, red, 8–10mm wide. Fl.6–8. Hummocks in moorland, in Sphagnum mosses, bogs; local. Britain, N. and C. Europe. Very similar species: **Small Cranberry**, *V. microcarpum* (Turcz.)Schmalh., but pedicels glabrous; flowers usually solitary; filaments pubescent outside; lvs 3–5mm long, widest at about $\frac{1}{4}$ the way up; bogs, moorland; rare. See also *Vaccinium, Arctostaphylos*, Pl.167.

6. Heather, *Calluna vulgaris*, and **7. Cross-leaved Heath**, *Erica tetralix*, see p.272.

Plate 112   Heath Family, Ericaceae

1. Dorset Heath, *Erica ciliaris* L., plant 30–80cm, similar to *E. tetralix* (Pl.111), but lvs ovate to ovate-oblong, 2–4mm long, in whorls of 3, glandular-hairy, long-ciliate; twigs spreading-hairy; flowers in elongated, secund racemes, 5–12cm long; corolla deep red, 8–10mm long, pitcher-shaped, slightly curved; calyx hairy; anthers without appendages, enclosed in the corolla (*E. tetralix* has anthers with narrow appendages). Fl.7–9. Moors, heaths, open pinewoods; calcifuge; local. W. Europe, Britain, southwards to Spain.

2. Bell Heather, *Erica cinerea* L., an erect, dwarf shrub, 20–60cm, with linear, dark green, glabrous lvs, 4–6mm long; lvs in whorls of 3, margins inrolled; flowers red-violet, whorled in dense racemes, 1–7cm long; corolla 5–6mm long, enclosing the anthers, which have short appendages; calyx glabrous. Fl.6–8. Moors, heaths, open, stony woods; calcifuge. Common, but rarer eastwards. W. and N. Europe, Britain, northwards to Norway and southwards to Spain.

3. Blue Heath, *Phyllodoce caerulea* (L.)Bab., dwarf shrub, 10–30cm; lvs dense, alternate, linear, evergreen, leathery, 6–10mm long, hairy below; leaf margin inrolled; sepals and corolla lobes 5; corolla pitcher-shaped, 7–8mm long, blue-violet to purple, shortly 5-lobed; flowers on slender, reddish, glandular-hairy stalks in racemes of 2–6. Fl.6–7. Heaths, rocky moorland, open, stony woods, peaty soils; calcifuge; rare. Mainly N. Europe, isolated occurrences in Scotland, the Pyrenees.

4. Cassiope, **Cassiope tetragona* (L.)D.Don, dwarf shrub, 10–30cm; lvs dense, overlapping, arranged in 4 rows, obtuse, 3–5mm long, the underside with a deep longitudinal furrow; flowers solitary, axillary; corolla campanulate, 5–7mm long, white, with pink-coloured lobes or pure white; pedicels glabrous; fruit a capsule. Fl.7–8. Heaths, tundra. Mainly the mountains of N. Europe, Arctic. Similar species: **Matted Cassiope**, **C. hypnoides* (L.)D.Don, but lvs acute, needle-shaped, alternate, not furrowed below; flowers terminal, white, hemispherical; pedicels shortly hairy; places where snow lies late in summer, tundra. N. Europe.

Primrose Family, Primulaceae

5. Brookweed, *Samolus valerandi* L., plant 10–50cm, with a basal leaf rosette; upper lvs alternate, obovate, entire, 1–8cm long; flowers tiny, 2–3mm across; corolla campanulate, white, with 5 rounded lobes. Fl.6–8. Salty, muddy soils, coastal, bottoms of dried-out ponds; widespread, but local or rare. Almost all of Europe. Protected!

---

6. Heather, *Calluna vulgaris* (L.)Hull (Pl.111), plant 30–100cm, shrub; lvs evergreen, in 4 rows on the stem, overlapping, linear-lanceolate, 1–4mm long; flowers in dense, secund racemes; flowers with a 4-merous epicalyx and 4 pale violet to whitish, free sepals; petals 4, united, the same colour as the sepals; stamens 8. Fl.8–10. Heaths, poor grassland, moors, pine and oak woods, bogs, on acid soils; common. Europe.

7. Cross-leaved Heath, *Erica tetralix* L. (Pl.111), plant 15–50cm; lvs needle-shaped, 3–5mm long, ciliate with stiff hairs, in whorls of 3–4; inflorescence head-like, 5–15 flowered; corolla red, 6–8mm long; stamens enclosed in the corolla. Fl.6–9. Moors, bogs and wet heaths, on peaty soils; common to rare. Mainly W. Europe, N. Germany, Britain to Spain and Portugal.

Plate 113   Primrose Family, Primulaceae

1. Bird's-eye Primrose, *Primula farinosa* L., plant 10–15cm; lvs in a basal rosette, dark green above, white-mealy below, usually irregularly toothed; flowers 10–15mm wide, pink or red-violet, in an erect umbel. Fl.5–7. Protected! Calcareous moorland, marshy meadows, stony flushes, stony alpine grassland. Widespread in the Alps and hinterlands, but rare, and has disappeared in many places because of drainage. N. Europe, Britain, Germany, Alps, Pyrenees, Carpathians.

2. Water-violet, *Hottonia palustris* L., water plant, 20–50cm; lvs submerged, in rosettes, pinnate to the middle nerve like the teeth of a comb; flower stalk erect, above the water; flowers white or reddish, yellow inside at the base. Fl.5–6. Moderately nutritious, still waters, over muddy, peaty soils, reservoirs, ditches, moorland lakes, often with Bladderwort (*Utricularia*, Pl.116) and other species with floating lvs; rare. Almost all of Europe, northwards to S. Scandinavia, Britain. Protected!

3. Tufted Loosestrife, *Lysimachia thyrsiflora* L., plant erect, 30–60cm; lvs lanceolate, sessile, decussate; flowers in dense, stalked racemes in the axils of the middle stem lvs; petals 5–6, narrowly lanceolate, 4–5mm long, yellow, red-spotted near the tip. Fl.5–6. On the banks of ponds or slowly flowing waters, on wet and frequently flooded, peaty soils; rare. Europe.

4. Yellow Loosestrife, *Lysimachia vulgaris* L., plant 50–150cm; stem roundish, villous; lvs opposite or in whorls of 3, ovate-lanceolate, up to 14cm long, dotted with glands; flowers in short, stalked racemes or panicles; corolla lobes oval, 7–12mm long, yellow; corolla lobes glabrous on the margins; calyx lobes with reddish edges. Fl.6–8. common or widespread. Europe. Similar species: **Dotted Loosestrife**, *L. punctata* L., but corolla lobes glandular-ciliate; calyx lobes green; stem 4-angled, softly glandular-hairy; lvs in whorls of 3–4, lanceolate-ovate, up to 10cm long. Fl.6–8. Cultivated plant and occasionally naturalised on river banks and in wet woods; native in E. and S.E. Europe, an escape elsewhere.

Plate 114 Bogbean Family, Menyanthaceae

1. Fringed Water-lily, *Nymphoides peltata* (Gmel.)Kunze (*Limnanthemum nymphoides* Link), plant 80–150cm; stem floating; lvs roundish-cordate, like small water-lily lvs; flowers yellow, 3–5cm across, bearded within, long-stalked, projecting above the surface of the water. Fl.7–8. Still or slowly flowing, nutrient-rich waters in a depth of 50–150cm, in communities of plants with floating lvs, usually in low-lying places which are warm in summer; rather rare. Britain, C. and S. Europe. Protected!

2. Bogbean, *Menyanthes trifoliata* L., plant 15–30cm, with a thick, long-creeping rhizome; lvs trifoliate, like clover; flowers white or reddish, bearded, in erect racemes. Fl.4–7. Bogs, fens, edges of lakes and ponds, often in shallow water; common in Britain, but rarer abroad; medicinal plant. N. and C. Europe, mountains of S. Europe. Protected!

Gentian Family, Gentianaceae

3. Marsh Gentian, *Gentiana pneumonanthe* L., plant 15–50cm, without basal leaf rosette; lvs linear, up to 5cm long, usually 1-veined; flowers in the axils of the uppermost lvs; corolla narrowly campanulate, 5-lobed, blue, with 5 green-dotted stripes within. Fl.7–9. Protected! Moorland pasture, level moors, wet heaths; rather rare. Europe, northwards to Scandinavia, Britain, only in the mountains in the south.

4. Marsh Felwort, *Swertia perennis* L., stem 15–40cm, simple, erect; lvs oval, the lower stalked, the upper sessile; flowers 2–3cm across, dirty blue, with darker spots, in racemes or panicles; pedicels 4-angled, winged. Fl.6–8. Protected! Base-rich moors, flushes; up to about 1500m; rare. Alps and hinterlands, mountains of C. and S. Europe.

Bedstraw Family, Rubiaceae

5. Common Marsh-bedstraw, *Galium palustre* L., plant 15–80cm; stems slender, rough with fine, hooked prickles; lvs narrowly ovate-lanceolate, widest towards the tip, obtuse, 1-veined, 4–6 to a whorl, becoming blackish on drying; inflorescence lax; corolla 4-lobed, white, 3–5mm across; anthers purple; fruit almost smooth. Fl.5–9. Reed beds, wet meadows, ditches, river banks, alder carr, fens and marshes; common to widespread. Almost all of Europe. Similar species: **Fen Bedstraw**, *G. uliginosum* L., but lvs linear-lanceolate, mucronate; anthers yellow; fruit finely rugose and rough; fens, moorland pasture.

6. Northern Bedstraw, *Galium boreale* L., plant 30–50cm; stem stiff, erect; lvs in whorls of 4, rough, lanceolate, obtuse, 3-veined; inflorescence pyramidal; corolla 4-lobed, white, 3–5mm across. Fl.6–8. Level moorland, poor pastures, water meadows, streamsides, rocky slopes, scree, also pine and oak woods; locally common or scattered. N. and C. Europe, Britain, mountains of S. Europe.

Plate 115 Borage Family, Boraginaceae

1. Water Forget-me-not, *Myosotis scorpioides* L. (*M. palustris* (L.)Hill), plant 20–100cm; stem weakly angled, densely leafy, with horizontally spreading or obliquely ascending hairs; lvs oblong-ovate, 2–10cm long, 5–20mm wide, hairy; inflorescence without lvs; corolla 4–8mm across, blue; calyx sparingly appressed-hairy, hairs never crisped. Fl.5–9. Wet meadows, ditches, river banks, by ponds; common or widespread. Europe.

2. Common Comfrey, *Symphytum officinale* L., plant 30–100cm; lvs narrowly lanceolate, the upper decurrent down the stem as far as the next leaf, stem therefore broadly winged; flowers red-violet or yellowish-white, with 5 acute coronal scales inside, between the stamens. Fl.5–7. Medicinal plant. Wet meadows, moorland pasture, river banks, wet woods; widespread. Almost all of Europe.

Mint Family, Labiatae or Lamiaceae

3. Skullcap, *Scutellaria galericulata* L., plant 10–40cm; lvs oval to lanceolate, shallowly toothed; flowers in pairs in the axils of the upper lvs, both pointing the same way; corolla 12–20mm long, blue, shorter than the bracts; corolla tube curved upwards; calyx with a projecting flap on the upper lip. Fl.6–9. Wet meadows, ditches, river banks, communities colonising river sediments; widespread. Europe. Also in upland pasture and on river banks and similar in appearance is the **Spear-leaved Skullcap**, *S. hastifolia* L., but lvs entire, the lower hastate; corolla 20–22mm long, longer than the bracts; rare. The **Lesser Skullcap**, *S. minor* Huds., has red-violet flowers 6–7mm long with a straight corolla tube; damp meadows, marshy woods, heaths and moorland, calcifuge; local or rare. Protected!

4. Gipsywort, *Lycopus europaeus* L., plant 20–80cm; lvs lanceolate, deeply and coarsely toothed, lower lvs pinnatisect; flowers in whorls in the axils of the upper pairs of lvs, sessile; calyx teeth 5, stiff, long-pointed; corolla 4–6mm long, white. Fl.7–8. River banks, ditches, alder carr, marshes and fens; rather common. Europe.

5. Water Mint, *Mentha aquatica* L., plant 20–80cm, aromatically scented, with rhizomes and long leafy stolons; lvs ovate, serrate; flowers in a terminal, head-like inflorescence and in whorls in the axils of the upper lvs below; corolla 5–7mm long, pink or lilac; calyx regularly 5-lobed (in **Pennyroyal**, *M. pulegium* L., calyx unequally 5-lobed, almost 2-lipped). Fl.7–9. River banks, ditches, wet meadows, reed beds, communities of large sedges, marshes and fens; common. Europe.

6. Horse Mint, *Mentha longifolia* (L.)Huds., plant 30–80cm; stem hairy; lvs oblong-lanceolate, 6–10cm long, sharply serrate, felted-hairy below; inflorescence spike-like, slender; calyx densely hairy, 5-toothed; corolla 3–4mm long, pink or reddish-lilac. Fl.7–9. River banks, wet meadows, associated with springs, roadsides and waste places; fairly common. C. and S. Europe, northwards to Britain. Similar species: **Round-leaved Mint**, *M. rotundifolia* (L.)Huds., but lvs roundish-ovate, crenate-serrate, felted below; corolla pale lilac or nearly white; ditches, river banks, damp arable fields, roadsides and waste places; rather rare. **Spear Mint**, *M. spicata* L., plant 30–80cm; lvs glabrous, or only hairy on the veins, green on both sides, oblong-lanceolate, sharply serrate; stem glabrous, often red-tinged; inflorescence slender; cultivated plant and occasionally naturalised, originally probably from S. Europe. See also *Stachys palustris*, Pl.30.

Plate 116 Figwort Family, Scrophulariaceae

1. Brooklime, *Veronica beccabunga* L., plant 20–60cm, fleshy; lvs broad elliptic to roundish, 1–4cm long, shortly stalked, irregularly, bluntly serrate; fls in opposite axillary racemes; corolla sky-blue. Fl.5–9. River banks, ditches, springs, marshes and wet meadows; widespread to common. Europe. Similar species: **Blue Water-speedwell**, *V. anagallis-aquatica* L., but lvs oblong-lanceolate, sessile with a weakly cordate base, almost entire; corolla pale lilac, with red-violet veins; by ponds and streams, ditches.

2. Alpine Bartsia, *Bartsia alpina* L., plant 5–15cm; stem with crisped hairs; lvs ovate. with cordate base, bluntly toothed, upper lvs a dirty violet colour; flowers solitary in the axils of the upper lvs; corolla 2-lipped, 15–20mm long, dark violet, glandular-hairy; calyx 4-lobed. Fl.6–8. Moorland flushes, dwarf shrub heaths, mountain meadows and rock ledges, in the Alps to over 2500m; widespread, but rare in the lowlands. Britain, N. Europe, mountains of C. Europe, the Alps, Pyrenees, Carpathians, mountains of the Balkan peninsula.

3. Marsh Lousewort, *Pedicularis palustris* L., plant 20–50cm; stems solitary, erect, branched, with flowers above only; lvs glabrous, deeply pinnatisect; calyx 2-lipped, 5-lobed; corolla pale purple, the upper lip darker, lower lip ciliate, as long as the upper lip. Fl.5–7. Wet heaths and meadows, moorland, peaty soils; common or local. Most of Europe, except Mediterranean region. Protected! Similar species: **Lousewort**, *P. sylvatica* L., but plant with several prostrate-ascending stems, with flowers below as well; lower lip glabrous, shorter than upper lip; damp heaths, marshes, bogs.

4. Moor-king, **Pedicularis sceptrum-carolinum* L., plant 30–100cm, the most impressive species of the genus; lvs 10–30cm long, pinnatisect to the midrib; inflorescence long, lax; corolla pale yellow, 3cm long; lower lip red-edged, upper lip falcate. Fl.6–8. Protected! Moorland pasture, damp grassy places; rare. N. and C. Europe.

See also *Melampyrum pratense* and *Scrophularia*, Pl.174.

Butterwort Family, Lentibulariaceae

5. Alpine Butterwort, **Pinguicula alpina* L., plant 5–15cm; lvs in a basal rosette, fleshy, 3–6cm long, inrolled at the margin, with sticky glands and digestive glands for capturing and digesting insects; flowers white, with a short, conical spur. Fl.5–7. Moors and bogs, damp, mossy rocks, from the lowlands up to 2700m; scattered to rare. Mainly the Alps and hinterlands, Pyrenees, Scandinavia. Protected!

6. Common Butterwort, *Pinguicula vulgaris* L., plant like *P. alpina*, but flowers blue-violet, with a slender, subulate spur. Fl.5–7. Moors and bogs, damp rocks; from the lowlands up to about 2000m; common in Britain but scattered or rare elsewhere in Europe. Mainly N. Europe and the C. European mountains. Protected!

7. Greater Bladderwort, *Utricularia vulgaris* L., plant usually free-floating, submerged, 10–200cm long, only the flowers above the water; lvs 2–8cm long, several times divided into filiform segments, bearing bladders 1–4mm long, for the capture of small animals (plankton) for additional nourishment; flowers 3–15 together in racemes, shining yellow, 13–20mm long. Fl.6–8. Still or slow-flowing, usually nutrient-rich waters, ditches; local to rare. Europe. Protected!

8. Intermediate Bladderwort, *Utricularia intermedia*, see p.282.

Plate 117   Plantain Family, Plantaginaceae

1. Shoreweed, *Littorella uniflora* (L.)Achers., plant 5–10cm, with grass-like, linear, usually erect lvs in a basal rosette, and slender stolons; flowers unisexual, scarious; male flowers solitary on a long stalk, with a 4-merous corolla, 6–8mm long; female flowers usually 2, sessile at the base of the stalk of the male flower. Fl.6–8. Sandy, flat river banks, in rather nutrient-poor lakes and ponds, down to about 2m depth; commoner in the north, rare in the south. Almost all of Europe. Protected!

Valerian Family, Valerianaceae

2. Common Valerian, *Valeriana officinalis* L., plant 30–150cm; lvs opposite, all pinnate, with 7–13 lanceolate, serrate or entire segments; inflorescence dense, corymbose; corolla 3–6mm across, pink. Fl.6–8. Species with numerous microspecies. Moorland pasture, river banks, ditches, wet woods, high-level shrub communities; widespread. The aggregate species in almost all of Europe, the microspecies usually confined to smaller regions.

3. Marsh Valerian, *Valeriana dioica* L., plant 10–30cm; basal lvs roundish to kidney-shaped; stem lvs pinnate, with a larger, oval, terminal segment; inflorescence corymbose; corolla of the female flower 1mm long, white, that of the male flower 3mm long, pink. Fl.5–6. Non-calcareous moorland pasture, fens and bogs; widespread. Almost all of Europe.

Teasel Family, Dipsacaceae. See *Succisa*, Pl.71.

Bellflower Family, Campanulaceae

4. Water Lobelia, *Lobelia dortmanna* L., plant 40–70cm; lvs submerged, linear, almost cylindrical, in a basal rosette; flowers 1–2cm long, whitish, with a bluish corolla tube; corolla 2-lipped, with a 2-lobed upper lip and a 3-lobed lower lip. Fl.7–8. Nutrient-poor, non-calcareous lakes; fairly common, to rare. Britain, N.W. Europe.

---

8. Intermediate Bladderwort, *Utricularia intermedia* Hayne (Pl.116), plant 10–20cm, with colourless rooting shoots, anchored in the soil, and with green, submerged, floating branches; bladders only on the rooting shoots; lvs of the floating branches roundish in outline, 5–20mm long, with strap-shaped segments, 0.5mm wide, which are abruptly pointed, and have 2–10 minute, bristle-tipped teeth on each side; flowers (rarely seen) with a cylindrical spur, 1cm long. Fl.7–8. Moorland hollows, pools and ditches, usually in shallow peaty water, 10–50cm deep; local. C. and N. Europe, Britain, southwards to N. Italy, Pyrenees. In similar habitats also grows the **Lesser Bladderwort**, *Utricularia minor* L., but leaf segments gradually pointed, without bristly teeth; flowers 2–6 together, pale yellow, brown-streaked; lateral margins of the lower lip incurved downwards; scattered; almost all of Europe.

Plate 118 Daisy Family, Compositae or Asteraceae

1. Trifid Bur-marigold, *Bidens tripartita* L., plant 15–100cm; stem brownish-red, with spreading, far-reaching branches; lvs dark green, 3–7 lobed, segments ovate-lanceolate, coarsely serrate, acute; teeth straight, directed forwards; flower heads 1–2cm wide, with or without yellow ray florets; disc florets brownish; the outermost bracts leaf-like, 5–8, yellow-brown; fruit smooth on its flat surfaces, rough with backwards-directed hairs on the margin. Fl.7–10. Pond margins, ditches, wet arable fields, also in waste places, on nitrogen-rich, clayey-muddy or marshy soils; widespread. Almost all of Europe. Similar species: **Greater Bur-marigold**, **B. radiata* L., but lvs pale green, leaf teeth curved inwards; flower heads larger, with 9–14 leaf-like bracts; stem pale green to faintly reddish, with erect branches; ditches, pond margins. **Beggarticks**, *B. frondosa* L. (*B. melanocarpa* Wieg.), lvs pinnate, the upper tripartite, long-stalked, the leaflets also stalked; flower heads long-stalked; fruit tuberculate on the faces and the angles upwardly ciliate; river banks, waste ground.

2. Nodding Bur-marigold, *Bidens cernua* L., plant annual, 10–100cm; lvs undivided, lanceolate, coarsely toothed, sessile; flower heads nodding, 2–4cm wide, often with strap-shaped ray florets; outer bracts 5–8, herbaceous, ciliate on the margin, much longer than the broad ovate, brown-striped inner bracts; fruit flat, 4-angled, with 4 barbed bristles. Fl.7–10. Village ponds, ditches, streamsides, on nutrient-rich muddy soils; fairly common to rather rare. Almost all of Europe, northwards to S. Scandinavia, Britain.

3. Sneezewort, *Achillea ptarmica* L., plant 20–100cm; lvs narrowly lanceolate, undivided, finely serrate, usually glabrous, shining; flower heads 12–17mm wide, from 5 to 30 per stem; ray florets 8–13, white, spreading, 4–6mm long; bracts felted-hairy, with scarious margins. Fl.7–9. Wet meadows, ditches, river banks, marshes; fairly common. Britain, S. Scandinavia, C. Europe, southwards to N. Spain, N. Italy.

4. Butterbur, *Petasites hybridus* (L.)G.M.Sch., plant 10–40cm at flowering time, female spikes at fruiting up to 100cm; lvs appearing at the end of the flowering period, roundish-cordate, up to 60cm wide, toothed, grey-felted beneath at first; scales of the inflorescence reddish; flower heads in dense racemes, with only tubular florets, reddish; male and female on different plants. Fl.3–5. River banks, wet meadows, alder scrub, copses; widespread. Britain, N. and C. Europe and the mountains of S. Europe. See also *P. albus*, Pl.180.

5. Marsh Ragwort, *Senecio aquaticus* Huds. ssp. *aquaticus*, plant 15–60cm; lvs pinnate, yellow-green, with narrow, forward-directed lateral segments and a large terminal segment; basal lvs often undivided; flower heads 2–3cm wide, in corymbose panicles, with a short outer involucre of 1–3 bracts; ray florets 10–12mm long, yellow. Fl.6–8. Wet meadows, moorland pasture, ditches, marshes; common. Almost all of Europe. (See also *S. erucifolius* and *S. aquaticus* ssp. *barbaraefolius*, Pl.74.) In plant communities colonising bare mud with reeds and sedges grows the **Fen Ragwort**, *S. paludosus* L., plant 80–150cm; lvs linear-lanceolate, sharply serrate, grey-felted beneath, 8–14cm long; flower heads 3–4cm across, yellow, 10–30 together in a corymbose panicle, each head with 10–20 ray florets; scattered to rare. Protected!

Plate 119   Daisy Family, Compositae or Asteraceae

1. Brook Thistle, *Cirsium rivulare* (Jacq.) All. (*C. salisburgense* (Willd.)G.Don), plant 30–100cm; stem leafless above; lvs green on both sides, shortly hairy, auriculate, amplexicaul, deeply pinnatisect, lobes lanceolate, usually undivided; flowers purple; involucre usually red-tinged. Fl.5–7. Wet meadows and upland pasture, ditches, wet stony places. Widespread in the Alps and hinterlands, but rather rare. Mainly the C. European mountains, Alps, Pyrenees, Carpathians. Similar species: **Tuberous Thistle**, *C. tuberosum* (L.)All. (*C. bulbosum* DC.), but flower heads solitary; lvs thinly arachnoid-woolly beneath, pinnate, with lobed or deeply serrate lobes; root thickened in the shape of a spindle; moorland pasture, chalk downs, scrub; rather rare. Britain, W. and C. Europe. Protected! Likewise single-headed is the **Meadow Thistle**, *C. dissectum* (L.)Hill, but root not thickened; lvs grey and arachnoid-woolly below; stem lvs only slightly pinnatisect; fens and bog margins, on wet peaty soil; local or rare. W. Europe, from Spain to Britain and N.W. Germany. Protected!

2. Cabbage Thistle, *Cirsium oleraceum* (L.)Scop., plant 50–150cm, with weak prickles; stem distantly leafy; lvs weak, scarcely spiny, pale green, the lower pinnatisect, the upper undivided, ovate, amplexicaul; flower heads several together, surrounded by pale, broadly ovate bracts; flowers yellowish-white, involucre not sticky. Fl.7–9. Wet meadows and woods, stream banks, scrub communities; in the Alps to 2000m; widespread. S. Scandinavia, C. Europe, only occasionally established in Britain, southwards to the W. Alps, N. Apennines, Bulgaria. The **Yellow Melancholy Thistle**, *C. eristhales* (Jacq.)Scop., likewise has pale yellow flowers, but heads not surrounded by bracts; involucre sticky; lvs soft, pinnatisect to the midrib, with lanceolate, prickly toothed side lobes and a very narrow end lobe; high-level shrub communities, deciduous mixed woods, wood margins; Alps, Apennines, the Swiss Jura, Carpathians, mountains of the Balkan peninsula.

3. Melancholy Thistle, *Cirsium helenoides* (L.)Hill (*C. heterophyllum* (L.)Hill), plant 50–100cm; stem densely leafy; lvs green above, snow-white felted beneath, most lvs undivided but some of them sometimes pinnatisect all on the same plant, lobes narrowly lanceolate; upper lvs amplexicaul; flower heads 1–3 together; flowers purple. Fl.7–8. Non-calcareous wet meadows, by streams, in wet shrub communities; local to rare. Mainly N. Europe, Britain and Scandinavia, the Alps and C. European mountains.

Plate 120 Daisy Family, Compositae or Asteraceae

1. **Marsh Thistle**, *Cirsium palustre* (L.)Scop., plant 50–150cm; stem prickly winged, leafy to the top; lvs lanceolate, deeply divided into narrow, spiny lobes, woolly-felted below; stem lvs decurrent; flower heads clustered, ovoid, shortly stalked, purple, 10–15mm long; fruit 2–3mm long, pappus 7–10mm long, white. Fl.5–6. Moorland pasture, marshes, wet grassland, wet woods; common to widespread. Europe. Similar species: **Grey Marsh Thistle**, *C. canum* (L.)All., but flower heads solitary; lvs less decurrent; stem almost leafless above, woolly-hairy; root bulbously thickened; damp meadows, moors; rare; C. and S.E. Europe.

2. **Saw-wort**, *Serratula tinctoria* L., plant 20–100cm; branched above, with numerous flower heads, stem leafy to the top; lower lvs ovate, simple, finely serrate, upper lvs deeply pinnatisect; flower heads narrowly ovoid, 15–20mm long, purple; bracts imbricate, appressed, with purple-red tips; fruit 5mm long, greenish, pappus yellowish. Fl.7–8. Moorland grazing, ditches, water meadows, open deciduous woods; local. Europe. Protected!

3. **Viper's-grass**, *Scorzonera humilis* L., plant 10–40cm; stem and involucre woolly; lvs radical, lanceolate, entire; stem lvs absent or scale-like; stem unbranched, flower heads solitary; flowers pale yellow, twice as long as the involucre; outer bracts half as long as the inner; fruit 5–7mm long, with brown hairs. Fl.5–6. Non-calcareous moorland pasture, poor grassland, heaths, pinewoods; rare. S. Scandinavia, Britain, C. Europe, in the south only in the mountains. Protected! Similar species: **Small-flowered Viper's-grass**, *S. parviflora* Jacq., but stem and involucre glabrous; flowers pale yellow, as long as the bracts; grassland on saline soils; rare; Germany to Spain, and S.E. Europe.

4. **Calycocorsus**, *Calycocorsus stipitatus* (Jacq.)Rauschert (*Willemetia stipitata* (Jacq.)Cass.), plant 15–40cm; stem glabrous below, spreading black-hairy above, with 1–2 small lvs; rosette lvs glabrous, faintly bluish-green, ovate, sinuate-dentate to entire; flower heads 1–5, yellow; bracts in 2 rows; fruit beaked, with a pappus 6–7mm long. Fl.6–8. Moors and bogs, stream banks; scattered. Mountains of C. and S. Europe, Alps and hinterlands, Bavaria.

5. **Marsh Hawk's-beard**, *Crepis paludosa* (L.)Moench, plant 40–80cm, laxly branched above; basal lvs narrowly ovate to lanceolate, irregularly sinuate-dentate; stem lvs cordate or hastate, amplexicaul, bluish beneath; flower heads yellow; involucre glandular-hairy; fruit 5mm long, 10-ribbed, scarcely narrowed above; pappus yellowish-white, brittle. Fl.6–8. Wet meadows, fens, streamsides, wet copses; widespread. Almost all of Europe. Similar species: **Leafy Hawk's-beard**, *C. pyrenaica* (L.) Greuter (*C. blattarioides* (L.)Vill.), stem densely leafy up to the flower heads; lvs ovate-lanceolate, sinuate-dentate, hastate and amplexicaul; basal lvs withered by flowering time; flower heads 3–4cm wide; involucre not glandular, rough with blackish hairs; pappus snow-white, soft, flexuous; high-level scrub communities, damp meadows of the Alps at 1000–2400m; scattered; mountains of C. and S. Europe.

6. **Meadow Fleabane**, *Inula britannica* L., plant 20–80cm, branched above; lvs lanceolate, woolly-villous below, lower narrowed into a short stalk, upper sessile with a cordate base; flower heads 3–5cm across, yellow, solitary or in lax corymbs; ray florets longer than disc florets; bracts spreading or reflexed. Fl.7–9. Wet meadows, ditches, river banks, salt tolerant; rather rare. S. Scandinavia, C. Europe to C. Italy. Rare or extinct in Britain. Protected! See *Pulicaria*, Pl.37, *Sonchus palustris*, Pl.40.

Plate 121   Yew Family, Taxaceae

1. Yew, *Taxus baccata* L., tree, to 20m, often with several trunks; bark at first red-brown, then grey-brown, scaly; lvs evergreen, needle-shaped, flat, all in one plane (ranked), dark green above, pale green below; plant dioecious; male flower consisting of 6–14 sporophylls (like stamens) in a catkin; female flowers solitary in the leaf axils, consisting of a single ovary; seed surrounded by a red, fleshy structure (aril). Flowering begins after 20 years. Fl.3–4. Poisonous! (The seed especially.) Deciduous woods, beech-with-fir woods, mixed mountain forests where winters are mild; local to rare; rejuvenation threatened by the increase of deer, to whom the lvs are non-poisonous. Britain, S. Scandinavia, C. Europe, mountains of S. Europe.

Pine Family, Pinaceae

2. Common Silver Fir, *Abies alba* Mill., tree, up to 50m, crown pyramidal, flattened out at the top in old trees, as the side branches exceed the main shoot; bark smooth, pale grey; needles 2–3cm long, flat, notched at apex, dark green, with disc-like, green, broadened stalks; lvs keeled below, with 2 white waxy stripes; eventually the needles fall away to leave a flat leaf scar. Begins to flower after 60–70 years. Fl.5–6. Forests, especially mixed mountain woods with beech and spruce in humid places, frost sensitive; widespread; rejuvenation threatened by excessive numbers of deer. Mountains of C. and S. Europe, planted elsewhere.

3. Norway Spruce, *Picea abies* (L.)Karst. (*P. excelsa* Link), tree to 50m, crown acute; bark red-brown; needles 4-angled, acute, with a brown, persistent, peg-like stalk; cones drooping, falling off as a whole on ripening; flowering begins after 30–60 years. Fl.5–6. Woods; common; originally only forming stands above 800m; in the Alps to 2000m; now widely spread by planting and replacing other tree species. N. and C. Europe, planted elsewhere. As a cultivated species, for forestry or decoration, the **Colorado Spruce**, *P. pungens* Engelm., has grey-green or silvery-white, hard, spiny needles. Origin N. America.

4. European Larch, *Larix decidua* Mill. (*L. europaea* DC.), tree to 40m, with grey-brown, flaking bark; needles pale green, 1–3cm long, golden-yellow and falling off in autumn, soft, clustered 15–30 together on short shoots; cones ovoid to globose, at first red, then brown when ripe, 2–5cm long; flowering begins after 30–60 years. Fl.4–5. Mountain woods in the Alps, Carpathians, Croatia; needs plenty of light; in the Alps up to 2500m; widespread; repeatedly planted in lowland areas for forestry and ornament.

See also *Juniperus*, Pl.41.

Plate 122 Pine Family, Pinaceae

1. Scots Pine, *Pinus sylvestris* L., tree to 40m; crown conical at first, later flat; bark dark brown in the lower half of the trunk, rusty-red in the upper half and in the crown; needles clustered in 2s on short shoots, grey- or blue-green, pointed, 4–7cm long; cones spherical to ovoid, 3–7cm long, distinctly stalked, pendent; begins to flower after 30–70 years; seeds first ripe in the second year after flowering. Fl.5–6. Woods, dunes, moors, stony, dry, sandy soils, grows where other trees are unable to compete; common. Europe. Similar species: **Austrian Pine**, *P. nigra* Arnold, but needles 8–15cm long, blackish-green, with yellowish tips; trunk and branches dark grey; S.E. Europe, but much planted elsewhere.

2. Mountain Pine, *Pinus mugo Turra (*P. montana* Mill.), prostrate shrub or a tree to 20m high; crown conical; bark grey- to black-brown; needles in pairs on short shoots, dark green, rather blunt, 2–5cm long; cones 2–5cm long, almost sessile. Fl.6–7. (Several microspecies are distinguished by the growth form and the shape of the cone, but these distinctions are not yet clear.) Mountain woods, in the Alps to about 2500m (Mountain Pine or 'krummholz' zone), high-level moors; widespread. Mountains of C. and S. Europe.

3. Arolla Pine, *Pinus cembra L., tree to 25m; crown conical or cylindrical; bark brown, young twigs densely, reddish-yellow hairy; needles clustered in 5s on short shoots, stiff, dark green, 6–10cm long, 3-angled, 1.5mm wide; cones ovoid, 5–8cm long, obliquely erect or spreading, first ripening 2 years after flowering; scales of cone bluish, with brown tips. Fl.6–7. Coniferous forests at high altitudes and on the tree limit in the Alps and Carpathians; common in the C. Alps, rare in the N. Alps, between about 1200 and 2600m.

4. Weymouth Pine, *Pinus strobus* L., tree to 40m; crown conical; bark grey- to dark-brown; young twigs slender, greenish, finely hairy at first; needles clustered in 5s on short shoots, soft, slender, blue-green, 8–15cm long, with 2 wax-stripes on the underside; cones cylindrical, narrow, drooping, often slightly curved, 10–15cm long; cone scales narrow, cuneate, with a yellow-brown thickening at the apex; begins to flower after about 30 years. Fl.5–6. Gardens, parks, planted for forestry; origin N. America.

Plate 123   Grass Family, Gramineae or Poaceae

1. Giant Fescue, *Festuca gigantea* (L.)Vill., plant 60–120cm; lvs flat, 5–15mm wide, shining, with clasping auricles at the base; ligules 1mm long, obtuse; panicle 10–30cm long, lax, finally drooping; spikelets (without awn) 10–15mm long; lemma with a sinuous awn 10–15mm long. Fl.7–8. Wet woods of alder or ash, wet woodland rides, mixed deciduous woods; common. S. Scandinavia, Britain, C. Europe, southwards to C. Italy. Similar species: **Wood Fescue**, *F. altissima* All., but spikelets 7–8mm long, lemma unawned; ligule 3mm long; rocky woods, often beside streams, beech-with-fir woods. See also *F. ovina*, Pl.42.

2. Wood Meadow-grass, *Poa nemoralis* L., plant 20–60cm; lvs spreading horizontally, 1–2mm wide, bluish-green; ligule very short or absent; panicle few-flowered, lax, 5–10cm long; spikelets often 1-flowered. Fl.6–7. Mixed deciduous woods, hedges, wood margins; common. Europe.

3. Wood Melick, *Melica uniflora* Retz., plant 30–50cm; lvs 3–6mm wide; leaf sheaths glabrous, with a ligule-like projection, up to 4mm long, opposite the blade; panicle lax; spikelets erect, 5–6mm long, 1-flowered. Fl.5–6. Deciduous mixed woods, beech woods; widespread. S. Scandinavia, Britain, C. and S. Europe.

4. Mountain Melick, *Melica nutans* L., plant 30–60cm; lvs 3–8mm wide; ligules very short, without an appendage opposite the blade; spikelets nodding, 6–10mm long, 2-flowered, in lax, one-sided panicles, up to 10cm long. Fl.5–6. Deciduous mixed woods, carr, open coniferous forests, usually on limestone; local to common. Almost all of Europe.

5. Blue Moor-grass, *Sesleria albicans* Kit. (*S. caerulea* (L.)Ard.), plant densely tufted, 10–40cm; lvs mid-green, 2–3mm wide, with a prominent midrib and a blunt tip; ligules 0.5mm long; spike ovoid to cylindrical, with small, scale-like, scarious bracts at the base; lemmas bluish, with an awn-like tip. Fl.3–5. Open pine and beech woods, dry or dryish calcareous grassland; local to rather rare, widespread in the Alps and hinterlands. Almost all of Europe, in the mountains in the south.

6. Wood Barley, *Hordelymus europaeus* (L.)Jessen (*Elymus europaeus* L.), plant 50–120cm; lower stem sheaths and nodes villous-hairy; lvs 6–10mm wide, green, soft, shortly hairy, with falcate auricles at the base, the midrib whitish and prominent beneath; ligule almost lacking; spikes 4–8cm long; spikelets 10-flowered, at times in 2s or 3s, making an interrupted spike, shortly stalked; glumes with an awn 2–3cm long. Fl.6–7. Deciduous mixed woods, beech woods, on calcareous soils; scattered. Britain, S. Scandinavia, C. and S. Europe.

Plate 124   Grass Family, Gramineae or Poaceae

1. Tor-grass, *Brachypodium pinnatum* (L.)Beauv., plant 50–100cm, yellow-green, with long rhizomes; stem stiff, shortly hairy on the nodes only; lvs 4–6mm wide, bristly-ciliate on the margins, rarely glabrous; ligules truncate, 1–2mm long; spike stiff, erect; spikelets 2–3cm long, 8–20 flowered; lemma with a short glume. Fl.6–7. Open, dry woods and wood margins, poor grassland, meadows, on chalk and limestone; widespread. Europe. Similar species: **False Brome**, *B. sylvaticum* (Huds.)Beauv., but plant in loose tufts; lvs dark green, flaccid, with a prominent whitish midrib below; spikes drooping; awn as long or longer than the corresponding lemma; woods and hedges, with deciduous trees, sometimes in grassland; common. Europe.

2. Tufted Hair-grass, *Deschampsia cespitosa* (L.)P.B., plant 30–120cm, forming tufts or tussocks; lvs flat, 2–4mm wide, very rough, deeply channelled above, with about 7 ribs; ligule 6–8mm long, often split; panicle green, 10–30cm long, silvery- or reddish-shining; main axis rough; spikelets 4–5mm long, usually 2-flowered; awn of the lemma usually hidden within the floret. Fl.6–7. Wet woods and clearings, damp meadows, moorland, marshes; common. Europe.

3. Wavy Hair-grass, *Deschampsia flexuosa* (L.)P.B., plant loosely tufted, 30–50cm; lvs inrolled, bristle-like; ligules 2–3mm long; panicle lax, with slender, sinuous branches; spikelets pale, brownish, violet-tinged like the branches of the panicle; awn of the lemma geniculate, exceeding the spikelet. Fl.6–8. Deciduous and coniferous woods on acid soils, heaths, poor pasture; calcifuge; widespread. Europe. See also *Agrostis capillaris*, Pl.42.

4. Wood Small-reed, *Calamagrostis epigejos* (L.)Roth, plant 80–150cm, with long rhizomes; lvs 5–10mm wide, flat or inrolled, hard, very rough on both sides; ligule up to 9mm long, truncate; panicle up to 30cm long, branches clustered and lobed in outline; spikelets with a tuft of hairs at the base of the lemma; lemmas awned on the back, awn much exceeding the lemma, almost reaching the tip of the glume, but not reaching out of the spikelet. Fl.6–8. Wood clearings, open woods, river banks, gravel pits, ditches and fens; local to common. Europe. Similar species: **Hairy Small-reed**, **C. villosa* (Chaix)Gmel., lvs 4–5mm wide, flaccid, with 2 tufts of hairs at the base; awn on the back of the lemma not exceeding the lemma; spruce forests and spruce moorland, oak woods on acid soils, dwarf shrub communities; mountains of C. and S.E. Europe. **Mountain Small-reed**, **C. varia* (Schrad.) Host, lvs dark green, 3–8mm wide, rough on both sides; awn geniculate, inserted on the lemma about $\frac{1}{4}$ the way up, exceeding the lemma and the glume and reaching beyond the spikelet; mountain woods, especially with Scots Pine, stony soils, a pioneer on landslides; mountains of C. and S. Europe.

Plate 125 Grass Family, Gramineae or Poaceae

1. Creeping Soft-grass, *Holcus mollis* L., plant 30–80cm, blue- to grey-green, with rhizomes; stem densely hairy only on the nodes (in contrast to *H. lanatus*, see Pl.77); lvs 3–8mm wide, rough; glumes glabrous; awn geniculate, projecting beyond the spikelet. Fl.7–8. Oakwoods on acid soils, heaths, wood clearings. Almost all of Europe.

2. Wood Millet, *Milium effusum* L., plant 60–100cm, blue-green, glabrous; lvs 10–15mm wide, keeled below; ligule up to 7mm long; panicle lax, 10–25cm long; panicle branches horizontal, later reflexed; spikelets 1-flowered, oval, 3–4mm long; glumes shining, glabrous, becoming hard after flowering. Fl.5–7. Mixed deciduous or coniferous woods, open spruce-fir-beech forests; widespread. Europe. See also *Dactylis glomerata* ssp. *aschersoniana*, Pl.77.

Sedge Family, Cyperaceae

3. Quaking-grass Sedge, **Carex brizoides* Jusl., plant 30–70cm, forming patches, with rhizomes; lvs flaccid, 2–3mm wide, longer than the stem; stem triquetrous, slender, bent over in fruit; inflorescence 2–3cm long, comprising 5–8 curved, similar-looking spikes, which are male below and female above; fruiting utricles 2–2.5mm long, narrowly winged, yellow-green; styles 2. Fl.5–6. Damp woods, mixed deciduous woods, carrs; common, formerly used for stuffing cushions. C. and S. Europe, southwards to Pyrenees, N. Italy.

4. Remote Sedge, *Carex remota* L., plant 30–60cm; lvs 2mm wide, weak; stem triquetrous, leafy throughout, later drooping; spikes 5–10, all alike, 5–8mm long, the lower flowers male, the upper female, the lower spikes considerably distant from each other (up to 8cm), hence the whole inflorescence 10–15cm long; bracts leaf-like, flaccid, very long; fruiting utricles 3–4mm long, whitish-green, flat on the inside, convex on the outside; styles 2. Fl.6–7. Wet woods, shady bogs, damp woodland rides, mixed deciduous woods; common. Almost all of Europe.

5. Glaucous Sedge, *Carex flacca* Schreb. (*C. glauca* Scop.), plant 20–60cm; basal sheaths red-brown; lvs 2–5mm wide, stiff, ribbed, blue-green; spikes of different appearance, the male slender, 2–4, terminal, the female cylindrical, 2–3cm long, on pedicels 3–8cm long, finally pendent; bracts leaf-like, the lowest reaching the tip of the inflorescence; fruiting utricle brown to black, convex on both sides, with an indistinct beak; styles 3. Fl.5–7. Open woods, calcareous grassland, fens, marshes; common. Europe.

Plate 126   Sedge Family, Cyperaceae

1. Pendulous Sedge, *Carex pendula* Huds., plant 50–150cm, forming tufts; lvs 7–15mm wide, shining; stem triquetrous, 2–4mm wide, regularly leafy throughout; male spike 1, terminal; female spikes 2–5, long-stalked, curved and pendulous, 7–15cm long, dense-flowered; bracts leaf-like; fruiting utricles 3–4mm long, narrowed into a short beak, shining; styles 3. Fl.5–6. Damp deciduous woods, shady marshes, stream banks; common to local. Britain, C. and S. Europe.

2. Small White Sedge, **Carex alba* Scop., plant 10–30cm, with long rhizomes, forms patches; basal leaf sheaths brown; lvs 1–1.5mm wide, stiff, glabrous; stem bluntly triquetrous; male spike 1, terminal, female spikes 2–4, lax-fruited, 5–10mm long, on erect stalks 0.5 to 3cm long, the uppermost female spike often exceeding the male spike; fruiting utricles globose, yellowish, glumes white and hyaline; styles 3. Fl.5–6. Likes warm places. Deciduous woodland, especially with beech, pinewoods; scattered. S. Germany, Alps, Pyrenees, Carpathians, mountains of the northern Balkan peninsula.

3. Fingered Sedge, *Carex digitata* L., plant 10–30cm; basal leaf sheaths red-brown, fibrous; lvs 2–4mm wide, dark green, flaccid; male spike 1, terminal, female spikes slender, 15–20mm long, 5–10 flowered, arranged digitately, the lowest somewhat distant, 1–2cm long, stalked; fruiting utricle yellow-green to pale brown; styles 3. Fl.3–5. Mixed woods; rare in Britain, commoner elsewhere. Most of Europe.

4. Bird's-foot Sedge, *Carex ornithopoda* Willd., plant 5–15cm, very similar to *C. digitata*, but all spikes arising from almost the same point; female spikes 6–10mm long, 2–6 flowered, finally curved like claws. Fl.4–5. Woods, scrub, dryish grassland; local. Almost all of Europe.

5. Wood Sedge, *Carex sylvatica* Huds., plant tufted, 30–70cm; basal leaf sheaths brown; stem leafy to the top; lvs 4–8mm wide, flaccid, shining green; male spike 1, terminal; female spikes 2–5, linear, 3–5cm long, lax-fruited, on slender, nodding stalks 3–10cm long; fruiting utricles 5–6mm long, gradually narrowed into the long, bifid beak; styles 3. Fl.5–7. Deciduous and coniferous mixed woods, on clayey soils; common to widespread. Britain, S. Scandinavia, C. and S. Europe.

See also *C. montana, C. pilulifera, C. pallescens*, Pl.43; and *Scirpus sylvaticus*, Pl.94.

Rush Family, Juncaceae

6. Hairy Wood-rush, *Luzula pilosa* (L.)Willd., plant 15–30cm, tufted; basal lvs 5–10mm wide, hairy, stem lvs narrower; leaf sheaths dark red; inflorescence umbellate; flowers solitary, on long pedicels; perianth segments 6, acute, brown, 3–4mm long; flower pedicels later reflexed. Fl.4–5. Woods, hedgebanks; widespread. Europe.

7. Snow Wood-rush, *Luzula nivea* (L.)DC., plant loosely tufted, 40–90cm; lvs 3–5mm wide, ciliate; flowers in clusters of 6–20; bracts as long as or longer than the inflorescence; inflorescence contracted; perianth segments 6, pure white. Fl.6–7. Deciduous and coniferous woods; scattered. Alps and hinterlands, Pyrenees, Cevennes, Apennines, elsewhere a garden escape. Similar species: **White Wood-rush**, *L. luzuloides* (Lam.) Dandy et Willm. (*L. albida* (Hoffm.)DC.), but inflorescence spreading; flowers in clusters of 2–6, yellowish-white; on acid soils in woods, poor grassland and dwarf shrub communities.

See also *L. multiflora*, Pl.43.

Plate 127 Rush Family, Juncaceae

1. Great Wood-rush, *Luzula sylvatica* (Huds.) Gaud., plant 30–90cm, with short, thick stolons; lvs 10–15mm wide, long-ciliate on the margins, shining, dark green; flowers clustered 3–4 together on the inflorescence branches; bracts shorter than the inflorescence; perianth segments brown or red-brown, with green midribs, the inner longer than the outer. Fl.5–6. Woods on acid soils, heaths, moorland; common. Almost all of Europe. Similar species: **Sieber's Wood-rush**, **L. sieberi* Tausch, with a smaller inflorescence and lvs 4–5mm wide; subalpine spruce forests, Alpenrose shrub communities; the Alps.

Arum Family, Araceae

2. Lords-and-ladies, *Arum maculatum* L., plant 15–40cm; lvs sagittate, 10–20cm long, green, rarely spotted; flowers small, sessile on a spadix, with the female flowers below, then the male flowers, and above that the flowers are transformed into a fringe which closes the mouth of the spathe; spadix flowerless above, violet, surrounded by a yellow-green, trumpet-shaped spathe, 10–25cm long; berries red. Fl.4–6. Poisonous! Wet woods, mixed deciduous woods, hedges; common. Pollinated by small flies which fall through the mouth of the spathe. Britain, C. and S. Europe.

Lily Family, Liliaceae

3. Yellow Star-of-Bethlehem, *Gagea lutea* (L.)Ker-Gawl. (*G. sylvatica* Loud.), plant 10–30cm, with 1 bulb only, auxiliary bulbs absent; with 1 basal leaf 6–12mm wide; inflorescence umbellate; perianth segments free, yellow, star-like. Fl.4–5. Damp woods, deciduous woods, scrub, orchards, pastures; widespread. Most of Europe.

4. Southern Wild Tulip, **Tulipa sylvestris* L. ssp. *australis* (Link)Pamp., plant 20–40cm; lvs usually 2, linear, 1cm wide; flowers campanulate, 2–3cm long; perianth segments narrowly lanceolate, pointed, yellow inside, red-tinged outside; capsule about as long as wide. Fl.4–6. Mountain pastures, open mountain woodland, rock outcrops; scattered. W. Alps, Apennines, mountains of S.W. Europe. Similar is the **Wild Tulip**, *T. sylvestris* ssp. *sylvestris*, but lvs usually 3, narrowly lanceolate, 1–2cm wide; flowers 4–6cm long, yellow, hairy inside at the base; capsule twice as long as wide. Fl.4–5. Protected! Vineyards, scrub, damp woods, meadows; rare. Germany, France, S. Europe, and naturalised in N. Europe, Britain.

5. Butcher's-broom, *Ruscus aculeatus* L., evergreen shrub, 30–100cm; lvs scale-like, soon falling, and the function of lvs taken by leaf-like, flattened lateral shoots (cladodes); cladodes in 2 rows, broadly lanceolate, leathery, evergreen, 2–3cm long, with spiny points; flowers small, greenish, 3mm wide, in 1s or 2s, with scaly bracts, on the upper side of the cladodes; the fruit is a spherical, red berry. Fl.2–4. Scrub, oakwoods, dry, stony slopes; likes warm places; scattered. S. Alps, S. Europe, France, Britain; also cultivated and becoming naturalised.

6. Bluebell, *Hyacinthoides non-scriptus* (L.)Chouard (*Endymion non-scriptus* (L.)Garcke), plant 15–30cm; lvs 5–6, linear, shining, 5–10mm wide; flowers blue, 4–20 in a secund raceme on a leafless, herbaceous stem; corolla lobes united at the base, campanulate; pedicels shorter than the bracts and bracteoles. Fl.4–5. Protected! Deciduous woods and scrub, hedgebanks; common in Britain but rarer elsewhere; France, Germany, W. Europe.

Plate 128   Lily Family, Liliaceae

1. Dog's-tooth Violet, *Erythronium dens-canis* L., plant 10–25cm; stem with 2 ovate to ovate-lanceolate, opposite, dark green lvs, 6–10cm long and brown-spotted; flowers 1 (rarely 2), nodding; perianth segments 6, free, curved outwards, 3cm long, pink to red-violet. Fl.2–4. Wood margins, scrub, mountain meadows; scattered. Mountains of S. Europe, Alps, Caucasus.

2. Martagon Lily, *Lilium martagon* L., plant 30–100cm; lvs oblong-spathulate, almost whorled in the middle of the stem, otherwise alternate; flowers nodding, 3–8 together in a lax raceme; perianth segments 6, free, rolled back, flesh-red, with dark spots, 3–6cm long. Fl.6–7. Protected! Mixed mountain woods, beechwoods, high-level scrub communities, mountain meadows, in the Alps up to about 2500m; scattered to rare. Mountains of C. and S. Europe, naturalised in Britain, Scandinavia.

3. Orange Lily, *Lilium bulbiferum* L., plant 50–100cm; lvs linear-lanceolate, alternate, up to 10cm long; often with bulbils in the axils of the upper stem lvs; flowers erect, orange, with black spots, solitary or 2–5; perianth segments not curved outwards. Fl.6–7. Protected! Meadows by woodland, wood margins, scrub, mountain meadows; often cultivated and occasionally naturalised, original distribution obscure, probably the mountains of C. and S. Europe (eastern).

4. Alpine Squill, *Scilla bifolia* L., plant 10–20cm; bulbs usually with 1 round stem and 2 amplexicaul lvs, up to 10cm long; inflorescence 2–8 flowered; perianth segments 6, free, spreading like a star, 6–12mm long, 2–3mm wide, pale blue, rarely white or pink. Fl.3–4. Protected! Deciduous woods, wet woods, scrub; rather rare. C. and S. Germany, Alps and their hinterlands, S. and S.E. Europe. The **Autumn Squill**, *S. autumnalis* L., has 3–6 grass-like lvs, which appear in spring and have withered by flowering time in autumn; flowers blue-violet, in 10–20 flowered racemes; perianth segments 4–5mm long and 1.5mm wide. Fl.8–10. Dry grassland; likes warm places; mainly S. Europe and W. Europe to France, Britain.

See also *Colchicum*, Pl.79.

Plate 129   Lily Family, Liliaceae

1. Ramsons, *Allium ursinum* L., plant 20–50cm, with a smell of garlic; bulbs oblong; lvs radical, usually in pairs, with an ovate-lanceolate blade 2–5cm long and a stalk 5–15cm long; inflorescence umbellate, 5–20 flowered, without bulbils; perianth segments 6, free, white; stamens 6, about half as long as the perianth. Fl.5–6. Wet woods, mixed deciduous woods and mountain woods, shady places; rather common. Almost all of Europe. Further broad-leaved species: **Few-flowered Leek**, *Allium paradoxum* (M.Bieb.)G.Don, lvs sessile, lanceolate, 1–2cm wide; flowers 1–3 together, with greenish bulbils; wet woods, parks; rare. **Black Leek**, **A. nigrum* L. (*A. multibulbosum* Jacq.), with basal lvs 3–5cm wide and greenish-white flowers in a many-flowered corymb, bulbils absent; vineyards, arable fields; rare. **Alpine Leek**, **A. victorialis* L., stem leafy; lvs 2–5cm wide, elliptic-lanceolate, shortly stalked; flowers greenish-yellow, in spherical corymbs; high-level scrub communities, alpine meadows; in the Alps to 3000m; Alps, Jura, Black Forest, Vosges, mountains of S. Europe.

2. May Lily, *Maianthemum bifolium* (L.)F.W.Schmidt, plant 5–15cm; basal lvs usually withered by flowering time; stem lvs usually 2, ovate-cordate, acute, stalked, 4–8cm long; stem with stiff, white hairs above; flowers in 8–15 flowered racemes, white; perianth segments 4, free, reflexed, 2–3mm long; berries red, 6mm wide. Fl.5–6. Deciduous and coniferous woods; rare or local. Scandinavia, Britain, mainly N. and C. Europe, southwards to the Apennines, northern Balkan peninsula.

3. Angular Solomon's-seal, *Polygonatum odoratum* (Mill.)Druce (*P. officinale* All.), plant 15–40cm; stem sharp-angled, curved and hanging down at the tip; lvs broadly elliptic, in 2 rows; flowers 1–2 together, rather inflated, 5–7mm wide, up to 2cm long; perianth segments 6, united in a tube, white, greenish at the free tips, glabrous; filaments glabrous. Fl.5–6. Poisonous! Wood margins, scrub, oak and pine woods; scattered. Europe.

4. Solomon's-seal, *Polygonatum multiflorum* (L.)All., plant 30–70cm, similar to *P. odoratum*, but stem terete; lvs arranged in more or less 1 plane; flowers 3–5 together, widened in a funnel at the tip, lobes hairy; filaments pubescent; berries dark blue, 8–10mm. Fl.5–6. Poisonous! Oak, beech and mixed coniferous woods; local. Almost all of Europe, northwards to S. Scandinavia, Britain.

5. Whorled Solomon's-seal, *Polygonatum verticillatum* (L)All., plant 30–70cm; stem erect, angled; lvs 3–7 in whorls, narrowly lanceolate, 5–15cm long; flowers 1–4 together, 6–10mm long; berries first red, then dark blue. Fl.5–6. Poisonous! Mountain woods, high-level shrub communities, in the Alps up to about 2000m; rare to local. Scandinavia, Germany, Britain, European mountains.

Plate 130   Lily Family, Liliaceae

1. Lily-of-the-valley, *Convallaria majalis* L., plant 10–25cm; lvs usually 2, broadly lanceolate, 10–20cm long; flowers nodding, campanulate, in a long-stalked, secund raceme; perianth segments 6, united, tips curved outwards; fruit a red berry. Fl.5–6. Poisonous! Protected! Oak and beech woods, pinewoods, stony screes; in the Alps up to 2200m; local or widespread. Almost all of Europe, in the south only in the mountains.

2. Herb Paris, *Paris quadrifolia* L., plant 10–40cm; stem glabrous, with 4 (or rarely 5 or 6) sessile, elliptic, net-veined lvs, arranged in a whorl at the top, lvs up to 10cm long; flower solitary, terminal, long-stalked, green and usually 4-merous; outer 4 perianth segments ovate-lanceolate, 2–3cm long, up to 5mm wide; inner 4 perianth segments rather shorter and narrower; stamens 8 (more rarely 10 or 12); fruit a black, pruinose berry, 1cm wide. Fl.5–6. Highly poisonous! Wet woods, oak and beech woods, mixed coniferous woods; scattered. Europe.

See also *Anthericum*, Pl.44.

Daffodil Family, Amaryllidaceae

3. Snowdrop, *Galanthus nivalis* L., plant 10–20cm; lvs 2, radical, grass-like, fleshy, 5–8mm wide, blue-green; stem with 1 drooping flower; outer 3 perianth segments pure white, 14–18mm long, the 3 inner only half as long, emarginate, with a green mark at the tip; fruit a fleshy, ovoid capsule. Fl.2–3. Protected! Wet woods, damp mixed deciduous woods, by streams; local; commonly planted, sometimes naturalised. Mainly S. Europe, naturalised in the north.

4. Spring Snowflake, *Leucojum vernum* L., plant 10–30cm; lvs linear, 20–30cm long, 1cm wide, pale green; flowers 1 or 2, nodding, campanulate, white, with a yellow margin; all perianth segments equally long, oval, suddenly narrowed into a blunt, yellow tip. Fl.2–4. Poisonous! Protected! Scrub, wet woods and woods in ravines, meadows, river banks; cultivated plant; scattered. Mainly C. Europe, rare in Britain, southwards to C. Italy. Similar species: **Summer Snowflake**, *L. aestivum* L., but stem 3–7 flowered, 30–60cm tall; lvs 30–50cm long and 1.5cm wide. Fl.4–5. Meadows, wet woods, willow thickets; C. and S. Europe, naturalised in the north.

See also *Narcissus*, Pl.45.

Yam Family, Dioscoreaceae

5. Black Bryony, *Tamus communis* L., herbaceous, anti-clockwise-twining plant, 1.5–3m; lvs very variable, ovate-cordate, long-pointed, stalked, 3–10cm long, dark green, shining; flowers unisexual, the male with a campanulate tube and star-like spreading perianth segments above, yellowish, clustered; the female with free perianth segments, solitary or in few-flowered, drooping racemes; berries red, shining, 12mm wide. Poisonous! Fl.4–6. Hedges, wood margins, open mixed oakwoods; common in Britain, but local or rare abroad. Mainly W. and S. Europe.

3   1   4

2   5

Plate 131 Orchid Family, Orchidaceae

1. Lady's-slipper, *Cypripedium calceolus* L., plant 15–50cm; lvs ovate to oblong, 6–12cm long, amplexicaul, pale green, finely hairy on the margins and veins; flowers 1–2, showy, large; perianth segments red-brown, lanceolate, the 2 lower four-fifths united; lip slipper- or shoe-shaped, yellow with darker veins and markings, 3–4cm long and up to 3cm wide; stamens 2 (other genera have only 1). Fl.5–7. Protected! Open, grass- and herb-rich deciduous and coniferous woods, scrub, on calcareous soils; in the Alps up to about 1600m (rarely higher); rare; exterminated in many localities by collecting. Mainly N. and C. Europe, very rare in Britain, absent from the Mediterranean region.

2. Red Helleborine, *Cephalanthera rubra* (L.)Rich., plant 20–50cm; stem densely glandular-hairy above; lvs ovate-lanceolate, acute, 6–12cm long; flowers 4–12 in lax spikes; bracts as long or longer than the ovary; perianth segments 15–20mm long, acute, connivent in the shape of a bell, pink or purple, the lip usually concealed; forward portion (epichile) of the lip with a red-violet margin and violet tip and with crested, yellow, longitudinal ridges; ovary hairy. Fl.5–7. Protected! Beech, oak and pine woods, also spruce-fir woods; scattered. C. and S. Europe, northwards to S. Scandinavia, very rare in Britain.

3. White Helleborine, *Cephalanthera damasonium* (Mill.)Druce (*C. alba* Simk., *C. grandiflora* S.F.Gray), plant 20–50cm; lvs ovate, acute, up to 10cm long, not folded lengthways, with 5–10 veins; inflorescence 3–8 flowered; perianth segments yellowish-white, blunt, 15–20mm long; flowers at most 4 times as long as wide; bracts over half the length of the ovary; lip reddish-yellow inside. Fl.5–6. Protected! Beech and beech-fir woods, more rarely oak and pine woods; scattered. C. and S. Europe, northwards to England.

4. Narrow-leaved Helleborine, *Cephalanthera longifolia* (L.)Fritsch, plant 20–50cm; stem glabrous; lvs lanceolate, folded, distichous, up to 10cm long; inflorescence 10–20 flowered; perianth segments pure white, acute, 10–15mm long; flowers at least 10 times as long as wide; middle and upper bracts shorter than the glabrous ovary. Fl.5–6. Protected! Open oak and beech woods, mixed pinewoods, scrub; likes warm places; rather rare. C. and S. Europe, northwards to S. Scandinavia, Britain.

Plate 132   Orchid Family, Orchidaceae

1. Dark-red Helleborine, *Epipactis atrorubens* (Hoffm.)Schult. (*E. rubiginosa* (Crantz) Gaud., *E. atropurpurea* Raf.), plant 30–60cm; stem reddish, often pubescent; lvs broad lanceolate, almost amplexicaul, distichous, pointed, longer than the stem internodes; flowers brown-red, smelling of cocoa; outer 3 perianth segments lanceolate, 6–8mm long, inner 2 paler coloured; lip somewhat shorter, forward portion of the lip with 2 strongly wrinkled basal bosses; ovary pubescent. Fl.6–8. Protected! Open, dry oak-pine woods, grassland with pines, dryish grassland, limestone rocks and screes; local. Almost all of Europe. Similar species: **Small-leaved Helleborine**, **E. microphylla* (Ehrh.) Sw., but lvs lanceolate, shorter than the stem internodes, grey-green, tinged violet; flowers greenish and violet tinged; all 5 perianth segments campanulate-connivent; herb-rich mixed deciduous woods; rare.

2. Broad-leaved Helleborine, *Epipactis helleborine* (L.)Crantz (*E. latifolia* All.), plant 25–60cm; lvs ovate, acute, 6–15cm, 2–8cm wide, dark green, amplexicaul, much longer than the stem internodes; flowers pale green; lip often tinged violet or pink or with purple markings. Fl.6–8. Protected! Oak and beech woods, mixed coniferous woods; local or fairly common. Almost all of Europe. Similar species: **Violet Helleborine**, *E. purpurata* Sm. (*E. sessiliflora* Peterm., *E. violacea* (Dur.)Bor.), but lvs smaller, lanceolate, 5–8cm long, 1–2.5cm wide, middle stem lvs scarcely longer than the internodes; inflorescence dense-flowered; flowers yellow-green and pink tinged; Fl.8–9. Mixed deciduous woods; rare.

3. Violet Limodore, **Limodorum abortivum* (L.)Sw., plant 20–50cm, blue to violet, without chlorophyll; lvs scale-like; flowers violet, erect, 3–4cm long, in lax spikes with 4–8 flowers; spur directed downwards, 15–25mm long; perianth segments connivent in the shape of a helmet, lip ovate, toothed on the margin. Fl.5–7. Woods of Downy Oak, pine, oak and beech; likes warm places; rare. S. Germany, Austria, France, C. and S. Europe. Protected!

4. Common Twayblade, *Listera ovata* (L.)R.Br., plant 20–50cm; stem with 2 opposite, broadly ovate lvs, 5–10cm long; inflorescence many-flowered; perianth segments green, connivent, 3–4mm long; lip yellow-green, 2-lobed, lobes obtuse, not spreading. Fl.5–6. Mixed deciduous woods, wet woods, scrub, mountain meadows, dunes; rather common. Almost all of Europe.

5. Lesser Twayblade, *Listera cordata* R.Br., plant 5–15cm, dainty; stem fragile, pale green, with 2 cordate, shining, green lvs in the middle; raceme 6–12 flowered; perianth segments yellow-green to reddish-brown, 2mm long; lip oblong, red, 3–4mm long, deeply bifid, lobes long-pointed and spreading. Fl.5–8. Mossy fir woods, mountain pinewoods, peaty moors, especially under heather; common in the north and rare in the south. N. Europe, N. Germany, Britain, Alps, Pyrenees, Jura, Vosges, Black Forest, Carpathians, mountains of the Balkan peninsula.

Plate 133   Orchid Family, Orchidaceae

1. Bird's-nest Orchid, *Neottia nidus-avis* (L.)Rich., plant 20–50cm, without chlorophyll, pale brown; stem thick, with sheathing scale-lvs; inflorescence many-flowered; perianth segments connivent, obtuse, 4–6mm long, pale brown; lip deeply bifid. Fl.5–6. Protected! Mixed deciduous woods, especially beechwoods; saprophyte, i.e. lives on decaying wood and dead organic material; fairly common. Almost all of Europe, northwards to C. Scandinavia, Britain.

2. Creeping Lady's-tresses, *Goodyera repens* R.Br., plant 10–30cm; basal lvs ovate to ovate-oblong, 1–3cm long, upper stem lvs narrowly lanceolate; stem densely glandular-hairy; inflorescence with 10–15 flowers, secund; flowers small, white or creamy-yellow, greenish outside, sweetly scented; perianth segments 4mm long; lip undivided, furrowed and curved downwards like a beak. Fl.7–8. Protected! Mossy coniferous woods, up to over 2000m in the Alps; rather rare. N. and C. Europe, Britain, Alps, Pyrenees, Balkan peninsula.

3. Ghost Orchid, *Epipogium aphyllum* Sw., plant 10–20cm, without chlorophyll, yellowish to brown, tinged purple above; stem tubular, with sheathing scale lvs; bracts oval, thin, whitish; flowers 2–4, pale yellow, with reddish streaks; perianth segments spreading, narrowly lanceolate, 10–14mm long; lip directed upwards, with 2 lateral, roundish lobes at the base, central lobe oval, with an undulate margin; spur directed upwards. Fl.7–8. Protected! Mossy mixed woods; rare. N. and C. Europe, Britain, in the south in the mountains (Pyrenees, Alps, Apennines, Olympus).

4. Coralroot Orchid, *Corallorhiza trifida* Chatel., plant yellowish to brown, 8–20cm, without green lvs; stem lvs scale-like; flowers yellow-green, without a spur, 4–9 together; perianth segments 6, 3–6mm long, lip undivided or weakly 3-lobed, whitish, red-spotted. Fl.6–7. Protected! Shady, mossy woods, especially pine, birch or alder; rather rare. N. and C. Europe, Britain, the Alps, Pyrenees, Apennines, mountains of the Balkan peninsula.

5. Lesser Butterfly-orchid, *Platanthera bifolia* (L.)Rich., plant 20–40cm; lvs usually 2, near the base of the stem, ovate, 5–15cm long, 2–5cm wide; upper stem lvs small, lanceolate; inflorescence lax, 5–15cm long; flowers white, strongly scented, 11–18mm wide, outer 3 perianth segments lanceolate, spreading, the inner 2 shorter and narrower, directed upwards; lip strap-shaped, 6–10mm long; pollinia (pollen masses) parallel; spur thread-like, pointed, 15–22mm long. Fl.5–7. Protected! Open deciduous and coniferous woods, especially pinewoods, wood margins, poor grassland, damp meadows; scattered. Europe.

6. Greater Butterfly-orchid, *Platanthera chlorantha* (Gust.)Rchb., very similar to *P. bifolia*, but flowers greenish-white, scentless, pollinia not parallel, diverging below; lip 10–16mm long; spur club-shaped, blunt, 20–40mm long. Fl.5–6. Protected! Mixed coniferous woods, boggy moorland pastures; rather uncommon. Almost all of Europe.

7. Calypso, *Calypso bulbosa* L., plant 10–25cm, with 1–3 scarious leaf sheaths at the base which surround the roundish bulb, and a broadly ovate, green, long-stalked leaf; stem with 1–2 sheathing lvs and a single, terminal flower; the 5 upper perianth segments linear, purple, spreading; lip slipper-shaped, 15–25mm long, whitish, pink or yellow flecked, without spur. Fl.5–6. Coniferous woods, marshes, moors. N. Europe.

See also *Ophrys insectifera*, Pl.47.

Plate 134 Orchid Family, Orchidaceae

1. Pale-flowered Orchid, *Orchis pallens* L., plant 15–40cm; lvs broad-ovate, blunt, up to 4cm wide, shining; inflorescence dense-flowered, ovoid, short; bracts scarious, 1-veined; flowers pale yellow; outer perianth segments spreading or reflexed, inner connivent; lip weakly 3-lobed. Fl.4–5. Protected! Open deciduous woods, scrub, dryish grassland; rare. S. Germany, Thuringia, Alps and hinterlands, Austria, S. Europe. Similar species: **Elder-flowered Orchid**, *Dactylorhiza sambucina* (L.)Sóo, but flowers dark red or yellow; lip undivided, undulate on the margin, with purple spots; siliceous grassland, dry margins of scrub; rare. Protected!

2. Early-purple Orchid, *Orchis mascula* L., plant 15–50cm; lvs lanceolate, the lower with a spreading blade, the upper sheathing and clasping the stem; inflorescence narrow, cylindric, 5–15cm long, usually lax-flowered; bracts scarious, 1-veined, violet-tinged; flowers purple; perianth segments lanceolate, acute, the 2 lateral spreading or reflexed; lip deeply 3-lobed, dark spotted, with spreading lateral lobes; spur club-shaped, about as long as the ovary, usually directed upwards. Fl.5–6. Protected! Mountain pastures, dryish grassland, open mixed deciduous woods; common or local. Europe.

3. Lady Orchid, *Orchis purpurea* Huds., plant 30–80cm; lvs oblong-ovate; inflorescence usually dense-flowered, 5–15cm long, cylindrical; bracts short, hyaline, scale-like; all 5 perianth segments connivent, forming a globose-ovoid helmet which is brownish-red outside; lip pale red, dark-spotted, with linear lateral lobes and 2 broad, finely toothed terminal lobes, often with a small tooth between them; spur half as long as the ovary. Fl.5–6. Protected! Likes warm places. Woods, especially oakwoods, scrub, dryish grassland; rather rare. From Britain and Denmark southwards, C. and S. Europe.

4. Common Spotted-orchid, *Dactylorhiza fuchsii* (Druce)Sóo, plant 20–70cm; lower lvs broadly elliptic to obovate, obtuse, usually with spots which are broadest in the direction across the leaf; flowers in dense-flowered, conical (later cylindrical) spikes, pale lilac or whitish; lateral perianth segments spreading; lip deeply 3-lobed, the central lobe pointed, rather longer than the laterals; spur conical, about as long as the ovary. Fl.6–7. Protected! Open woods, grassland, fens, heaths, on calcareous soils; fairly common to rare. Almost all of Europe. Similar species: **Heath Spotted-orchid**, *D. maculata* (L.)Sóo, but lvs lanceolate to linear, acute, with roundish spots; lip of flower weakly 3-lobed, its central lobe smaller and shorter than the laterals; on acid soils, heaths, woods, marshes; fairly common to rare.

Plate 135   Bog Myrtle Family, Myricaceae

1. Bog Myrtle, *Myrica gale* L., shrub 50–200cm, usually dioecious; lvs lanceolate to obovate, shortly stalked, weakly serrate, 2–5cm long, with a strong aromatic smell; flowers in erect catkins, 5–15mm long, appearing before the lvs. Fl.4–5. Heather moorland, damp scrub; scattered or fairly common. Britain, N. Europe. Protected!

Willow Family, Salicaceae

2. Aspen, *Populus tremula* L., tree 5–20m, with bark yellowish-brown, later dark grey, smooth; lvs almost circular, bluntly toothed, on long, slender, sideways-compressed petioles, glabrous, fluttering in the slightest breeze; flowers in catkins 5–10cm long; male flowers with 7–17 stamens; female flowers with purple stigmas. Fl.3–4. Open woods, wood margins; widespread. Europe.

3. White Poplar, *Populus alba* L., tree 15–35m, with smooth, pale grey bark and a wide-spreading crown; young twigs, buds and the underside of the lvs white-felted; lvs ovate, sinuate-lobed on the margins; flowers in catkins 5–10cm long; bracts of the flowers hairy; female flowers with yellow stigmas. Fl.3–4. Wet woods, parks, roadsides, waste places; rather common. Britain, C. and S. Europe. Similar species: **Grey Poplar**, *P. canescens* (Ait.)Sm., but older lvs grey-felted beneath.

4. Black Poplar, *Populus nigra* L., tree 15–30m, with deeply fissured bark; younger branches roundish; lvs broadly ovate to rhomboid, finely serrate, glabrous, 5–10cm long; bracts of the flowers glabrous; male flowers with 20–30 stamens. Fl.3–4. Wet woods, river banks, parks; widespread. Britain (rare in the pure form), C. and S. Europe. Various hybrids are often planted, such as the **Italian Poplar**, *P. × canadensis* Moench, with the young branches with corky ridges; many different cultivars.

5. White Willow, *Salix alba* L., tree or shrub, up to 20m; twigs yellow-brown to red-brown; lvs lanceolate, 5–8cm long, with a finely serrate margin, the underside densely appressed-hairy (silky-hairy on both sides when young); catkins appearing with the lvs; catkin scales uniformly coloured; ovary glabrous, almost sessile; male flowers with 2 stamens. Fl.4–5. Banks of lakes and rivers, reservoirs, wet woods; common. Britain, C. and S. Europe. Similar species: **Crack Willow**, *S. fragilis* L., but twigs easily broken at the junctions; lvs soon entirely glabrous and pale below, usually asymmetric at tip. Same places as *S. alba*. Most of Europe.

6. Almond Willow, *Salix triandra* L., shrub 2–5m; twigs yellow-green to red-brown; the bark of older trunks coming away in flakes; lvs lanceolate, glabrous, 5–10cm long, up to 2cm wide, both sides green and almost the same colour; catkin scales uniformly coloured; female catkins slender; ovaries glabrous, with a stalk of about $\frac{1}{4}$ to $\frac{1}{2}$ their length; male flowers with 3 stamens. Fl.4–5. Wet thickets, banks of rivers and streams; widespread. Almost all of Europe.

---

1. Osier, *Salix viminalis* L. (Pl.136), shrub, more rarely a tree, 4–10m; twigs greenish-yellow to blue-green, brush-like; lvs narrowly lanceolate, almost entire, 8–15cm long, the margins usually inrolled and undulate; catkins appearing before the lvs, cylindrical, densely silky-hairy, 10–15mm wide; catkin scales 2-coloured, pale at the base, blackish-brown at tip; ovaries almost sessile, hairy; male flowers with 2 stamens. Fl.3–4. Willow thickets, wet woods; widespread; planted and pollarded and used to obtain wicker for basket making. Almost all of Europe.

Plate 136   Willow Family, Salicaceae

1. Osier, *Salix viminalis*, see p.318.

2. Purple Willow, *Salix purpurea* L., shrub 2–6m; young twigs purple-tinged; lvs 4–12cm long, dark green above, blue-green below, glabrous; catkins appearing before the lvs, slender, dense-flowered; catkin scales 2-coloured, densely hairy; ovaries sessile, felted-hairy; filaments 2 but united up to the anthers, purple at first, yellow when they open, finally brown. Fl.4–5. Willow thickets, shingle banks in rivers; common. Britain, C. and S. Europe.

3. Eared Willow, *Salix aurita* L., shrub 1–3m; branches spreading at a wide angle; twigs and buds glabrous; young wood with raised longitudinal striations under the bark; lvs 2–5cm long, roundish-ovate, rugose, undulate at the margin, irregularly toothed, with large stipules; catkins 2cm long; catkins scales 2-coloured; ovaries felted-hairy, stigmas globose; stamens 2, anthers yellow. Fl.4–5. Willow thickets, on the edges of moorland, lakesides, marshy woods; widespread. Britain, N. and C. Europe, southwards to the Pyrenees, Apennines.

4. Grey Willow, *Salix cinerea* L., shrub 2–5m; young twigs and buds grey-felted; young wood with raised striations under the bark like *S. aurita*; lvs broadly ovate, 4–10cm long, grey-green, finely serrate or irregularly crenate; stipules up to 5mm long; catkins 4–9cm long; ovaries felted; stigmas deeply bifid, spreading; stamens 2, anthers orange-red at first. Fl.4–5. Like *S. aurita*; common. Europe.

5. Goat Willow, *Salix caprea* L., shrub 3–9m; twigs red-brown, glabrous like the buds; lvs elliptic, 3–10cm long, the underside grey-green to whitish, densely white-felted; stipules small; catkins dense, cylindrical, 4–10cm long, sessile at first, later stalked; catkins scales 2-coloured, with long white hairs; ovaries felted; stamens 2. Fl.3–4. Thickets, wood margins, gravel pits, wet woods; common. Almost all of Europe.

6. Dark-leaved Willow, *S. nigricans* Sm. (*S. myrsinifolia* Salisb.), shrub 2–5m; twigs brown to blackish-brown, dull; lvs 3–8cm long, ovate, serrate, dark green above, blue-green below, but pure green near the tip below; stipules large, kidney-shaped; catkins appearing almost at the same time as the lvs, ovoid, 2–3cm long; catkin scales 2-coloured; villous; ovaries glabrous, stalked; stamens 2. Fl.4–5. Woods, moors, shingle, by water; scattered. Most of Europe, mainly in the mountains.

7. Bay Willow, *Salix pentandra* L., shrub to 12m, rarely a tree, with glabrous, olive-green to red-brown twigs, which shine as though varnished; lvs broadly lanceolate, finely serrate, 5–15cm long; female catkins lax-flowered; catkin scales uniformly coloured, yellow-green; stamens usually 5; lvs and catkins appearing at the same time. Fl.5–6. Wet or marshy woods, moors; scattered. Britain, C. and N. Europe, southwards to the Pyrenees, N. Italy, the Balkan peninsula.

Walnut Family, Juglandaceae

8. Walnut, *Juglans regia* L., tree to 25m, with a rounded crown and grey, vertically fissured bark; lvs pinnate with 7–9 entire, ovate leaflets, up to 15cm long; male flowers in hanging catkins 5–15cm long; female flowers 2–3 together, terminal; ovary inferior, stigmas 2; fruit (a drupe) with a green, fleshy skin and a brown shell enclosing the nut. Fl.5. Deciduous woods in places with a mild climate which are safe from late frosts; cultivated. S. Europe, introduced into the rest of Europe.

Plate 137   Birch Family, Betulaceae

1. Silver Birch, *Betula pendula* Roth (*B. verrucosa* Ehrh.), tree to 25m, the lower part of the trunk (up to about 10m) fissured, bulging and blackish, with white patches, and the upper part smooth and white; lvs triangular-rhomboid, long-pointed, viscid when young, glabrous; twigs pendulous, with small, resinous warts when young; male catkins 3–6cm long, female 1.5–3cm long; scales of fruit 3-lobed (from the combination of a bract and 2 bracteoles), falling when ripe; fruit a winged nut; wing 2–3 times as wide as the nut. Fl.4–5. Woods, moors, heaths, quarries; common. Europe. Similar species: **Downy Birch**, *B. pubescens* Ehrh., but bark usually smooth, white, rarely grey to black; young twigs and lvs softly hairy; lvs ovate or rhomboid, shortly pointed; branches spreading or ascending; wing of fruit scarcely wider than the nut; moors and marshy woods, oakwoods on acid soils.

2. Grey Alder, *Alnus incana* (L.)Moench, tree to 20m, with smooth, pale grey bark (even on old trees); lvs ovate-elliptic, gradually pointed, doubly serrate, with 7–12 pairs of lateral veins, young twigs hairy; female catkins sessile. Fl.2–4. Wet woods, mountain streams and rivers; particularly in calcareous districts; local or common. The tree enriches the soil with nitrogen via nitrogen-fixing fungi, which live symbiotically with the alder in nodules on the roots. N. and C. Europe (but planted in Britain), southwards to C. Italy, the Balkan peninsula.

3. Alder, *Alnus glutinosa* (L.)Gaertn., tree to 20m; bark blackish, fissured; lvs roundish-ovate, up to 10cm long, obtuse or emarginate, with 5–8 pairs of lateral veins, weakly serrate; buds and young lvs viscid; male and female catkins on the same twigs, the female stalked, becoming woody and forming cones. Fl.3–4. Wet woods, marshy woods, river banks; common. Europe.

4. Green Alder, **Alnus viridis* (Chaix)DC., shrub 2–3m tall; lvs ovate, acute, green on both sides, serrate; teeth longer than wide; buds rather acute, sessile (stalked in the other species); male catkins pendulous, yellow, up to 6cm long, shedding pollen as the lvs unfold; female catkins below the male catkins, ovoid, 10–15mm long, green, later becoming woody and forming brown cones. Fl.5–7. Shady, damp and steep slopes in the higher parts of mountains, in the Alps at about 1400–2400m, lower in the mountains of C. Europe; scattered. Mountains of C. and S. Europe.

Hazel Family, Corylaceae

5. Hornbeam, *Carpinus betulus* L., tree to 20m, with smooth, grey bark, often with a network of ridges; lvs ovate, 5–8cm long, usually asymmetric, doubly serrate, pale green; male inflorescences appearing at the same time as the lvs; male flowers without perianth segments, solitary in the axils of bracts of cylindrical catkins; female flowers with inconspicuous involucre of 3-lobed bracts in lax catkins; fruit a nut, 5–10mm long, enclosed in the green involucre. Fl.4–5. Deciduous woods, hedges, in low-lying districts; widespread. Mainly C. and S.E. Europe, northwards to S. Sweden, England.

6. Hazel, *Corylus avellana* L., shrub 2–6m; lvs ovate, acute, 5–12cm long, doubly serrate; male flowers in catkins 2–8cm long, 1–4 together; female flowers 2–6, surrounded by bud scales, so that at flowering time only the red, filiform styles project; fruit a hard-shelled, brown nut, surrounded by the irregularly lobed involucre. Fl.2–4. Scrub, wood margins, deciduous woods, hedges; widespread. Europe.

Plate 138 Beech Family, Fagaceae

1. Beech, *Fagus sylvatica* L., tree to 40m, with a smooth, grey bark; deep-rooting; buds narrow, long-pointed, brown; lvs broadly ovate, up to 10cm long, entire, ciliate; male flowers in spherical, long-stalked catkins; female flowers in 2s on a long stalk, surrounded in common by a weakly spined fruiting involucre (cupule); fruit a 3-angled nut, falling out of the cupule in autumn; flowering begins after 40–60 years. Fl.4–5. Deciduous mixed woods, sensitive to late frosts; widespread. Almost all of Europe where the climate is of oceanic type (high rainfall and relatively mild winters), northwards to S. Scandinavia, Britain, in the south mainly in the mountains, where it often forms the tree-limit zone.

2. Sweet Chestnut, *Castanea sativa* Mill., tree to 30m, with fissured bark; deeply rooted; lvs oblong-lanceolate, with spiny teeth, leathery, up to 25cm long; male flowers in erect catkins, up to 20cm long; female flowers usually in 3s at the base of the male inflorescence; perianth segments 6, only free at the tip, styles 6, stiff, thread-like; fruiting involucre globose, 5–7cm across, with hard spines, enclosing usually 3 dark-brown, smooth, hemispherical nuts; flowering begins after 20–30 years. Fl.6. Deciduous woods in areas with warm summers; non-calcareous to acid soils. Introduced from the Mediterranean region in Roman times. S. Europe, S. Alps, Austria, Germany and France, planted and naturalised in Britain. The **Horse Chestnut**, *Aesculus hippocastanum* L., belongs to a different family (Hippocastanaceae); lvs palmate with 5–7 leaflets; inflorescences erect with flowers white, spotted with yellow or pink; fruit is the familiar 'conker'. Commonly planted in Britain, native of S.E. Europe.

3. Pedunculate Oak, *Quercus robur* L., tree to 40m, with a dark brown, fissured bark and gnarled, wide-spreading branches; lvs very shortly stalked, with a cordate, auriculate base, asymmetric, with the largest lobes at the tip of the leaf blade; male flowers in lax, drooping spikes; female flowers 2–5, axillary; involucre 6-merous; fruiting inflorescence on a stalk 3–8cm long; fruit (acorn) an ovoid, smooth nut, enclosed in the cup-like fruiting involucre; flowering begins after 40–80 years. Fl.4–5. Mixed deciduous woods, hedgerows; widespread. Almost all of Europe. Similar species: **Sessile Oak**, *Q. petraea* (Mattuschka)Liebl., tree to 40m; lvs with stalks 1–3cm long, cuneate-based, the largest lobes in the middle of the blade; fruit on a stalk of at most 1cm long; Fl.4–5. Mixed deciduous woods, usually on acid soils; widespread. A frequently planted forestry or decorative tree is the **Red Oak**, *Q. rubra* L., with lvs 15–20cm long, with pointed lobes and narrow sinuses; origin N. America.

4. Downy Oak, **Quercus pubescens* Willd., shrub or tree, 5–20m; branches widely spreading; young twigs and lvs pubescent, becoming glabrous later (except for the underside of the leaf); lvs 5–10cm long, on stalks 1–2cm long; scales of the fruiting involucre hairy. Fl.4–5. Woods, dry, sunny slopes; rare in the north, common in the south. S. Germany, warm valleys of the Alps, Austria, S. Carpathians, S. Europe.

5. Turkey Oak, *Quercus ilex* L., shrub or tree, to 20m, with oval, evergreen lvs, 3–6cm long, undulate on the margins and spiny-toothed (5a) to entire, and densely grey-hairy beneath; fruiting involucre with blunt, appressed scales. Fl.4–5. Dry stony slopes; characteristic tree of the hardwood forests of the Mediterranean. Principally S. Europe, but commonly planted and sometimes naturalised elsewhere.

Plate 139 Elm Family, Ulmaceae

1. Small-leaved Elm, *Ulmus minor* Mill. (*U. carpinifolia* Gleditsch, *U. campestris* L.), tree, 10–30m, with ridged, grey-brown bark; lvs smooth, 6–10cm long, sharply serrate, with 8–12 lateral nerves on each side, almost glabrous, widest at the middle, with stalks up to 1.5cm; flowers hermaphrodite, almost sessile, in clusters; fruit ovate to roundish, up to 2cm long; seed closest to the upper margin of the wing. Fl.3–4. Wet woods, deciduous woods, woods of Downy Oak, hedgerows; likes warm places; widespread. C. and S. Europe, probably introduced in Britain. Similar species: **European White Elm**, **U. laevis* Pall. (*U. effusa* Willd.), but lvs with 12–29 lateral veins on each side, hairy below, with stalks up to 1cm long; flowers long-stalked, pendulous; wing of fruit villous-ciliate; seed in centre of fruit or below; rare. Mainly C. and E. Europe. **English Elm**, *U. procera* Salisb., has lvs with 10–12 pairs of veins, hairy below, widest at the middle; seeds two-thirds the way down fruit. Hedges and roadsides. England and France only, much reduced by Dutch Elm disease.

2. Wych Elm, *Ulmus glabra* Huds. (*U. scabra* Mill.), tree to 30m, with vertically furrowed bark; lvs asymmetric, 8–15cm long, rough, sharply serrate; lvs with 12–20 lateral veins on each side, hairy below, widest in the upper third, with stalk 3–7mm long; flowers almost sessile; seeds in the centre of the fruit. Fl.3–4. Deciduous woods, thickets, woods in river gorges, hedges; in the Alps to 1400m; scattered. Europe.

Hemp Family, Cannabaceae

3. Hop, *Humulus lupulus* L., plant 2–6m; stem twining clockwise; lvs opposite, usually deeply 3-lobed, toothed; plant dioecious; male inflorescences in panicles; perianth segments 5, free, pale green, stamens 5; female inflorescence cone-like, in a short spike (the 'hop', used to flavour beer); fruit a nut 3mm long. Fl.7–8. Wet woods, scrub, hedges; cultivated; common. Europe.

Sandalwood Family, Santalaceae

4. Bavarian Bastard Toadflax, **Thesium bavarum* Schrank (*T. montanum* Ehrh.), plant 20–60cm, without stolons; lvs broadly lanceolate, 2–7mm wide, 3–5 veined, bluish-green; flowers usually 5-merous, white, with 1 larger and 2 smaller bracts; perianth segments at fruiting time inrolled to the base. Fl.6–9. Sunny woodland margins, pinewoods, dryish grassland; rare. C. and S.E. Europe. See also *Thesium*, Pl.49.

Mistletoe Family, Loranthaceae

5. Mistletoe, *Viscum album* L., plant shrubby, up to 1m, a semiparasite on trees (deriving mineral salts and water from the host plant); twigs branched dichotomously; lvs leathery, evergreen; flowers unisexual, inconspicuous, in the fork between the branches; berries white or yellowish. Fl.2–5. The ssp. *album* on deciduous trees, the **Fir-mistletoe**, ssp. *abietis* (Wiesb.)Abrom., on Silver Fir, and the **Pine-mistletoe**, ssp, *austriacum* (Wiesb.)Vollm. (*V. laxum* Boiss. et Reut.), on pines; scattered. Britain, S. Scandinavia, C. and S. Europe.

Nettle Family, Urticaceae. See *Urtica dioica*, Pl.4.

Birthwort Family, Aristolochiaceae

6. Asarabacca, *Asarum europaeum*, see p.328.

4

5

6

2

3

1

Plate 140 Pink Family, Caryophyllaceae

1. Wood Stitchwort, *Stellaria nemorum* L., plant 20–50cm, with far-creeping stolons; stem terete, softly glandular-hairy above; lvs opposite, ovate-cordate, acute, ciliate on the margins, the lower ones stalked; flowers 5-merous; calyx 4–6mm long, petals twice as long, bifid almost to the base, white. Fl.5–7. Herb-rich mixed woods, wet woods, mountain woods, high-level scrub communities, in the Alps to 2200m; local to common. Europe.

2. Greater Stitchwort, *Stellaria holostea* L., plant 10–30cm; stem 4-angled, glabrous below, with scattered hairs above; lvs stiff, narrowly lanceolate, sessile; petals bifid to halfway, twice as long as the calyx, which is 6–8mm long. Fl.4–6. Herb-rich mixed woods, scrub, hedges; common. Almost all of Europe, northwards to S. Scandinavia, Britain. The **Long-leaved Stitchwort**, **S. longifolia* Muhl. (*S. diffusa* Schldl.), has likewise narrowly lanceolate, sessile lvs, 2–3cm long, but petals divided almost to the base, scarcely longer than the calyx; stem rough upwards, 4-angled below; plant loosely tufted, 10–20cm. Fl.6–8. Coniferous woods on moorland; rare; ice-age relict, in the natural distribution zone of the Norway Spruce; mainly northern C. Europe, Alps, Carpathians, N.E. Europe.

3. Berry Catchfly, *Cucubalus baccifer* L., plant much-branched, clambering, shortly hairy, 50–150cm; lvs ovate, acute, shortly stalked; flowers in lax inflorescences, with unequally 5-lobed, campanulate, inflated calyx and deeply bifid, greenish-white petals; styles 3; fruit a black berry, 6–8mm long. Fl.7–9. Wet woods, thickets, nutrient-rich alluvial soils in river valleys; rather rare. C. and S. Europe, northwards to S. England.

4. Three-nerved Sandwort, *Moehringia trinervia* (L.)Clairv., plant ascending, 10–30cm; stem solitary, not tufted, shortly hairy; lvs ovate, acute, usually 3-nerved, shortly hairy, the lower stalked, the upper almost sessile; flowers 5–6mm wide, white, solitary, axillary or several together at the ends of branches; petals 5, shorter than the calyx; sepals 5, broadly hyaline-margined, 3-veined, shortly hairy. Fl.5–7. Deciduous or coniferous mixed woods, woodland clearings, forest rides, scrub; widespread. Almost all of Europe, in the south only in the mountains.

See also Pl.49, *Dianthus*, Pl.50 and *Silene dioica*, Pl.80.

Buttercup Family, Ranunculaceae

5. Shining Meadow-rue, **Thalictrum lucidum* L., plant 60–120cm; lvs 2-pinnate; leaflets oblong, cuneate or linear, dark green above, shining; flowers clustered at the ends of the panicle branches; flowers and stamens erect, yellowish. Fl.6–7. Wet woods, upland pasture; rare. C. Europe (Germany, Alps), S.E. Europe.

---

6. Asarabacca, *Asarum europaeum* L. (Pl.139), plant 5–10cm; stem creeping, with whitish or brownish scale leaves at base; lvs roundish to kidney-shaped, dark green, 3–10cm wide, often evergreen; flowers solitary, shortly stalked, symmetric; perianth 3-merous, campanulate, brown-red, 10–15mm long. Fl.3–5. Deciduous and coniferous woods, wet woods. Medicinal plant. Common in C. and E. Europe, rare in Britain. See also *Aristolochia*, Pl.4.

Dock Family, Polygonaceae. See *Rumex sanguineus*, Pl.4.

Plate 141   Buttercup Family, Ranunculaceae

1. Greater Meadow-rue, *Thalictrum aquilegifolium* L., plant 40–120cm, lvs 3-pinnate; leaflets roundish, coarsely and bluntly toothed, blue-green; inflorescence a panicle, many-flowered; perianth segments inconspicuous, soon dropping; filaments thickened above, pale violet, rarely white; achenes pendulous on long stalks, 3-angled, winged. Fl.5–7. Wet woods, high-level scrub communities, in the Alps up to over 2000m; scattered. Mainly C. Europe, southwards to N. Spain, Pyrenees.

2. Lesser Meadow-rue, *Thalictrum minus* L., plant 30–100cm; lvs on the stem distributed regularly, 3–4 pinnate; leaflets roundish or roundish-cuneate, blue-green, pale green below; stem furrowed; inflorescence a panicle, often with spreading branches; perianth segments and filaments yellowish, hanging; achenes sessile, spindle-shaped, ribbed lengthwise. Fl.5–8. Light, dry woods, dry scrub, dry grassland on wood margins, limestone rocks and cliffs, quarries, dunes; local or rather rare. (Variable species with several subspecies.) Mainly C. and S. Europe, northwards to S. Scandinavia, Britain. Protected! Similar species: **Rock Meadow-rue**, *Thalictrum minus* ssp. *saxatile* Sch. et Kell. (*T. saxatile* DC.), lvs blue-green on both sides, with prominent nerves beneath, lvs distributed in the lower half or in the middle of the stem only; dry grassland, open scrub, pinewoods, in the Alps up to about 2000m; rare. **Small Meadow-rue**, *T. simplex* L., but leaflets oblong to linear; flowers greenish; moorland pasture, calcareous grassland; rare. Protected!

3. Stinking Hellebore, *Helleborus foetidus* L., plant 30–60cm; lvs evergreen, palmate with 3–9 lobes, up to 30cm in diameter, segments toothed; inflorescence much branched; flowers campanulate, 2cm wide, 5-merous, greenish-yellow, reddish on the margins, smelling unpleasantly; achenes with a hooked beak, 3–8 together, about 2cm long. Fl.2–4. Poisonous! Oak and beech woods, wood margins, dry, calcareous bushy slopes; scattered. Mainly S.W. Europe, S. Germany, northwards to Britain.

4. Christmas Rose, *Helleborus niger* L., plant 10–30cm; lvs over-wintering, 7–9 lobed, segments only serrate towards the tip; stem usually 1-flowered, with 1–2, oval, entire stem lvs in the upper part only; flowers 5–10cm across, white or pink, later turning green; perianth segments spreading; nectaries yellow or yellow-green. Fl.1–4. Mixed deciduous and pine woods; scattered. Alps and S.E. European mountains. Poisonous! Protected!

5. Green Hellebore, *Helleborus viridis* L., plant 10–30cm; lvs not over-wintering, with 7–11 lobes, segments toothed to the base; stem simple or branched, with divided, serrate stem lvs; flowers 2–3 together, drooping, 4–6cm wide; perianth segments spreading, broadly ovate, green; nectaries green. Fl.3–4. Protected! Open woods, scrub; rare; once cultivated as a medicinal plant and occasionally naturalised. C. Europe, southwards to Spain and N. Italy, northwards to Britain, Germany.

6. Winter Aconite, *Eranthis hyemalis* (L.)Salisb., plant 5–15cm; basal lvs usually appearing after the first flowering time, roundish, 5–7 lobed to the base, segments 2–3 lobed; stem 1, red-brown, with a whorl of lvs underneath the yellow flowers, which are 2–3cm wide; achenes up to 15mm long. Fl.2–3. Damp deciduous woods of S.E. Europe, in C. and N. Europe in gardens, parks, vineyards, scrub; scattered; originally from S. and S.E. Europe.

Plate 142   Buttercup Family, Ranunculaceae

1. Monk's-hood, *Aconitum napellus* L., plant 50–150cm; lvs palmately divided almost to the base in 5–7 lobes, lobes with narrow, more or less linear teeth; flowers usually in a simple or sparingly branched, dense raceme, blue-violet; helmet of the flower wider than high; filaments usually hairy. Fl.6–8. Poisonous! Protected! (Very variable species with several subspecies.) High-level shrub communities, thickets, grey alder woods, shady stream banks, in the Alps to up over 2000m; scattered. Alps, Pyrenees, mountains of C. Europe, S. Britain. Similar species: **Variegated Monk's-hood**, **A. variegatum* L., flowers violet and white-checked, on long eglandular pedicels; helmet higher than wide; Grey Alder woods, subalpine scrub communities. Poisonous! Protected!

2. Panicled Monk's-hood, **Aconitum paniculatum* Lam., plant 50–150cm; pedicels and the upper part of the stem glandular-hairy, sticky and pubescent; inflorescence branched; raceme lax, short; helmet of flower as wide as high; flowers blue-violet. Fl.6–8. Poisonous! Protected! Mountain and river ravine woods, high-level shrub communities; in the Alps to 2400m; scattered. Mountains of C. and S. Europe (eastern).

3. Wolf's-bane, **Aconitum vulparia* Rchb. (*A. lycoctonum* L.), plant 50–150cm; lvs palmately 5–7 lobed, with broad lobes; flowers yellow or whitish, with a tall, narrow helmet, in a simple or branched raceme. Fl.6–8. Poisonous! Protected! Wet woods, woods in river ravines, damp mixed deciduous woods, high-level shrub communities, in the Alps to 2400m. Mountains of C. and S. Europe.

4. Yellow Anemone, *Anemone ranunculoides* L., plant 10–20cm; basal lvs not appearing until after flowering; stem lvs 3, almost whorled, shortly stalked or almost sessile, 4–8cm long, 3-lobed nearly to the base, lobes coarsely toothed or cut; flowers 2cm wide, usually 2; perianth segments 5, yellow, hairy outside; achenes densely bristly- and short-hairy. Fl.4–5. Wet woods, damp mixed deciduous woods; rather rare. Almost all of Europe, northwards to S. Scandinavia (garden escape in Britain), southwards to N. Spain, C. Italy, Albania.

5. Wood Anemone, *Anemone nemorosa* L., plant 10–25cm; lvs in 3s, almost whorled, stalked, divided to the base, 3–6cm long, lobes deeply 2–5 divided and coarsely toothed; flowers solitary, 2–4cm wide, perianth segments 6–8, white, or pink outside, glabrous. Fl.3–4. Deciduous and coniferous woods, mountain meadows, orchards; widespread. Almost all of Europe.

6. Snowdrop Windflower, **Anemone sylvestris* L., plant 15–40cm; basal and stem lvs roundish, palmately 2–5 lobed, 4–10cm wide, sparsely hairy, lobes 2–3 divided, coarsely toothed at the tip; flowers usually 1, 4–7cm across; perianth segments 5–6, white, silky-hairy outside. Fl.4–6. Wood margins, slopes, gorges, pinewoods, likes warm places; rare. C. Europe, northwards to N. Russia, Siberia, southwards to N. Spain, S. Alps, mountains of the Balkan peninsula. Protected!

7. Hepatica, **Hepatica nobilis* Mill. (*Anemone hepatica* L.), plant 5–15cm; lvs radical, over-wintering, 3-lobed, green above, the undersides often red-brown or violet, with 1 flower each; perianth segments 6–10, blue, more rarely pink or white, with 3 green bracts directly under the flower, looking like a calyx. Fl.3–4. Beech, oak and mixed coniferous woods; scattered. Most of Europe (but not Britain).

See also *Pulsatilla*, Pl.51.

1
2
3
4
5
6
7

Plate 143   Buttercup Family, Ranunculaceae

1. Baneberry, *Actaea spicata* L., plant 30–60cm; basal lvs absent; stem lvs long-stalked, twice ternate to pinnate, with oval, irregularly and coarsely toothed segments; leaf blades 20–30cm long; flowers white, small, 4-merous, in many-flowered racemes; fruit a green berry, becoming black later. Fl.5–6. Mixed deciduous woods, river ravine woodland, limestone pavements; scattered. Almost all of Europe, mainly in the mountains.

2. Columbine, *Aquilegia vulgaris* L., plant 30–80cm; basal lvs long-stalked, twice ternate, often blue-green, leaflets roundish, 3-lobed; uppermost lvs sessile, 3-lobed; flowers blue-violet, 3–5cm long, with a long spur; stamens numerous, scarcely reaching out of the flower. Fl.5–7. Protected! Deciduous woods, scrub, wood margins, dry grassland, fens; likes warm places; scattered. Almost all of Europe.

3. Dark Columbine, **Aquilegia atrata* Koch, plant 30–70cm, very similar to the last species, but flowers brown-violet; stamens far exceeding the flower. Fl.6–7. Protected! Open coniferous woods, wood margins, moorland pasture; scattered. S. Germany, the Alps and hinterlands, Apennines.

4. Traveller's-joy, *Clematis vitalba* L., plant 3–8m, twining and climbing, with a woody stem; lvs with 3 or 5 long-stalked, ovate or cordate, coarsely and asymmetrically toothed leaflets; inflorescence in panicles, both lateral and terminal; flowers 5-merous, 2–3cm wide, white; perianth segments pubescent on both sides; style bearded in fruit. Fl.6–7. Poisonous! Wet woods, thickets, wood margins, hedges; rather common. C. and S. Europe, northwards to England. Similar species: **Virgin's Bower**, *C. flammula* L., but leaflets entire; perianth segments white, densely felted-hairy only on the margins. Mediterranean region, and naturalised elsewhere. Widespread in the E. Mediterranean is also the **Purple Clematis**, *C. viticella* L., plant 3–6m, twining; lvs 2-pinnate, leaflets ovate, entire or unequally 2–3 lobed; flowers blue-purple to crimson, drooping, 3–4cm across; perianth segments crisped at the tip; cultivated and occasionally naturalised.

5. Erect Clematis, **Clematis recta* L., plant 50–150cm, erect, not climbing or twining, herbaceous; stem rarely woody, angled, hollow; lvs with 5, 7 or 9 entire, ovate, acute leaflets; flowers in a terminal panicle, white; perianth segments densely felted-hairy only on the margins. Fl.5–7. Dry, bushy, stony slopes, mixed deciduous woods; likes warm places; rare. S. Germany, Elbe district, Thuringia, S. Alps, Pyrenees, S. and E. Europe.

See also *C. alpina*, Pl.185.

Plate 144   Buttercup Family, Ranunculaceae

1. Lesser Celandine, *Ranunculus ficaria* L. (*Ficaria verna* Huds.), plant 5–20cm, creeping, often rooting at the nodes; lvs roundish-cordate, distantly and bluntly toothed, very shiny; flowers yellow, 2–3cm broad; sepals 3, petals 8–12, ovate-oblong, yellow. Fl.3–5. Wet woods, mixed woods, orchards, hedges, meadows, grassy banks; common or widespread. Europe.

2. Woolly Buttercup, **Ranunculus lanuginosus* L., plant 30–70cm, densely spreading-hairy; basal lvs palmately divided, lobes with broadly ovate, toothed segments; flower stalks terete, not furrowed; flowers orange-yellow, 2–2.5cm wide; achene with a hooked, curved or inrolled beak; receptacle glabrous at fruiting time. Fl.5–7. Mixed and wet woods, woods in river ravines; scattered. C. and S. Europe (eastern).

3. Wood Buttercup, **Ranunculus nemorosus* DC. (*R. breyninus* auct.), plant 20–80cm; basal lvs dark green, pentagonal in outline, and 3-lobed almost to the base; leaf lobes cuneate, themselves lobed and toothed; flower stalks deeply furrowed, sparsely hairy; inflorescence many-flowered; flowers shining yellow; petals 15–20mm long; achenes with a beak 1.5mm long, which is inrolled at the tip. Fl.5–7. Mixed deciduous woods, mountain meadows, dry grassland; widespread. (Aggregate species with several subspecies.) C. and S. Europe, northwards to Denmark, introduced elsewhere.

4. Many-flowered Buttercup, **Ranunculus polyanthemos* L., plant 30–60cm, with 5-lobed basal lvs; leaf lobes deeply cut, with linear to linear-lanceolate segments; flower stalks weakly furrowed, appressed-hairy; petals yellow, 7–14mm long; achenes with a slightly curved, not inrolled beak, 0.5mm long. Fl.5–7. Open oakwoods, wood margins, sunny thickets; local or rare. C. and N. Europe.

5. Goldilocks Buttercup, *Ranunculus auricomus* L., plant 15–50cm; stem and basal lvs very different; basal lvs roundish to kidney-shaped in outline, glabrous, shining, almost undivided to deeply cut in 5 lobes, with broadly cuneate, lobed segments; stem lvs divided to the base in lanceolate segments; flowers yellow, 1–3cm across; pedicels terete, not furrowed; achenes hairy; receptacle at fruiting time glabrous or hairy. Fl.4–6. Mixed deciduous and wet woods, meadows. Almost all of Europe. (Aggregate species with almost innumerable microspecies; these often confined to very small areas.)

Plate 145   Barberry Family, Berberidaceae

1. Barberry, *Berberis vulgaris* L., deciduous shrub, up to 3m; main shoots with 1–7, usually trifid thorns, in whose axils the short shoots appear, bearing clusters of obovate lvs with prickly margins and 2–6cm long; flowers 6-merous, in many-flowered, pendulous racemes; sepals and petals yellow; berries oblong, red, 8–11mm long, Fl.5–6. Hedges, wood margins, wet woods, open deciduous and pine woods; scattered. Intermediate host of wheat rust. C. and S. Europe, Britain, S. Scandinavia.

Fumitory Family, Fumariaceae

2 Bulbous Corydalis, *Corydalis bulbosa* (L.)DC. (*C. cava* (L.)Schw. et Koerte), plant 15–30cm, with a globose, hollow bulb; lvs twice ternate, blue-green; flowers 18–28mm long, in 6–20 flowered, erect racemes, purple or white; bracts ovate, entire; outer petals 2, the upper with a backwards-pointing spur and a broadened tip (upper lip), and the lower broadened at tip also (lower lip), the inner petals 2, united at the tip. Fl.3–5. Deciduous woods, wet woods, orchards, scrub; scattered. Mainly C. Europe, northwards to S. Sweden, garden escape in Britain.

3. Solid-tubered Corydalis, *Corydalis solida* (L.)Clairv., plant 10–20cm, similar to *C. bulbosa*, but bulb solid; upper lvs of the inflorescence palmately divided; stem with pale scale lvs at the base. Fl.3–5. Mixed deciduous woods, hedges; rather rare. S. Sweden, C. Europe, Pyrenees, in the south in the mountains, in Britain a garden escape.

Cabbage Family, Cruciferae or Brassicaceae

4. Honesty, *Lunaria rediviva* L., plant 30–140cm; lvs cordate, stalked, serrate, the lower almost opposite; flowers 4-merous, pleasantly scented; petals 12–20mm long, pale violet, lilac or white; sepals 5–6mm long; siliquae elliptic to broadly lanceolate, pointed at both ends, 4–8cm long. Fl.5–7. Mountain woods, shady, humid slopes; rather rare in the wild. Most of Europe. Also cultivated and often escaping.

5. Trifoliate Bitter-cress, *Cardamine trifolia* L., plant 10–20cm; stem leafless or with 1 leaf; basal lvs 3-lobed, long-stalked, over-wintering; leaflets roundish-rhomboid, crenate; petals 4, 8–11mm long, white; anthers yellow; siliquae 15–25mm long and 2mm wide, stalk about the same length. Fl.4–6. Herb-rich beech or mixed beechwoods, mountain woods; scattered. Mountains of C. and S. Europe (eastern).

6. Tower Cress, *Arabis turrita* L., plant 10–70cm; basal lvs narrowly ovate, narrowed into the stalk, irregularly serrate; stem lvs oblong, with cordate, auriculate base, auricles clasping; inflorescence leafy below; petals 4, 6–8mm long, yellowish-white; siliquae curved and twisted downwards, 8–12cm long, 2–3mm broad. Fl.4–6. Open woods, thickets, stony slopes; likes warm places; rare. S. Germany (Black Forest, Jura), Swiss Jura, Alps, S. Europe. Rare introduction in Britain.

7. Tower Mustard, *Arabis glabra* (L.)Bernh. (*Turritis glabra* L.), plant 60–120cm; basal lvs sinuate-dentate to pinnatifid, withered by flowering time; stem lvs entire, sessile and clasping the stem, glabrous, bluish pruinose; corolla 5–8mm long, yellowish-white; siliquae 4-angled, 4–7cm long, appressed to the stem. Fl.5–7. Wood margins and hedges, open oakwoods, dry banks, cliffs and rocks; local. Europe.

See also *Alliaria*, Pl.15.

Plate 146   Cabbage Family, Cruciferae or Brassicaceae

1. Coralroot, *Cardamine bulbifera* (L.)Crantz (*Dentaria b.*L.), plant 30–50cm; basal lvs and lower stem lvs pinnate, stalked, with 5–7 ovate-lanceolate, serrate leaflets, upper stem lvs undivided; leaf axils with small, brown-violet bulbils; corolla pale violet, pink or white. Fl.5–6. Herb-rich beech and maple mixed woods; rare. Mainly the mountains of C. Europe, northwards to S. Scandinavia, Britain, southwards to Greece.

2. Five-leaflet Bitter-cress, *Cardamine pentaphyllos* (L.)Crantz (*D. pentaphyllos* L.), plant 25–50cm; rhizome fleshy, with tooth-like scale lvs (as in all species of the genus *Dentaria*); stem lvs alternate, palmately 3–5 lobed, leaflets ovate-lanceolate, serrate; petals 4, pink or violet, 12–18mm long. Fl.4–6. Herb-rich beechwoods or beech-fir woods, woods in river gorges; scattered. Mountains of C. and S. Europe (western). Similar species: **Seven-leaflet Bitter-cress**, *C. heptaphylla* (Vill.)Schultz (*Dentaria heptaphylla* L.), but lvs 7-lobed; corolla white or pale lilac. Fl.4–5. Deciduous and mixed woods; rare; C. and S. European mountains.

3. Drooping Bitter-cress, *Cardamine enneaphyllos* (L.)Crantz (*Dentaria enneaphyllos* L.), plant 20–30cm; stem lvs in 3s, almost whorled; corolla yellowish-white. Fl.4–5. Beech and beech-fir woods; rare. Mountains of C. and S. Europe (eastern).

Saxifrage Family, Saxifragaceae

4. Alternate-leaved Golden-saxifrage, *Chrysosplenium alternifolium* L., plant 5–20cm, forming patches, with filiform stolons; basal lvs roundish-reniform, 2–5cm wide, deeply cordate at the base, crenate, long-stalked; stem lvs similar, alternate; stem 3-angled; inflorescence cymose, with yellowish bracts; sepals simple, 4, greenish-yellow, 5–6mm wide; stamens 8. Fl.3–5. Wet woods, woods in river ravines, stream banks, springs, wet rocks; local to common. Almost all of Europe, southwards to C. Italy, northwards to Britain, Norway. Similar species: **Opposite-leaved Golden-saxifrage**, *C. oppositifolium* L., but stem 4-angled, lvs opposite. Fl.4–5. Similar habitats; scattered.

5. Round-leaved Saxifrage, *Saxifraga rotundifolia* L., plant 20–70cm; lvs long-stalked, pale green, roundish, coarsely crenate; flowers in lax panicles, white, with red spots at the base. Fl.6–9. Mixed mountain woods, high-level shrub communities, stream banks, in the Alps to over 2200m; widespread. Mountains of C. and S. Europe.

Currant Family, Grossulariaceae

6. Mountain Currant, *Ribes alpinum* L., shrub 80–150cm, without spines; lvs 3(5)-lobed, coarsely toothed, 2–4cm long, both sides with red-tipped glandular hairs; leaf stalk ½ the length of the blade; plant usually dioecious; male flowers in 10–30 flowered, erect racemes; female flowers 2–5 similarly; flowers 5-merous, greenish-yellow; petals shorter than the calyx; berries red. Fl.4–5. Mixed mountain woods, in river gorges, alpine pastures, cliffs, in the Alps to 2000m; cultivated plant; local. N. and C. Europe, Britain, south to N. Spanish mountains, Apennines, mountains of Balkan peninsula.

7. Gooseberry, *Ribes uva-crispa* L. (*R. grossularia* L.), shrub 60–120cm, spiny; lvs 3–5 lobed, 3cm wide, pubescent-hairy, with blunt teeth; flowers 1–3 together, greenish; calyx densely hairy, sepals oval, spreading, 2–3 times as long as the whitish petals; berries ovoid, 1–2cm wide, reddish, green or yellowish, glabrous or bristly. Fl.4–5. Wood margins, wet and river gorge woods, hedges, ruins of old buildings; cultivated in many forms; rather common. Almost all of Europe, in the south only in the mountains.

1

2

3

4

5

6

7

Plate 147 Currant Family, Grossulariaceae

1. Black Currant, *Ribes nigrum* L., shrub 80–150cm, without spines, with a peculiar, strong smell; lvs 3–(5) lobed, up to 10cm long, with yellowish glands on the underside; young twigs pubescent-hairy; flowers greenish, in hanging racemes; petals erect, much shorter than the reflexed, densely hairy sepals; fruit a black, edible berry, 10–15mm across. Fl.4–5. Alder carr and wet woods; fairly common to rather rare. Cultivated for its fruit and escaping. Protected! Mainly C. and N. Europe, Britain, Scandinavia. Similar species: **Red Currant**, *R. rubrum* L., but lvs not strong-smelling, without glands; berry red or white, edible, 6–10mm. Similar places, sometimes an escape. Britain, W. Europe.

Rose Family, Rosaceae

2. Goat's-beard Spiraea, *Aruncus dioicus* (Walter)Fern. (*A. sylvestris* Kost, *A. vulgaris* Raf.), plant 80–150cm, dioecious; lvs 2–3 times ternate; leaflets ovate, sharply double-serrate; inflorescence a panicle, up to 50cm long; flowers small, 2–4mm across, male flowers yellowish-white, with over 20 stamens, female flowers pure white, very numerous; petals and sepals 5. Fl.4–7. Woods in river gorges, mountain streams, high-level shrub communities; scattered. Mountains of C. Europe, Alps, Pyrenees, Apennines, mountains of the Balkan peninsula. See also *Filipendula*, Pl.56.

3. Wild Cotoneaster, *Cotoneaster integerrimus* Med., shrub 50–200cm; lvs ovate, 2–4cm long, upper surface glabrous, green, lower surface grey-felted; flowers 1–5 together; sepals 5, glabrous outside, 2mm long; petals 5, white to bluish-red; fruit globose, red, 6–8mm across. Fl.4–5. Thickets on rocky places, sunny, stony slopes, dry oak-pine woods; rather rare. Mainly C. and S. Europe (eastern), northwards to S. Scandinavia, Britain (very rare). Similar species: **Hairy Cotoneaster**, *C. nebrodensis* (Guss.)Koch (*C. tomentosus* (Ait.)Lindl.), but lvs hairy above, white-felted beneath, 2–6cm long; calyx felted; fruit felted-hairy; dry woods and scrub, limestone pavement; rare; mountains of C. and S. Europe.

4. Rowan, *Sorbus aucuparia* L., shrub or tree with smooth bark, 5–15m; lvs imparipinnate; leaflets ovate, 4–6cm long, sharply serrate; flowers in many-flowered, umbellate panicles; sepals 5; petals 5, 4–5mm long, white, styles 2–4; fruit almost spherical, red, 8–10mm long. Fl.5–6. Open deciduous and coniferous woods, up to the tree limit, mountain rocks; common to widespread. Europe. Similar species: **Service-tree**, *S. domestica* L., but bark rough; leaflets serrate only in the upper half; inflorescence only 6–12 flowered, styles usually 5; fruit pear-shaped, reddish-yellow; likes warm places; deciduous woods; Mediterranean region, elsewhere planted or naturalised.

5. Common Whitebeam, *Sorbus aria* (L.)Crantz, shrub or tree, 2–10m; lvs ovate, irregularly toothed, 8–14cm long, dark green, the underside grey- to white-felted, with 10–14 pairs of veins; flowers white; styles 2; fruit globose or ovoid, orange to red. Fl.5–6. Sunny deciduous woods, limestone cliffs, mountain woods; scattered. (An aggregate species with many microspecies.) Mountains of C. and S. Europe, Britain, often planted and naturalised. Similar species: **Vosges Whitebeam**, *S. mougeotii* Soy.-Will. et Godr., but lvs cut to $\frac{1}{4}$ the leaf width and with 8–10 pairs of veins; lobes obtuse, directed forwards, becoming smaller from the tip downwards; likes warm places; mixed deciduous woods, mountain woods, scrub; rare; mountains of C. and S. Europe.

2

5

4

3

1

Plate 148   Rose Family, Rosaceae

1. Wild Service-tree, *Sorbus torminalis* (L.)Crantz, tree or shrub 3–15m; lvs lobed, underside grey-felted, becoming glabrous later, with 4–5 prominent vein pairs; leaf lobes acute, the lower large, patent; flowers in corymbose panicles, white; fruit ovoid, brown. Fl.5–6. Oak-hornbeam woods, Downy Oak scrub; local. Britain, France, Germany, C. and S. Europe.

2. Swedish Whitebeam, *Sorbus intermedia* (Ehrh.)Pers., tree or shrub, 3–15m, like *S. torminalis*, but lvs not becoming glabrous below, with 3–8 veins on each side and obtuse, rounded lobes. Fl.5–6. Deciduous woods, N. Europe and North Sea area; in Britain planted, and sometimes naturalised.

3. Crab Apple, *Malus sylvestris* (L.)Mill. ssp. *sylvestris* (*Pyrus malus* L.), tree 8–10m; twigs usually thorny; lvs roundish to ovate, 3–5cm long, finely serrate, the underside hairy only on the veins, which are distinctly prominent; flowers 5-merous, 3–4cm wide, in few-flowered racemes; petals white above, red-tinged on the underside; anthers yellow; fruit hard and sour. Fl.4–5. Wet woods, hedges, scrub; scattered. Almost all of Europe. The **Cultivated Apple**, *Malus sylvestris* ssp. *mitis* (Wallr.)Mansf., is rarely thorny, the lvs are persistently pubescent all over below, and the fruit is larger and sweeter; cultivated and often naturalised. The **Wild Pear**, *Pyrus pyraster* Burgsd. (*P. achras* Gaertn., *P. communis* auct.), has petals which are pure white both sides, red anthers and round lvs with indistinct veins beneath; stony scrub, hedges, wet woods; likes warm places; cultivated and naturalised.

4. Midland Hawthorn, *Crataegus laevigata* (Poir.)DC. (*C. oxyacantha* L.), shrub, similar to the next species, but lvs shallowly lobed, with cuneate base; styles 2–3; fruit with 2–3 stones. Fl.5–6, variable. Deciduous woods, hedges; widespread. Almost all of Europe, northwards to Britain, S. Scandinavia.

5. Hawthorn, *Crataegus monogyna* Jacq., shrub 1–8m; lvs deeply pinnatisect, usually 5-lobed, base broad; leaf lobes entire or toothed at the tip; flowers in corymbs, 5-merous, 10–15mm across, white; style 1; fruit globose, red, 6–10mm wide, with 1 stone. Fl.5–6. (Variable, with several subspecies.) Sunny scrub, woods, rocks, hedges; common. Almost all of Europe, northwards to Scotland.

6. Raspberry, *Rubus idaeus* L., plant 60–120cm; stem biennial, woody, erect or drooping at tip, with red prickles; lvs pinnate, with 5 or 7 serrate, pale green leaflets, which are white-felted below; inflorescence nodding, paniculate; petals narrowly ovate, white, 5mm long; sepals reflexed after flowering; fruit red, glabrous, coming away from the receptacle. Fl.5–6. Wood clearings, wet woods, ravines, heaths; common. Europe.

7. Dewberry, *Rubus caesius* L., plant 30–80cm; stem roundish, bluish-pruinose, prickly; lvs trifoliate, coarsely and unequally serrate; stipules lanceolate; inflorescence corymbose; flowers 1.5–2cm wide; petals 5, white (rarely pink); sepals grey-felted; fruit blue-pruinose, consisting of 5–20 large drupelets. Fl.5–6. Wet woods, wood margins, hedges, river banks, dry grassland and scrub; common. Almost all of Europe.

Plate 149   Rose Family, Rosaceae

1. Stone Bramble, *Rubus saxatilis* L., plant 10–30cm; stem annual, not woody, with weak prickles or unarmed; lvs trifoliate, long-stalked, green on both sides, coarsely double-serrate; flowers 3–10 together; petals white, 5mm long; fruit pale red (with taste of currant). Fl.5–6. Mixed woods, scrub, shady rocks, in the Alps to 2000m; local to rare. Europe, in the south in the mountains.

2. Blackberry, *Rubus fruticosus* L., aggregate, variable with very numerous microspecies, 50–200cm; stem arched or procumbent or erect or drooping, prickly; the armature of the stem varies enormously, according to the presence or absence of long hairs, felt, sessile and stalked glands, bristles and pricklets; lvs 3–5 lobed, digitate or pedate, often over-wintering, leaflets ovate, elliptic, obovate to roundish, acute, serrate, sessile to long-stalked; stipules filiform; flowers white to pink, in panicles; fruit black, shining, formed from 20–50 small drupelets, remaining attached to the receptacle. Fl.5–8. Woods, hedges, clearings, heaths; common. Almost all of Europe.

3. Snowy Mespil, **Amelanchier ovalis* Med., shrub 1–3m; lvs elliptic, 2–4cm long, dark green and glabrous above, grey-felted below, but later glabrous and grey-green, finely serrate; flowers in few-flowered racemes, 5-merous; petals narrowly lanceolate, 15–20mm long, white or yellowish-white, villous outside; stamens 20; fruit blue-black. Fl.4–5. Scrub over rocks, limestone pavement, sunny oak or pine woods; scattered. Mountains of C. and S. Europe.

4. Wood Avens, *Geum urbanum* L., plant 30–60cm; basal lvs long-stalked, pinnate, with a large terminal leaflet, all leaflets coarsely toothed and cut; stipules large, leaf-like; stem with several flowers; flowers 5-merous, erect to spreading, 1–2cm wide; petals yellow, as long as the inner sepals; style curved and hooked, glabrous below, feathery above. Fl.5–9. Mixed deciduous woods, scrub, walls, shady hedgerows; widespread. Almost all of Europe. See also *G. rivale*, Pl.106.

5. Agrimony, *Agrimonia eupatoria* L., plant 30–100cm; stem usually branched above, densely rough-hairy, with sessile glands, lvs opposite, 10–15cm long, white-felted below, pinnate, with 5–9 pairs of coarsely toothed lateral lobes and a large terminal lobe, with smaller lobes in between; flowers 5-merous; petals yellow, scarcely emarginate; fruiting receptacle grooved, with spreading hooked spines, which become harder later as the fruit ripens. Fl.7–9. Hedges, wood margins, poor grassland, dryish grassland; widespread. Almost all of Europe.

6. Rock Cinquefoil, *Potentilla rupestris* L., plant 30–50cm; basal lvs pinnate, long-stalked; leaflets oval, 1–3cm long, serrate; upper lvs trifoliate; flowers 5-merous, 1–2cm across; petals white, oval, 1–2 times as long as the lanceolate sepals; filaments and achenes glabrous. Fl.5–7. Sunny, open oak and pine woods, walls, rocks; likes warm places; rare. Protected! Britain (very rare), C. and S. Europe.

Plate 150 Rose Family, Rosaceae

1. White Cinquefoil, *Potentilla alba* L., plant 5–20cm; lvs 5-lobed digitate, green above, silvery-white and silky-hairy beneath; leaflets lanceolate, 3–6cm long, with 1–4 small teeth on each side; stem hairy, 1–5 flowered; petals 5, white, cordate, emarginate, little longer than the sepals; filaments glabrous. Fl.4–6. Wood margins, open oak and oak-pine woods in warm and dry districts; rare. C. and E. Europe.

2. Barren Strawberry, *Potentilla sterilis* (L.)Garcke, plant 5–10cm, with stolons which form leaf rosettes at their tips; basal lvs 3-lobed, like those of the Strawberry, each leaflet with 4–6 teeth on each side, the upperside sparsely hairy, the underside densely hairy; flowers 10–15mm wide; petals white, not overlapping; filaments glabrous, getting more slender upwards. Fl.3–5. Scrub, herb-rich deciduous woods, wood margins, hedge banks; common to widespread. C. Europe, northwards to Scotland, S. Scandinavia, southwards to C. Spain, Yugoslavia. Similar species: **Pink Barren Strawberry**, *P. micrantha* Ram., but plant without stolons, leaflets with 6–11 teeth on each side; calyx and corolla reddish at base; filaments ciliate below, broadened and strap-shaped. Mixed woods, wood margins; likes warm places; rare. S.C. and S. Europe. Protected!

3. Trailing Tormentil, *Potentilla anglica* Laich., plant 15–30cm, stem prostrate, often rooting at the nodes; basal lvs 3–5 lobed, with leaflets with 4–6 teeth on each side; stem lvs with stalks 1–2cm long, stipules usually entire; flowers 4(–5)-merous, yellow; petals 5–7mm long. Fl.6–8. Woodland rides, damp woods, marshy meadows, heaths, hedgebanks; local to rare. Derived from, but distinct from, the hybrid *P. erecta* × *P. reptans*, which also occurs, but is sterile. Britain, N. and C. Europe. Similar to the Tormentil, Pl.107.

4. Wild Strawberry, *Fragaria vesca* L., plant 5–20cm, with long stolons; lvs trifoliate, leaflets ovate, sessile, serrate, sparsely hairy, silky-hairy beneath; flowers 1–1.5cm wide; petals 5, white; sepals horizontally spreading or reflexed in fruit (as in *F. viridis*, Pl.56), calyx staying behind when the ripe, soft, red fruit is picked. Fl.5–6. Wood margins, clearings and where woods are clear-felled; common. Europe.

Plate 151   Rose Family, Rosaceae

1. Field Rose, *Rosa arvensis* Huds., shrub 50–150cm, prostrate or clambering, with slightly curved prickles which are all alike; lvs imparipinnate with 7 leaflets; leaflets ovate, 1–3cm long, serrate, dark green, pale green below; flowers solitary, long-stalked, 4cm across, white; sepals entire, rarely pinnate, falling away before the fruit is ripe; styles united in a column, which clearly exceeds the receptacle (hypanthium) and is at least as long as the inner stamens; fruit ellipsoid to spherical, dark red, 1–1.5cm long. Fl.6–7. Wood margins, open deciduous woods, hedges, scrub; common. Britain, C. and S. Europe.

2. Alpine Rose, *Rosa pendulina* L., shrub, 0.5 to 2m; stem and twigs with few, straight prickles, or without prickles; lvs with 9–11 dark green, doubly serrate leaflets; flowers usually solitary; pedicels 1–3cm, with glandular hairs; sepals entire, erect after flowering and persistent; petals pink to dark red; fruit flask-shaped, with glandular hairs and prickly bristles; orange to red. Fl.6–7. Mixed mountain woods, high-level shrub communities, rock outcrops, Mountain Pine communities; scattered. Mountains of C. and S. Europe.

3. Dog Rose, *Rosa canina* L., shrub, 1–3m, with hooked prickles; lvs with 5–7 leaflets, glabrous on both sides, non-glandular, leaflets serrate; peduncles with falcate prickles; sepals reflexed after flowering, falling off before the fruit is ripe, the outermost sepals pinnatisect, with narrow appendages; flowers pink; styles free, short, only just exceeding the receptacle; fruit ovoid, red. Fl.6. Woods and wood margins, waysides, hedges, open woodland, scrub; common or widespread. Europe.

4. Cinnamon Rose, *Rosa majalis* Herrm. (*R. cinnamomea* L.), shrub 0.5–2m; flowering branches with curved, hooked prickles, usually arranged in pairs, all like one another, lower branches and stem also with many acicular prickles and stiff bristles; lvs with 5–7 leaflets, dark green to blue-green and hairy above, grey-green and pubescent to felted-hairy below; sepals entire, erect after flowering, persistent, longer than the petals, which are pink to dark red; fruit spherical, glabrous, orange to red. Fl.5–6. Sunny hedges, stony slopes, wet woods; rare. N. and C. Europe (eastern).

5. Blackthorn, *Prunus spinosa* L., shrub to 4m, much-branched and very thorny; lvs ovate-lanceolate, 2–4cm long, dark green, finely serrate; flowers solitary or in pairs, but closely packed, appearing before the lvs; petals 5–8mm long, white; sepals irregularly and finely serrate; stamens about 20; fruit spherical, blue-pruinose, 1–1.5cm wide (called a 'sloe'). Fl.4–5. Wood margins and waysides, sunny hedges, poor pastures; common. Europe.

6. St. Lucie Cherry, *Prunus mahaleb* L. (*Cerasus mahaleb* (L.)Mill.), shrub 1–6m; lvs ovate-roundish, with a cuspidate tip, dark green, with short, obtuse teeth on the margin, 4–8cm long; flowers 4–10 in corymbose racemes, white, 1cm across; fruit ovoid, black. Fl.4–5. Sunny hedges, scrub, stony slopes, woods of Downy Oak; likes warm places; rare, commoner in the south. S. Germany, Austria, S. Europe, otherwise planted.

Plate 152   Rose Family, Rosaceae

1. Wild Cherry, *Prunus avium* L. (*Cerasus avium* (L.)Moench), tree to 25m, bark red-brown to black and peeling off in strips around the trunk; lvs ovate, 6–15cm long, serrate; flowers in clusters, without bracts; pedicels 3–5cm long, with only bud-scales at the base; petals 10–15mm long, oval, white; stamens up to 20; fruit spherical, 10–15mm across, red to blackish-red, with a smooth stone. Fl.4–5. Mixed deciduous or coniferous woods, wood margins, hedges; fairly common or local. Closely related to garden cherries. Mainly C. and S. Europe, also Britain, Scandinavia. Similar species: **Dwarf Cherry**, *Prunus cerasus* L. (*C. vulgaris* Mill.), shrub or tree, up to 10m; flower stalks 2–4cm long, with bud-scales at the base and with 1–2 bracts; petals almost circular; open woods, hedges; often planted, local; originally from S.E. Europe.

2. Bird Cherry, *Prunus padus* L. (*Padus avium* Mill.), shrub or tree to 12m; lvs soft, broad-lanceolate, 5–10cm long, finely serrate; flowers 10–20 together in first erect, then hanging racemes; pedicels 10–15mm long; petals 5–10mm long, white; sepals fringed, glandular; fruit globose, shining, black, 7–8mm wide, stone furrowed. Fl.4–5. Wet woods, wood margins, scrub; local or fairly common. Almost all of Europe. Similar species: **Rum Cherry**, *Prunus serotina* Ehrh. (*Padus serotina* (Ehrh.)Borkh.), but lvs leathery, very shiny above; pedicels 3–6mm long; calyx still present on the ripe fruit, which has a smooth stone. Fl.6–7. Wood margins, cultivated tree; originally from N. America.

Pea Family, Papilionaceae or Fabaceae

3. Mountain Zigzag Clover, *Trifolium alpestre* L., plant 10–30cm, growing in patches; lvs trifoliate, leaflets narrow-lanceolate, 3–5cm long, dark green, not spotted; stipules sagittate, over 3cm long; flower heads globose to ovoid, usually in pairs; calyx tube 20-veined, hairy; corolla 10–15mm long, purple. Fl.6–7. Wood margins, open, dry woods, scrub, dry, calcareous grassland; likes warm places; rather rare. C. and S. Europe.

4. Red Trefoil, *Trifolium rubens* L., plant 30–60cm, similar to the last species, but calyx tube glabrous outside (calyx teeth are ciliate, however), 20-veined; flower heads in 2s, oblong-cylindrical, finally 3–7cm long, with an involucre below; leaflets finely serrate; stipules glabrous, longer than the petioles and longer than the leaflets. Fl.6–7. Sunny wood margins, open oak and pine woods; likes warm places; rare. C. Europe, southwards to N. Spain and C. Italy. In the similar **Zigzag Clover**, *T. medium* L., the calyx tube is 10-veined, glabrous (but the teeth are ciliate); flower heads globose, usually solitary, without involucre; leaflets oblong-elliptic, usually spotted; stipules ciliate; wood margins, scrub, dryish grassland; widespread.

5. Greater Badassi, *Dorycnium pentaphyllum* Scop. ssp. *germanicum* (Gremli) Gams, plant 15–30cm, woody at base; lvs 3–5 lobed; leaflets 2–4mm wide, appressed silky-hairy; flower heads 6–14 flowered; pedicels 1–2mm long, shorter than the calyx tube; corolla 5–7mm long, white, with a violet tip to the keel; wings with pocket-like longitudinal folds; pods ovoid, 3–5mm long. Fl.7–8. Open, dry pinewoods, poor calcareous grassland; rare. Hinterlands of the Alps, E. Alps, S.E. Europe. Similar species: **Herbaceous Badassi**, *D. pentaphyllum* ssp. *herbaceum* (Vill.)Rouy, but leaflets 4–6mm wide, with long spreading hairs (or becoming glabrous); flowers 3–5mm long, 15–35 together; pedicels as long or longer than the calyx tube; sunny scrub, dryish grassland; very rare; mainly S. and S.E. Europe. Protected!

Plate 153   Pea Family, Papilionaceae or Fabaceae

1. Black Broom, *Lembotropis nigricans* (L.) Griseb. (*Cytisus nigricans* L.), plant 50–150cm, becoming black on drying; twigs appressed short-hairy, brush-like; lvs trifoliate, glabrous above, hairy beneath; flowers in terminal, leafless, many-flowered racemes; corolla 10–12mm long, yellow; pedicels 1–2 times as long as the short and campanulate calyx; pods 2–3cm long. Fl.6–8. Open woods, sunny wood margins; likes warm places; rare. S. Germany, S. Alps, E. Europe.

2. Broom, *Cytisus scoparius* (L.)Link (*Sarothamnus scoparius* (L.)Koch), shrub 50–200cm, with an erect, branched stem and brush-like, 4–5-angled, green twigs; lower lvs trifoliate, upper simple, lanceolate, 5–20mm long; flowers 1–2 together, axillary; calyx glabrous; corolla 20–25mm, yellow; style very long, rolled up in a spiral; pods 3–5cm long, hairy. Fl.5–6. Wood margins, clearings, oakwoods on acid soils, heaths; calcifuge; common. S. Scandinavia, Britain, C. and S. Europe (western).

See also *Chamaecytisus* and *Genista*, Pls.56 and 57.

3. Bladder-senna, *Colutea arborescens* L., shrub 2–5m; lvs imparipinnate, leaflets obovate, emarginate at tip, 1–3cm long; flowers 3–6 together, yellow, 15–25mm long; calyx campanulate, with 5 unequal teeth; pods bladder-like, inflated, 6–8cm long, 3cm thick. Fl.6–8. Oakwoods, scrub, sunny wood margins, waste places; rare. Poisonous! S. Germany, Austria, S. Europe, naturalised in N. Europe, Britain.

4. Bush Vetch, *Vicia sepium* L., plant 30–60cm; lvs with 8–14 ovate, pinnate leaflets and a terminal tendril; flowers 2–5 together in shortly stalked racemes; calyx teeth of unequal length; corolla 12–15mm long, brownish-violet; pods 2–3.5cm long, shortly hairy at first, then glabrous, 3–6 seeded. Fl.5–8. Scrub, wood margins, mixed deciduous woods, also meadows, hedges; common or widespread. Almost all of Europe. See also *V. cracca*, Pl.81.

5. Danzig Vetch, *Vicia cassubica* L., plant 30–60cm; stem simple or branched, shortly hairy; lvs with 16–28 ovate leaflets, which are 5–7mm wide, with sparse short hairs on both sides and a branched terminal tendril; stipules entire, sagittate to hastate; flowers 10–15 together in stalked racemes; racemes shorter than the subtending lvs; corolla 9–13mm long, with a red-violet and dark-veined standard, whitish wings and a whitish keel with a violet tip; pod glabrous, 1–3 seeded. Fl.6–7. Open, dry woods, sunny wood margins, scrub; rare. C. Europe, S. Scandinavia, casual in Britain, southwards to C. Spain. Protected!

6. Purple Wood Vetch, *Vicia dumetorum* L., plant 60–200cm; lvs with 6–10 broadly ovate leaflets, 15–40mm long and 10–25mm wide and a branched terminal tendril; stipules toothed; flowers 4–12 together in stalked racemes, which are as long as the subtending bract; corolla 13–18mm long, red-violet, dirty yellow after flowering; pods glabrous, 4–10 seeded. Fl.6–8. Herb-rich deciduous woods, woodland rides, thickets; rather rare. C. Europe, northwards to S. Sweden, southwards to C. Italy. Similar species: **Pea Vetch**, *Vicia pisiformis* L., but flowers pale yellow, racemes shorter than the bracts; dry woods, sunny wood margins, rather rare. Mainly C. and E. Europe, southwards to Bulgaria.

Plate 154   Pea Family, Papilionaceae or Fabaceae

1. Wood Vetch, *Vicia sylvatica* L., plant 50–200cm, prostrate or climbing; lvs with 12–24 oblong leaflets and a branched tendril; flowers in 10–20 flowered, long-stalked racemes; corolla 13–17mm long, white, with violet veins, tip of the keel violet. Fl.6–8. Herb-rich mixed woods, open woods in river ravines, mountain woods, cliffs, in the Alps up to over 2000m; scattered. Almost all of Europe, mainly in the mountains.

2. Scorpion Senna, *Coronilla emerus* L., shrub 50–200cm; stem erect, woody; twigs green, angled; lvs shortly stalked, with 5–9 oval leaflets, which are 1–2cm long, and rounded or emarginate at tip; flowers 2–3 together; corolla yellow, 16–22mm long, the standard often red-striped; calyx short, ⅓ the length of the corolla, with short, triangular calyx teeth; pod 5–10cm long, narrowly cylindrical, almost terete. Fl.4–5. Sunny thickets, open oakwoods and mixed coniferous woods, stony slopes; likes warm places; rare. In scattered localities in S. Scandinavia, S. Germany, Austria, valleys of the N. Alps, C. and S. Alps, S. Europe.

3. Scorpion Vetch, *Coronilla coronata* L. (*C. montana* Scop.), plant 30–50cm, erect; lvs imparipinnate, with 7–13 oval leaflets, which are rather fleshy, and blue-green below; stipules filiform, soon falling off, the lower united, the upper free; flowers in 12–20 flowered umbels; corolla yellow, 7–10mm long; pods segmented, 2–3cm long, segments with 4 obtuse angles. Fl.5–7. Open woods, scrub, dry sunny slopes; calcareous soils, likes warm places; rare. C. and S. Europe (eastern), mainly in the mountains. The **Small Scorpion Vetch**, *C. vaginalis* Lam., is prostrate, 5–15cm; umbel 4–10 flowered, flowers yellow; leaflets 7–9, rather thick; stipules ovate, united, persistent; open pinewoods, rocks, dry grassland; mountains of C. and S. Europe.

4. Wild Lentil, *Astragalus cicer* L., plant 20–60cm; stem and lvs appressed-hairy; lvs with 17–25 oblong-lanceolate leaflets; flowers 8–25 together in dense racemes; corolla 12–16mm long, pale yellow; pods inflated, roughly hairy. Fl.6–8. Scrub, wood margins, calcareous soils, likes warm places; rather rare. C. and S. Europe.

5. Wild Liquorice, *Astragalus glycyphyllos* L., plant 50–150cm; stem and lvs glabrous; lvs with 11 or 13 ovate leaflets, 2–5cm long; flowers 8–30 together; corolla 13–15mm long, yellow-green; pods glabrous, linear, slightly curved. Fl.6–7. Wood margins, scrub, stony slopes, rough grassy places; widespread. S. Scandinavia, Britain, C. Europe, rare in S. Europe.

6. Round-leaved Restharrow, *Ononis rotundifolia* L., plant 15–30cm, glandular-hairy, woody at the base; lvs trifoliate; leaflets ovate to roundish, 2–3cm long, sinuate-dentate, the middle leaflet long-stalked; flowers 1–3 together in the axils; corolla 15–22mm long, pink. Fl.5–7. Open pinewoods, stony slopes; scattered. C. and S. Alps, Pyrenees, Cevennes, Abruzzi.

Plate 155   Pea Family, Papilionaceae or Fabaceae

1. Black Pea, *Lathyrus niger* (L.)Bernh., plant 30–80cm, becoming black when dried; stem erect, not winged; lvs pinnate with 8–12 elliptic to ovate leaflets, which have an awn-like point at the tip; inflorescence long-stalked, 3–10 flowered; corolla 10–15mm long, purple, later violet; pod 8–14 seeded. Fl.6–7. Open oakwoods and oak-pine woods, sunny edges of scrub, likes warm places; rather rare. C. and S. Europe (eastern), northwards to S. Scandinavia, very rare in Britain.

2. Spring Pea, *Lathyrus vernus (L.)Bernh., plant 20–40cm; stem erect, unbranched, not winged; lvs pinnate with 4–8, broad ovate, long-pointed leaflets, 3–7cm long and 1–3cm wide, each with an awn-like point at the tip; raceme 3–7 flowered, as long as the bract; corolla purple, later blue-green. Fl.4–5. Beechwoods, beech-and-fir woods and oak-hornbeam woods; widespread. Almost all of Europe, but not reaching Britain, in the south only in the mountains. The **Sword Pea**, **L. bauhinii* Genty, has 4–8 linear-lanceolate leaflets, 2–4mm wide; raceme much longer than its bract; corolla purple or blue-violet; open pinewoods; very rare; Germany, Switzerland, S.E. Europe.

3. Bitter Vetch, *Lathyrus montanus* Bernh. (*L. linifolius* (Reichard) Bässler), plant 15–30cm, prostrate to ascending; stem winged; lvs pinnate with 4–8 lanceolate to linear, blue-green leaflets with an awn-like point at the tips; leaf stalk narrowly winged; inflorescence long-stalked, 3–6 flowered; corolla 10–15mm, pale purple at first, then deep blue; calyx teeth unequal in length; pod 3–4cm long, 6–10 seeded. Fl.4–6. Open oak and oak-beech woods, woodland rides, poor grassland, heaths, hedges and thickets; calcifuge; fairly common. Almost all of Europe.

4. Narrow-leaved Everlasting Pea, *Lathyrus sylvestris* L., plant 100–200cm, prostrate, ascending or climbing; lvs pinnate with 3-veined leaflets 5–14cm long and 5–15mm broad, with a branched, terminal tendril; stem and petiole broadly winged; the wing on the petiole narrower than that on the stem; inflorescence 3–6 flowered, peduncle about as long as the bract; corolla 12–18mm, yellowish-green, tinged with red; pod 5–7cm long. Fl.7–8. Wood margins, woodland rides, sunny hedges, screes; fairly common or local. Almost all of Europe, northwards to S. Scandinavia, Britain. Similar species: **Broad-leaved Everlasting Pea**, *L. latifolius* L., but leaflets 5-veined, 15–50mm wide; petioles as broadly winged as the stem; peduncles of the racemes much longer than the bracts; flowers carmine-red. Fl.6–7. Sunny hedges; rare; cultivated plant and occasionally naturalised (as in Britain); wild in S. Europe. In the **Various-leaved Everlasting Pea**, *L. heterophyllus* L., the lower lvs have 2 leaflets and the upper 4 or 6; stem and petioles broadly winged; corolla purplish-red; sunny hedges, screes; rare; mainly mountains of C. Europe, northwards to S. Sweden, as an escape in Britain.

Flax Family, Linaceae. See *Linum*, Pl.61.

Wood Sorrel Family, Oxalidaceae

5. Wood Sorrel, *Oxalis acetosella* L., plant 5–12cm; all lvs radical, long-stalked, tri-foliate and like clover lvs; flowers solitary, on long peduncles, which have 2 scale-like bracts in the middle; petals 5, white, more rarely pink or bluish, with purple veins, 10–15mm long; sepals oblong-ovate, 4–5mm long. Fl.4–5. Herb-rich coniferous and mixed deciduous woods, high-level shrub communities, scrub, shady rocks, can stand deep shade; in the Alps to over 2000m; common. Europe, in the south only in the mountains.

1

2

3

4

5

Plate 156   Crane's-bill Family, Geraniaceae

1. Shining Crane's-bill, *Geranium lucidum* L., plant annual, glabrous, 10–30cm; stem fragile, red-tinged; lvs shining, almost fleshy, roundish in outline, divided almost to the middle in broad, bluntly toothed segments; the underside often red-tinged; petals 7–9mm long, pinkish-red, rounded at tip; sepals with 3 sharply keeled veins. Fl.5–8. Scrub, wood clearings, shady walls and rocks; uncommon to rare. C. and S. Europe, scattered as far as S. Scandinavia, locally common in Britain.

2. Bloody Crane's-bill, *Geranium sanguineum* L., plant 15–60cm, prostrate or ascending; stem spreading-hairy, usually dichotomously branched; lower lvs roundish to kidney-shaped in outline, divided in 7 almost to the base, with linear segments; flowers solitary; petals 15–20mm long, rounded or irregularly emarginate at tip, purple-red. Fl.6–8. Sunny wood margins, dry scrub, open oak and pinewoods, calcareous grassland and dunes; likes warm places; local. S. Scandinavia, Britain, C. and S. Europe, only in the valleys in the N. Alps.

3. Herb Robert, *Geranium robertianum* L., plant 20–50cm, smelling unpleasantly; stem reddish, with scattered, spreading glandular hairs; lvs triangular to pentagonal in outline, 3–5 lobed to the base, consisting of 3–5 stalked leaflets which are pinnatifid almost to the midrib; inflorescence usually 2-flowered; petals 9–12mm long, pink, not emarginate at tip; sepals 6–7mm long; anthers red-brown. Fl.5–10. Woods, hedges, shady rocks and walls, shingle, waste places; common. Most of Europe.

See also *G. sylvaticum*, Pl.81.

Spurge Family, Euphorbiaceae

4. Dog's Mercury, *Mercurialis perennis* L., plant 15–30cm, dioecious; stem leafy above, with only scale lvs below; lvs oblong-lanceolate, 4–12cm long, bluntly toothed; petioles over 5mm long; flowers unisexual, small, with a 3-lobed, green calyx, in panicles or clusters. Fl.4–5. Herb-rich deciduous woods and mixed coniferous woods, thickets, mountain rocks; common. S. Scandinavia, Britain, C. and S. Europe. Similar species: **Broad-leaved Mercury**, **M. ovata* Sternb., but lvs broadly ovate, sessile or with stalks at most 2mm long; stem with small green lvs in the lower part; woods in warm places, oak-pine woods, sunny thickets; rare; Bavaria, S. Alps, S.E. Europe.

5. Sweet Spurge, *Euphorbia dulcis* L., plant 20–50cm; stems usually several; lvs alternate, oblong-ovate, narrowed into the short stalk, dark green above, blue-green below; inflorescence usually 5-rayed; involucral glands roundish-oval, yellow-green at first, later red-brown; bracts of the individual inflorescences not united; capsule covered with hemispherical warts. Fl.4–6. Herb-rich deciduous woods, mixed coniferous woods, hedgebanks; scattered. C. Europe, southwards to N. Spain, C. Italy, Balkan peninsula, naturalised in Britain.

6. Wood Spurge, *Euphorbia amygdaloides* L., plant 30–60cm, with many non-flowering, densely leafy stems, which become woody and give rise to flowering shoots in the second year; lvs alternate, ovate-lanceolate, 3–6cm long, over-wintering, densely imbricate under the inflorescence; inflorescence much-branched; involucral glands half-moon-shaped; the pair of bracts under the individual inflorescences united into 1 roundish leaf; capsule glabrous, finely punctate, somewhat trigonous. Fl.4–5. Herb-rich deciduous woods with beech, oak, hornbeam; hedgebanks; fairly common to local. Britain, C. and S. Europe.

Plate 157   Rue Family, Rutaceae

1. Burning Bush, *Dictamnus albus* L., plant 60–120cm, strong scented of lemon or cinnamon; lvs simply pinnate, with 7–11 ovate-lanceolate, finely serrate leaflets, 5–8cm long; flowers in an erect, terminal raceme; petals 5, pink with dark veins, 2–3cm long, 4 directed upwards, the fifth deflexed; sepals 5, 6–8mm long. Fl.5–6. Protected! Sunny, rocky slopes, oak scrub, open oak-pine woods; warm places; rare. C. and S. Europe.

Maple Family, Aceraceae

2. Sycamore, *Acer pseudoplatanus* L., tree to 30m; lvs over 10cm wide, 5-lobed, with acute sinuses and unequally and coarsely toothed lobes; flowers yellow-green, in drooping racemes 5–15cm long; sepals and petals 5, free; each half of the fruit with a wing 4–6cm long, the wings of the two halves making a narrow angle to each other. Fl.5–6. Woods in river gorges, mixed beechwoods; mountain woods, in the Alps often almost to the tree limit; often planted and naturalised and in plantations and hedges; common. C. and S. Europe, naturalised in N. Europe.

3. Norway Maple, *Acer platanoides* L., tree to 30m; lvs over 10cm across, 3–7 lobed, dark green on both sides, shining; leaf lobes long-pointed, with acute teeth and rounded sinuses; flowers yellow-green, appearing with the lvs, in erect, corymbose panicles; the wings of the two halves of the fruit at an obtuse angle to each other, or spreading horizontally. Fl.4–5. Herb-rich deciduous woods with oak, lime, hornbeam, elm; in the Alps rarely over 1000m; scattered. S. Sweden, C. Europe, southwards to the Pyrenees, Apennines, Greece. Planted and naturalised like Sycamore, but less commonly. Similar species: **London Plane**, *Platanus × hybrida* Brot. (in the Plane Family, Platanaceae), lvs palmately 3–5 lobed, lobes acute, entire or with large, acute, curved teeth, central lobe wider at the base than it is long; bark flaking off in patches; flowers in dense, spherical, unisexual, pendulous heads, 1cm across; planted in streets and parks; a hybrid between *P. occidentalis* (native in N. America) and *P. orientalis* (native in S.E. Europe and Asia). The **Oriental Plane**, *P. orientalis* L., has 5–7 lobed, toothed lvs; central lobe narrower at the base; marshy woods; frost sensitive; S.E. Europe.

4. Field Maple, *Acer campestre* L., tree or shrub 3–15m; lvs 5–8cm wide, (3–) 5-lobed, lobes obtuse, the central lobe with a large blunt tooth on either side, the lateral lobes usually entire; flowers in erect, corymbose inflorescences; pedicels and bracts densely hairy; fruit felted-hairy at first, the wings spreading horizontally. Fl.5–6. Wet woods, oak-hornbeam woods, hedges; in the Alps scarcely over 1000m; common. Britain, Scandinavia, C. and S. Europe. Similar species: **Italian Maple**, *A. opalus* Mill. (*A. opulifolium* All.), tree to 12m; lvs with 3 coarsely crenate, obtuse lobes, grey-green below; flowers in sessile corymbs, appearing before the lvs; oakwoods in warm places; Germany, Switzerland, Alps, Cevennes, Pyrenees, S.W. Europe; protected! **Montpelier Maple**, *A. monspessulanum* L., up to 5m; lvs with 3 entire lobes; flowers in erect, corymbose inflorescences; wings of the fruit overlapping or running parallel to each other; sunny oak scrub; rare; Germany, Swiss Jura, S. Europe.

Holly Family, Aquifoliaceae

5. Holly, *Ilex aquifolium* L., shrub 1–10m (in the Alps scarcely more than 2m, as twigs not covered by snow are killed by frost); lvs evergreen, shining (see p.364.)

Plate 158   Balsam Family, Balsaminaceae

1. Touch-me-not Balsam, *Impatiens noli-tangere* L., plant 30–80cm; stem branched; lvs alternate, ovate, bluntly serrate, 3–10cm long; flowers 1–4, drooping, 2–3cm long, with a bent spur, golden-yellow; capsule spindle-shaped, 2–3cm long, springing open when ripe. Fl.7–8. Wet woods, river gorge woodland, damp deciduous woods, by streams; generally common (but rare in Britain). Almost all of Europe, southwards to C. Italy.

2. Small Balsam, *Impatiens parviflora* DC., plant 30–60cm; lvs ovate, acuminate, sharply serrate; flowers 8–10mm long, in erect, 4–10 flowered racemes, pale yellow, with a straight spur; capsule club-shaped, 15–20mm long. Fl.6–9. Wood margins, hedges, deciduous woods on good soils, gardens, waste places; common. Originally from Asia, but since 1840 naturalised out of botanic gardens.

3. Indian Balsam, *Impatiens glandulifera* Royle, plant 50–200cm; lvs alternate, the upper whorled, ovate-lanceolate, sharply serrate, with stalked glands on the teeth and on the petioles; flowers white, pink or wine-red, 2–4cm long, with a bent spur, in long-stalked, 2–15 flowered racemes. Fl.7–8. River banks, wet woods, waste places; scattered. Originally from the Himalayas; garden plant and naturalised for 50 years.

Spindle-tree Family, Celastraceae

4. Spindle-tree, *Euonymus europaeus* L., shrub to 5m; young twigs 4-angled, green, later often with corky ridges; leaf buds shortly ovoid; lvs ovate, acute, finely serrate, 4–10cm long; inflorescence 2–6 flowered, long-stalked, axillary; flowers greenish, 8–10mm wide, 4-merous; fruit a red, 4-angled capsule; seeds white, enclosed in an orange-red aril which is visible when the capsule springs open on ripening. Fl.5–6. Poisonous! Hedges, scrub, deciduous woods; common. Almost all of Europe. Similar species: **Broad-leaved Spindle**, *E. latifolius* (L.)Mill., young twigs somewhat compressed, oval in cross-section; leaf buds long acuminate; flowers 5-merous; capsule usually 5-angled, winged on the angles; mixed deciduous woods; likes warm places; rare; Alps and hinterlands, Swiss Jura, mountains of S. Europe. The **Rough-stemmed Spindle**, *E. verrucosus* Scop., has all the twigs covered with black, corky warts; flowers 1–3; petals 4, yellow-green, with red spots; E. Alps, E. Europe.

Bladder-nut Family, Staphyleaceae

5. Bladder-nut, *Staphylea pinnata* L., shrub to 5m; lvs opposite, long-stalked, imparipinnate, with 5 or 7 ovate-lanceolate, finely serrate leaflets, 6–10cm long, blue-green below; flowers in drooping panicles, symmetric, 5-merous; petals 10–15mm long, yellowish, spathulate, connivent into a bell-shape; fruit a hyaline, inflated capsule, 3–4cm long and wide. Fl.5–6. Deciduous woods, wood margins; likes warm places; rare. S. Germany, Alsace, French Jura, Alps, S.E. Europe, rarely naturalised in Britain. Protected!

(from p.362), glabrous, leathery, spine-toothed or entire, 3–8cm long, with very small stipules; flowers unisexual, 4–5-merous; corolla white, 3mm long; female flowers 1–3; male flowers several together; fruit 6–8mm, red. Fl.5–6. Protected! (In Germany.) Deciduous woods, thickets; common to rare. S. Norway, Britain, N.W. Germany, Vosges, Black Forest, Lake Constance, Rhine valley, Alps and hinterlands, S. Europe.

Plate 159 Box Family, Buxaceae

1. Box, *Buxus sempervirens* L., shrub, 0.5 to 3m; lvs evergreen, ovate, leathery, opposite, almost sessile, 10–25mm long; flowers unisexual, in clustered, axillary inflorescences with 1 female flower and many male flowers; corolla absent; sepals 4–8, yellow-green, 2mm long; fruit ovoid, 8mm long, wrinkled. Fl.3–4. Deciduous woods, scrub; likes warm places and a mild winter climate; local or rare as native shrub, but cultivated and naturalised. Britain, France, Germany, Alps, Pyrenees, S. Europe.

Buckthorn Family, Rhamnaceae

2. Buckthorn, *Rhamnus catharticus* L., shrub 1–3m; twigs and lvs opposite; tips of the branches usually thorny; lvs roundish to broadly ovate, 4–6cm long, with 2–3 pairs of curved lateral veins; petiole much longer than the stipules; flowers 2–8 together in the leaf axils, yellow-green, 4-merous, 4–5mm wide; berries 6–8mm across, black. Fl.5–6. Hedges, open woods, poor grassland, fens; local. Britain, most of Europe.

3. Rock Buckthorn, **Rhamnus saxatilis* Jacq., shrub 50–100cm, thorny; lvs ovate-lanceolate, 1–3cm long, the underside softly hairy, narrowed into the short stalk, which is as long as the stipules; flowers and berries as in *R. catharticus*. Fl.4–5. Wood margins, sunny thickets; rather rare. S. Germany, Alps, S.E. Europe, southwards to N. Spain, C. Italy. The **Mediterranean Buckthorn**, *R. alaternus* L., has evergreen, leathery, oval lvs which are 3–5cm long, serrate or entire, and with the underside often blue-green; inflorescences 6–12 flowered, axillary, corymbose; berries 4–6mm wide, red; rocky, dry slopes; widespread in the Mediterranean region, naturalised in Britain.

4. Alder Buckthorn, *Frangula alnus* Mill. (*Rhamnus frangula* L.), shrub 1–4m; twigs thornless; lvs alternate, roundish to ovate, 2–5cm long, with 7–12 lateral veins on each side; flowers 5-merous, 2–10 together in the leaf axils, yellow-green, about 5mm across; fruit a red berry, 5–8mm across, turning dark blue later. Fl.5–6. Alder carr, birch moors, deciduous woodland, limestone scrub, fens; local or common. Most of Europe.

Lime Family, Tiliaceae

5. Small-leaved Lime, *Tilia cordata* Mill., tree to 25m; lvs cordate, suborbicular, 3–8cm wide, serrate, the upperside dull, dark green, the underside blue-green, with tufts of rust-coloured or yellowish hairs in the angles of the veins, otherwise glabrous; petioles glabrous, shoots glabrous; inflorescence 4–10 flowered, the whole with a wing-like bract 4–8cm long; flowers 5-merous; petals 4–8mm long, yellowish-white; young twigs and buds olive-green to reddish; fruit thin-walled, leathery, obscurely 4-angled. Fl.6–7. Mixed deciduous woods; in places with mild winters; scattered. S. Scandinavia, Britain, C. and S. Europe. The **Caucasian Lime**, **T. × euchlora* Koch, a hybrid, has dark green, shining lvs and mucronate leaf teeth; young twigs and buds yellow-green.

6. Large-leaved Lime, *Tilia platyphyllos* Scop. (*T. grandifolia* Ehrh.), tree to 30m; lvs 5–15cm wide, shortly hairy above, with dense tufts of white hairs in the angles of the veins below, otherwise shortly hairy below also; petioles hairy; shoots hairy; inflorescence 2–5 flowered, with a wing-like bract 5–12cm long; fruit thick-walled, woody, 5-angled. Fl.6. Deciduous woods, scattered. C. and S. Europe, possibly wild in Britain, also planted. Similar species: **Lime**, *T. × vulgaris* Hayne (*T. × europaea* L.), large tree; lvs broadly ovate, glabrous except for hair tufts below; fruit slightly ribbed. Planted, the hybrid between *T. cordata* and *T. platyphyllos*. Europe.

7. Wonder Violet, **Viola mirabilis*, see p.368.

Plate 160   Violet Family, Violaceae

1. Common Dog-violet, *Viola riviniana* Rchb., plant 5–30cm, with an erect aerial stem and a basal leaf rosette; stem and lvs glabrous; lvs long-stalked, broadly cordate to kidney-shaped, dark green, 3–5cm long, crenate on the margin; stipules lanceolate, fringed like the teeth of a comb; flowers in the axils of stem lvs; corolla pale violet, 14–22mm long; petals usually overlapping each other; spur whitish, thick, 3mm long, furrowed at the tip; appendages of the acute sepals 2–3mm long, often emarginate. Fl.4–6. Deciduous and mixed woods, hedgebanks, heaths, mountain rocks; common or widespread. Almost all Europe. See also *V. canina*, Pl.64.

2. Hairy Violet, *Viola hirta* L., plant 5–25cm, without stolons, with a basal leaf rosette, but without an erect, aerial stem; flowers in the axils of basal lvs; lvs cordate, acute, hairy, up to 8cm long; stipules broadly lanceolate, not or shortly fringed; flowers not scented, violet-blue; petals emarginate, spur reddish-violet, curved upwards at the tip; sepals obtuse. Fl.3–5. Wood margins, sunny scrub, open oak and pine woods, calcareous grassland; rather common. Almost all of Europe.

3. Early Dog-violet, *Viola reichenbachiana* Jord. (*V. sylvestris* Lamk.), plant 10–20cm, with an erect aerial stem and a basal leaf rosette, similar to *V. riviniana*, but corolla reddish-violet, 12–15mm long; petals not overlapping; spur dark violet, slender, 4–6mm long, with rounded tip; appendages of acute sepals are 1–2mm long. Fl.3–5. Mixed woods, hedges; widespread. Britain, S. Sweden, C. and S. Europe.

4. Sweet Violet, *Viola odorata* L., plant 5–10cm, with procumbent stolons which root at the tips and a basal leaf rosette, without an aerial stem; lvs cordate, with lanceolate-ovate, shortly fringed stipules 4–5mm wide; flowers scented; corolla dark violet, rarely white or pink; spur straight, the same colour; sepals blunt. Fl.3–4. Scrub, wood margins, plantations; common. Originally from S. Europe, now most of Europe.

Daphne Family, Thymelaeaceae

5. Mezereon, *Daphne mezereum* L., shrub 30–150cm, flowers before the lvs appear; twigs brush-like, leafy at the tip; lvs lanceolate, 5–12cm long, blue-green; flowers scented, pink, 10–14mm wide, 4-lobed, 1–3 together in the axils of fallen lvs of the previous year, sessile, arranged like spikes in the upper part of the twig; fruit red, spherical, 6–10mm across. Fl.3–4. Protected! Poisonous! Deciduous and mixed woods, shrub communities, stony screes; in the Alps to over 2000m; rare to widespread. Almost all of Europe. The **Spurge Laurel**, *D. laureola* L., has rough, evergreen lvs and yellow-green flowers in 5-flowered, axillary racemes; fruit black; likes warm places; deciduous woods; fairly common to very rare; Germany (rare), Swiss Jura, Alps, Britain (fairly common), W. and S. Europe. Protected!

6. Sea-buckthorn, *Hippophae rhamnoides*, see p.370.

Violet Family, Violaceae

7. Wonder Violet, **Viola mirabilis* L. (Pl.159), plant 10–25cm, with a basal leaf rosette; lvs broadly cordate to almost kidney-shaped, 5–10cm wide, with cordate base and petiole 10–18cm long; stem and petiole with a single line of hairs; basal flowers pale violet, scented, usually sterile, stem flowers without corolla, rudimentary, but fertile. Fl.4–6. Deciduous woods; scattered. Most of Europe (not Britain).

Plate 161   St. John's-wort Family, Hypericaceae

1. Pale St. John's-wort, *Hypericum montanum* L., plant 30–80cm, erect; stem terete; lvs ovate-oblong, 2–8cm long, dotted with glands on the margins, sessile and semi-amplexicaul; inflorescence in a head; flowers pale yellow; sepals acute, ciliate with black glandular hairs. Fl.6–8. Sunny deciduous woods, thickets; scattered. Most of Europe.

2. Slender St. John's-wort, *Hypericum pulchrum* L., plant 30–60cm, woody at base; lvs almost triangular, cordate, rather blunt, sessile, 1–2cm long, not gland-dotted on the margins; flowers golden-yellow; sepals bluntish, with sessile or shortly stalked black glands; petals often red-tinged, with black glands on margins. Fl.7–9. Heaths, moors, open woodland; calcifuge; widespread. Britain, C. Europe, S. Scandinavia.

3. Hairy St. John's-wort, *Hypericum hirsutum* L., plant 40–100cm; stem terete, densely hairy like the lvs; lvs ovate, 2–6cm long; flowers yellow; sepals and petals with black, stalked glands on the margins. Fl.6–8. Mixed deciduous woods, woodland rides, river banks and damp grassland; widespread. Europe.

See also *H. perforatum*, Pl.63.

Willow-herb Family, Oenotheraceae or Onagraceae

4. Rosebay Willow-herb, *Epilobium angustifolium* L., plant 50–150cm; stem bluntly angled; lvs all alternate, narrowly lanceolate, 8–12cm long and 1–2cm wide, glabrous, blue-green and with prominent lateral veins beneath; flowers 2–3cm across, in long, erect racemes, pink to purplish-red; petals shortly clawed; stigma 4-lobed. Fl. 7–8. Woodland clearings and rides, scrub, rocky places and scree, waste ground; common to local. Most of Europe. Similar species: **Rosemary Willow-herb,** *E. dodonaei* Vill., but lvs linear, 2–5mm wide, stiff, green on both sides, without prominent lateral veins below; petals not clawed; gravelly banks, open ground; rare; C. and S. Europe.

5. Broad-leaved Willow-herb, *Epilobium montanum* L., plant 10–80cm; stem simple or slightly branched above, with spreading curved hairs above; lvs ovate, closely toothed, subsessile, 4–7cm long, 15–35mm wide, grass-green; flowers 8–12mm long; stigma 4-lobed; calyx and fruit with spreading curved hairs. Fl.6–9. Mixed woods, hedges, clearings, garden weed; widespread. Europe. Similar species: **Hill Willow-herb,** *E. collinum* Gmel., but stem branched from base; lvs stalked, 1–4cm long, 5–15mm wide, grey-green, sinuate-dentate; flowers 4–6mm across; calyx and fruit appressed-hairy; rocks and walls; calcifuge; widespread, but not in Britain. See also Pl.109.

6. Alpine Enchanter's-nightshade, *Circaea alpina*, and **7. Enchanter's-nightshade,** *Circaea lutetiana*, see p.372.

Oleaster Family, Elaeagnaceae

6. Sea-buckthorn, *Hippophae rhamnoides* L. (Pl.160), thorny shrub up to 6m; lvs linear-lanceolate, 5–6cm long, 3–7mm wide, dark green above, silvery-white below, inrolled at the margins; flowers unisexual, small, brownish, males in head-like inflorescences, females in few-flowered racemes; fruit ovoid, 6–8mm long, orange-red. Fl.3–5. Protected! Woods on river banks, shingle banks, coastal dunes, open pine-woods; scattered to rare. Mountains of C. and S. Europe, coasts of the North Sea and the Baltic, Lake Constance, Rhine valley; cultivated.

Plate 162   Ivy Family, Araliaceae

1. Ivy, *Hedera helix* L., climbing shrub 6–20m; twigs and branches with adhesive roots; lvs alternate, evergreen, shining, 3–5 lobed on non-flowering shoots and ovate-lanceolate to rhomboid on flowering shoots, 5–10cm long; flowers 5-merous, in hemispherical umbels, green; petals 3–4mm long, fleshy, brown outside, green inside; fruit a dark blue berry, 8–10mm wide. Fl.9–11. Damp woods, deciduous woods with oak or beech, walls, rocks; tolerates shade but frost sensitive. Almost all of Europe.

Carrot Family, Umbelliferae or Apiaceae

2. Sanicle, *Sanicula europaea* L., plant 20–50cm; basal lvs evergreen, long-stalked, 5–8cm across, divided to the base into 5 palmate segments, segments coarsely toothed; stem lvs smaller, almost sessile; flowers in a terminal umbel; secondary umbels head-like, with stalked and sessile flowers; bracteoles 4–8; corolla white or yellowish, 3mm wide; fruit spherical, densely covered with prickles, 4–5mm across. Fl.5–6. Oak and beech woods, mixed coniferous woods; common. Most of Europe.

3. Great Masterwort, *Astrantia major* L., plant 30–100cm, dichotomously branched above; basal lvs long-stalked, 5–7 lobed in outline, 10–20cm wide, deeply palmately 5–7 lobed, lobes further divided 2–3 times and coarsely toothed, the lateral lobes partly united; stem lvs similar, smaller; flowers long-stalked, in simple, head-like umbels, surrounded by star-like, spreading, white or reddish bracts 11–30mm long and rough; corolla usually reddish; calyx teeth ovate-lanceolate, with sharp points. Fl.6–8. Woods by rivers and in ravines, thickets, mixed coniferous woods, mountain meadows, high-level shrub communities; in the Alps to over 2000m; widespread. Mountains of C. and S. Europe, naturalised in Britain. Similar species: **Bavarian Masterwort,** **A. bavarica* F. W. Schultz, plant 20–50cm, lateral leaf segments free nearly to the base; bracts thin, 10–15mm long; calyx teeth ovate, rather blunt or short-pointed; rare. **Lesser Masterwort,** **A. minor* L., plant 20–40cm; basal lvs 5cm wide; leaf segments narrowly lanceolate and toothed; bracts 5–11mm long.

4. Long-leaved Hare's-ear, **Bupleurum longifolium*, and **5. Sickle-leaved Hare's-ear,** *Bupleurum falcatum*, see p.374.

---

6. Alpine Enchanter's-nightshade, *Circaea alpina* L. (Pl.161), plant 5–15cm, curved-ascending; lvs shining, glabrous, broadly lanceolate, 1–3cm long, sinuate-dentate; leaf stalks winged; flowers in short racemes which elongate later, with bracts 1mm long which easily fall off. Fl.6–8. Woods in ravines and by streams, mixed coniferous woodlands, shady rocky places; rather rare. Britain, N. and C. Europe, Alps, Pyrenees, Apennines, Corsica, mountains of the Balkan peninsula.

7. Enchanter's-nightshade, *Circaea lutetiana* L. (Pl.161), plant 20–60cm; stem erect, with scattered hairs; lvs dull, broadly lanceolate, gradually tapered, stalked, 5–10cm long, rounded at the base, scarcely cordate, toothed and hairy on the veins; flowers in long terminal racemes; pedicels with spreading glandular hairs, without bracts; sepals 2; petals 2, deeply emarginate, 2–4mm long, white or reddish; fruit 3–4mm long, with hooked bristles. Fl.6–8. Mixed deciduous and coniferous woods, shady damp places; widespread. Almost all of Europe. Similar species: **Upland Enchanter's-nightshade,** *C. intermedia* Ehrh. (a hybrid of *C. alpina* and *C. lutetiana*), but lvs shining, rather abruptly pointed, clearly cordate at base; fruit up to 2mm long; woods in ravines and by streams, damp ash woods, shady rocky places; rather rare.

Plate 163   Carrot Family, Umbelliferae or Apiaceae

1. Austrian Ribseed, **Pleurospermum austriacum* (L.)Hoffm., plant 60–150cm; stem hollow, 2–3cm wide, furrowed; lvs dark green, shining, 2–3 pinnate, leaflets unequally coarsely toothed and dissected; umbel 12–20 rayed, up to 25cm across; bracts pinnately lobed, large; bracteoles lanceolate; flowers white; petals up to 3mm long, not emarginate; fruit 6–10mm long, with wing-like ribs. Fl.6–8. Woods by rivers, wood margins, river banks, high-level shrub communities; scattered; in the Alps to over 2000m. Mainly Alps and hinterlands, Jura, Rhône, northwards to S. Sweden, southwards to Dalmatia, Bulgaria, Romania. Similar species: **Southern Ribseed,** **Grafia golaka* (Hacquet) Rchb. (*Pleurospermum golaka* Rchb.), plant 50–100cm, glabrous; lvs rough, shining, 3-ternate or 3-pinnate; bracts broadly lanceolate, entire or with 2 teeth at the apex only; bracteoles usually 3, on the outer side of the secondary umbels; flowers white; petals 1mm long, emarginate; fruit oblong-ovoid, 8–13mm long, ribbed; stony slopes; S. E. Alps on limestone, Abruzzi, mountains of the Balkan peninsula.

2. Ground-elder, *Aegopodium podagraria* L., plant 50–100cm, with long creeping rhizomes; basal lvs twice ternate; leaflets ovate-oblong, 5–10cm long, toothed; petiole 3-angled, pithy (in *Angelica sylvestris* petiole hollow); umbel with 15–25 rays; bracts and bracteoles absent; corolla white; fruit oblong-ovoid, 3mm long, like caraway seed. Fl.6–8. Woods by rivers and in ravines, wood margins, river banks, and a weed in gardens; common. Almost all of Europe.

3. Wild Angelica, *Angelica sylvestris* L., plant 80–150cm; stem terete, often pruinose; lvs 2-pinnate, dark green; leaflets ovate to broadly lanceolate, 6–12cm long, finely serrate; leaf sheaths large, inflated; petioles channelled above, with a half-moon-shaped hollow in cross-section; umbel 20–40 rayed, the rays pubescent; bracts 0–3; bracteoles many; corolla white or reddish, greenish at first; fruit oval, 4–6mm long. Fl.7–9. River banks, wet woods and meadows, fens; widespread. Europe. Similar species: **Garden Angelica,** *A. archangelica* L., but petioles terete; lvs dark green, strong-smelling; leaflets partly pinnatisect; rays pubescent only above; flowers greenish; river banks, ditches, willow thickets; rare. **Marsh Angelica,** **A. palustris* (Besser) Hoffm., has a sharply angled, furrowed stem, petioles keeled beneath, cordate to triangular-ovate, crenate-serrate leaflets and glabrous rays to umbels; rare; E. Europe.

---

4. Long-leaved Hare's-ear, **Bupleurum longifolium* L. (Pl.162), plant 30–100cm; lvs oblong-ovate or broadly lanceolate, net-veined, up to 15cm long, the lowest narrowed into the winged stalk, the upper cordate and amplexicaul; inflorescence an umbel with 4–8 rays; bracts 3–4, roundish, often united at the base, 3–7 veined; flowers yellow-green; fruit 4–5mm long, almost black. Fl.5–6. Deciduous woods with oak, hornbeam or beech, thickets, wood margins; likes warm places; rare. Mountains of C. Europe, Alps and hinterlands, Carpathians, mountains of the Balkan peninsula. Similar species: **Thorow-wax,** *Bupleurum rotundifolium* L., but lvs roundish-ovate, the lowest sessile, the upper connate (joined at the base around the stem), 3–7cm long; bracts absent; fruit 3–4mm long; cornfields; rare. Protected!

5. Sickle-leaved Hare's-ear, *Bupleurum falcatum* L. (Pl.162), plant 20–100cm, woody at base, much-branched; lower lvs oblong-spathulate, with long, scarcely winged stalks, upper lvs lanceolate, 5–7 veined; bracts and bracteoles narrowly lanceolate, not united; fruit 3–4mm long, brown. Fl.6–9. Scrub, open oak and pine woods, dry grassland; warm places; rare. Britain (very rare), C. Europe, south to C. Spain, Albania.

1

2

3

Plate 164　Carrot Family, Umbelliferae or Apiaceae

1. Sermountain, *Laserpitium latifolium* L., plant 50–150cm, stock with abundant fibres; stem terete, finely ribbed, glabrous, up to 2cm across; lvs very large, up to 1m long, 1–2 times ternate, blue-green and glabrous; leaflets broadly ovate, toothed, 3–15cm long, stalked; umbel with 20–40 rays; bracts scarious; bracteoles thread-like, scarcely scarious; corolla white. Fl.7–8. Open, dry woods, scrub, high-level shrub communities, stony grassland; in the Alps to over 2000m; scattered. Most of Europe, but not Britain. Similar species: **Gaudin's Sermountain,** *L. krapfii* Crantz ssp. *gaudinii* (Moretti) Thell., but plant smaller, leaflets deeply 3-lobed; umbel 10–15 rayed, bracts usually absent; pinewoods, stony slopes; C. and S. Alps.

See also *Pimpinella major*, Pl.82 and *Heracleum*, Pl.83.

2. Alpine Sermountain, *Laserpitium siler* L., plant 30–120cm, aromatic, with a pronounced tuft of fibres at the base; stem finely ribbed; lvs 2–4 pinnate, up to 1m long, blue-green; leaflets linear-lanceolate, entire, glabrous, 1–5cm long, with a pale and swollen margin; umbel up to 25cm across, 20–40 rayed; bracts rough, lanceolate; bracteoles ovate-lanceolate, broadly scarious on the margins; corolla white; fruit 5–12mm long. Fl.6–8. Oak and pine woods, sunny thickets; likes warm places; scattered or rare. Alps and hinterlands, Jura, mountains of N. Spain, Pyrenees, Apennines, mountains of the Balkan peninsula. The **Prussian Sermountain,** *L. prutenicum* L., has a furrowed, usually stiffly hairy stem and 2–3 pinnate lvs, with lanceolate, ciliate segments; flowers yellowish-white; moorland pasture, open oak and pine woods; rare; C., E. and S. Europe. Protected!

3. Bulbous Chervil, *Chaerophyllum bulbosum* L., plant 80–180cm, annual or biennial, with bulbous or turnip-like thickened root; stem glabrous above, often pruinose, stiffly hairy below, with red spots, swollen at the nodes; lvs 3–4 pinnate, with narrowly lanceolate, acute segments; umbel 5–12 rayed; bracts absent; bracteoles 4–8, usually glabrous, unequal in length; corolla white, glabrous; fruit 4–6mm long. Fl.6–8. River banks, ditches, open woods by rivers; scattered. C., E. and S.E. Europe. Similar species: **Golden Chervil,** *Chaerophyllum aureum* L., but plant 60–120cm, perennial, root not thickened, but with a thick, branched rhizome which is difficult to pull out of the ground; lvs 3–4 pinnate, with finely acuminate segments, the underside usually softly hairy; bracteoles 5–10, lanceolate, with scarious margins, long-ciliate; corolla white, with inflexed lobes at the tip; fruit 8–11mm long, yellowish when ripe; ditches, river banks, hedges, wood margins, waste places, over-fertilised mountain meadows; Scotland, C. Europe southwards to the Pyrenees, Apennines. **Rough Chervil,** *C. temulentum* L., has a spindle-shaped root and is easy to pull out of the ground; stem stiffly hairy, red-spotted; lvs 2–3 pinnate with obtuse, ovate leaf segments and rounded lobes and teeth; shaded, weedy meadows, nutrient-rich hedges and wood margins; common in Britain; C. and S. Europe.

See also *Anthriscus*, Pl.82.

Plate 165 Carrot Family, Umbelliferae or Apiaceae

1. Hairy Chervil, *Chaerophyllum hirsutum* L., plant 50–100cm; lvs 2–3 pinnate, hairy; the two lowest leaf segments about as large as the rest of the leaf; segments further divided into irregular, coarsely toothed lobes; upper leaf sheaths 10–60mm long; umbel 10–20 rayed; bracts absent; bracteoles 5–10, lanceolate, with a bearded ciliate margin; petals ciliate, white or pink; fruit 8–12mm long. Fl.5–6. Mountain woods, meadows, high-level shrub communities; common in mountains of C. Europe, rare in N. Germany.

2. Mountain Chervil, *Chaerophyllum villarsii* Koch, plant 50–100cm, very similar to *C. hirsutum*, but the two lowest leaf segments much smaller than the rest of the leaf; segments regularly and more finely divided; lobes divided almost to the midrib; upper leaf sheaths 3–10mm long. Fl.5–8. Mountain meadows, alder thickets, high-level shrub communities, open mountain woods; widespread. Mountains of C. and S. Europe.

Dogwood Family, Cornaceae

3. Dogwood, *Cornus sanguinea* L. (*Thelycrania s.* (L.)Fourr.), shrub 2–5m; young twigs red; lvs opposite, ovate, acute, 5–8cm long, with 3–4 pairs of veins, green on both sides, becoming red in autumn; flowers in many-flowered corymbs, 4-merous, appearing after the lvs; petals 4–6mm long, white; fruit blue-black, spherical, 6–8mm long. Fl.5–6. Hedges, woods and scrub; common, also often planted. Almost all of Europe. Similar species: **White-fruited Dogwood,** *C. alba* L., but lvs grey-green below with 5–7 pairs of veins; fruit white or pale blue; cultivated shrub, occasionally naturalised; originally from Asia.

4. Cornelian Cherry, *Cornus mas* L., shrub 2–6m; young twigs green; lvs opposite, ovate-lanceolate, 5–8cm long, green on both sides, hairy in the angles of the veins beneath, with 3–4 pairs of veins; flowers yellow, appearing before the lvs, on short shoots, 10–20 together in spherical umbels, these with an involucre of 4 yellowish-green bracts; petals 2–3mm long; fruit oblong, 1–2cm long, red, juicy. Fl.3–4. Woods, scrub; often planted and naturalised, seldom native; Britain, C. and S.E. Europe.

5. Dwarf Cornel, *Cornus suecica* L. (*Chamaepericlymenum suecicum* (L.)Aschers. et Graeb.), herbaceous plant, 5–20cm; stem unbranched, erect, 4-angled, always with 1 terminal umbel of 8–25 dark red flowers, 2mm long, surrounded by 4 white, ovate bracts, 5–8mm long; lvs ovate, sessile, 3–5 veined, blue-green below. Fl.7–8. Moorland, dwarf shrub heaths, peaty soils; rare. Britain, N.W. Germany, principally N. Europe.

Wintergreen Family, Pyrolaceae

6. One-flowered Wintergreen, *Moneses uniflora* A. Gray (*Pyrola u.* L.), plant 5–10cm; lvs in basal rosettes, roundish-spathulate, up to 2cm long with a finely serrate margin; flowers solitary, white, 2cm across; petals spreading horizontally. Fl.5–7. Protected! Coniferous woods, oakwoods on acid soils; scattered. Scotland, N. and C. Europe, southwards to the Pyrenees, Corsica, Bulgaria, Caucasus.

Plate 166   Wintergreen Family, Pyrolaceae

1. Round-leaved Wintergreen, *Pyrola rotundifolia* L., plant 15–30cm; stem blunt-angled below, usually green, rarely red; lvs roundish-ovate; flowers 8–15 together in erect racemes, facing in all directions; corolla white, open, campanulate; calyx teeth lanceolate, acute, spreading. Fl.6–7. Protected! Coniferous woods, acid beech and oak woods, birch moorland and wet sand dunes; rather rare. Britain, N. and C. Europe. Similar species: **Pale-green Wintergreen,** **P. chlorantha* Sw., but stem sharply angled below, usually red; flowers 3–8 together; calyx teeth ovate, shortly acuminate, usually appressed to the greenish-white corolla; dry pinewoods; rare. Protected! Most of Europe, but rare in the west, absent from Britain.

2. Common Wintergreen, *Pyrola minor* L., plant 10–20cm; lvs roundish-ovate; flowers globose, closed, 5–20 together in a raceme, facing in all directions; style shorter than the flower, not thickened; calyx teeth appressed to the corolla. Fl.6–7. Protected! Coniferous woods, acid beech and oak woods, birch moorland; scattered. Britain, N. and C. Europe, southwards to C. Italy. Similar species: **Intermediate Wintergreen,** *P. media* Sw., but lvs almost circular; style longer than flower, thickened above; calyx teeth spreading. Protected! Pinewoods and moors; rare; Britain, N. and C. Europe.

3. Serrated Wintergreen, *Orthilia secunda* (L.)House (*Pyrola s.* L.), plant 5–25cm; lvs in the lower one-third of the stem (not in a basal rosette), ovate, acute; flowers in a many-flowered, secund raceme, 5–15cm long, greenish-white, campanulate to almost spherical; petals 3–4mm long. Fl.6–7. Protected! Coniferous woods, mixed deciduous woods on acid soils, damp rock ledges; scattered to rare; medicinal plant. Almost all of Europe.

4. Umbellate Wintergreen, **Chimaphila umbellata* (L.)Bart. (*Pyrola u.* L.), plant 5–20cm; lvs leathery, ovate-lanceolate, 3–5cm long, sharply serrate; flowers 3–7 in terminal umbels, nodding; petals 5–6mm long, overlapping, white to pale pink; style short, thick, with a disc-shaped stigma. Fl.6–8. Protected! Dry pinewoods, sandy soils; rare. C. and N. Europe.

5. Yellow Bird's-nest, *Monotropa hypopitys* L., plant 10–20cm, brown or yellowish, without chlorophyll; saprophytic (living from the decaying remains of other plants); leaf scales 1–2cm long, pale yellow; flowers 4–5-merous, campanulate, sessile, 8–15 together in dense racemes. Fl.6–7. Coniferous woods, also oak and beech woods, and with Creeping Willow on sand dunes; scattered. Almost all of Europe.

Heath Family, Ericaceae

6. Spring Heath, **Erica herbacea* L. (*E. carnea* L.), dwarf shrub 15–30cm; lvs evergreen, needle-shaped, acute, in whorls of 4; flowers in secund racemes; corolla pale to dark red, rarely white; anthers dark, projecting out of the oblong-urceolate corolla tube. Fl.2–5. Pinewoods, Mountain Pine scrub; scattered; in the Alps to 2700m. Mountains of C. and S. Europe.

Plate 167   Heath Family, Ericaceae

1. Bilberry, *Vaccinium myrtillus* L., dwarf shrub 15–50cm; twigs angled, green; lvs ovate, acute, finely serrate, pale green, 2–3cm long; flowers solitary, axillary; corolla globose, 4–5mm wide, greenish and red-tinged; fruit spherical, 5–8mm, blue-black. Fl.4–6. Woods on acid soils, dwarf shrub communities, heaths, moors; widespread; in Alps to 2500m. Britain, Europe, in mountains in the south.

2. Bog Bilberry, *Vaccinium uliginosum* L., dwarf shrub 20–100cm; twigs terete, brown; lvs obovate, obtuse, entire, blue-green, strongly net-veined below; flowers several together in a raceme, pink or whitish; fruit blue-black, 6–10mm wide. Fl.4–6. Pine and birch moorland, heaths, dwarf shrub communities, often with *V. myrtillus*; scattered; Alps to 2500m. N. and C. Europe, in the south only in mountains. Protected!

3. Cowberry, *Vaccinium vitis-idaea* L., dwarf shrub 10–20cm; lvs evergreen, leathery, obovate, inrolled at the margin, shining, pale green below, 1–3cm long; flowers in dainty racemes, campanulate, usually 4-merous, whitish or pink, 5–8mm long; fruit globose, 5–8mm wide, red. Fl.5–8. Coniferous woods, moors, heaths, dwarf shrub communities; widespread; in the Alps to about 2500m. Britain, N. and C. Europe, southwards to the Pyrenees, Apennines, mountains of the Balkan peninsula.

4. Bearberry, *Arctostaphylos uva-ursi* (L.)Spreng., dwarf shrub 20–60cm; lvs evergreen, glabrous, shining, leathery, entire, 1–3cm long, not inrolled at the margin, net-veined beneath; flowers 3–8 together; corolla ovoid, 5-merous, yellowish or pink; fruit red, 6–8mm wide. Fl.3–7. Pinewoods, heaths, moors, often on rocks; Alps to 2500m; local. Most of Europe. Protected! **Alpine Bearberry,** *A. alpinus* (L.)Spreng., has deciduous lvs which become red in autumn; lvs serrate, weakly ciliate, net-veined on both sides; fruit finally black; alpine dwarf shrub communities; local; Arctic and the mountains of Europe, Scotland. **Rhododendron,** *Rhododendron ponticum* L., an evergreen shrub up to 3m, has leathery leaves and conspicuous, campanulate purple flowers with 10 stamens, in terminal racemes. Woods on sandy or peaty soils. Introduced in Britain, native in S.W. and S.E. Europe.

Primrose Family, Primulaceae

5. Oxlip, *Primula elatior* (L.)Hill., plant 10–20cm; lvs basal, 10–20cm long, narrowed into winged stalk, irregularly toothed; flowers in many-flowered, secund umbel; calyx 8–13mm long, narrowed at base, slender, with lanceolate calyx teeth; corolla pale yellow, with flat limb. Fl.3–5. Protected! Herb-rich deciduous woods, woods by rivers, mountain meadows; rare in Britain, common in C. Europe.

6. Cowslip, *Primula veris* L. (*P. officinalis* L.), plant 10–20cm; leaf blade sharply demarcated from the winged petiole; calyx inflated, campanulate, with ovate teeth; corolla egg-yellow, with red spots at the throat and a campanulate limb. Fl.4–6. Protected! Wood margins, oakwoods, calcareous grassland and meadows; widespread. Britain, C. Europe. **Primrose,** *P. vulgaris* Huds., has lvs tapering to base; flowers solitary on long stalks from base; calyx cylindrical with triangular teeth; corolla yellow with greenish markings; woods and hedges; common. Britain, W. Europe.

7. Cyclamen, **Cyclamen purpurascens* Mill. (*C. europaeum* auct.), plant 5–15cm, with 1 spherical bulb; lvs cordate, glabrous, dark green above, with pale spots, reddish beneath; stem 1-flowered; corolla reddish-violet, with reflexed lobes. Fl.7–9. Protected! Mixed deciduous woods, mountain fir and beech woods; rare. C. and S. Europe.

8. Creeping Jenny, *Lysimachia nummularia*, see p.384.

Plate 168   Primrose Family, Primulaceae

1. Yellow Pimpernel, *Lysimachia nemorum* L., plant 10–30cm, prostrate to ascending; stem rooting in the lower part; lvs opposite, ovate, acute, 2–3cm long, shortly stalked; flowers solitary in the axils of the upper lvs, 10–15mm wide, yellow; calyx teeth subulate. Fl.5–8. Damp mixed deciduous woods, woodland rides, shady hedgebanks; common to scattered. Almost all of Europe.

2. Chickweed Wintergreen, *Trientalis europaea* L., plant 5–20cm; lvs lanceolate, entire, 2–4cm long, arranged in a whorl at the top of the stem; flowers solitary, axillary, long-stalked; corolla white, 10–15mm wide, lobes 7, acute, spreading horizontally, divided almost to the base; calyx 4–6mm long, with 7 lanceolate lobes. Fl.5–7. Pinewoods, birch moorland, deciduous woods on acid soils, heaths, poor grassland; rather rare. Britain, N. and C. Europe.

Olive Family, Oleaceae

3. Ash, *Fraxinus excelsior* L., tree to 40m, bark smooth when young, later with vertical fissures, blackish; lvs pinnate, with 9–13 ovate-lanceolate, finely serrate leaflets; leaf buds broad, black, opposite; flower panicle erect, appearing before the lvs; corolla and calyx absent; fruit a winged nut. Fl.4–5. Riverside woods and in ravines, herb-rich mixed woods, scrub and hedges; common or widespread. Almost all of Europe, northwards to S. Scandinavia, only in the mountains in the south.

4. Flowering Ash, **Fraxinus ornus* L., tree 5–10m, with a smooth bark; lvs with 5–9 elliptic, serrate, shortly stalked leaflets, with the underside paler green or whitish; buds grey-felted; flowers white, in many-flowered, pyramidal inflorescences, scented, appearing with the lvs; petals 4, linear, 10–15mm long; anthers with long filaments; fruit 3cm long, winged. Fl.4–5. Woods of Downy Oak, scrub; S. Alps, S. Europe; also planted as an ornamental tree.

5. Wild Privet, *Ligustrum vulgare* L., shrub 1–5m; lvs ovate-lanceolate, opposite, entire, glabrous; flowers small, in dense panicles, 4-merous, strongly scented; corolla deeply 4-lobed, white; fruit a black berry, inedible, 6–8mm long. Fl.6–7. Scrub, wood margins, mixed woods, hedges; likes warm places; prefers calcareous soils; common. Almost all of Europe, northwards to S. Scandinavia.

Milkweed Family, Asclepiadaceae

6. Vincetoxicum, **Vincetoxicum hirundinaria* Med. (*V. officinale* Moench, *Cynanchum v.* (L.)Pers.), plant 30–120cm; stem hollow, pubescent; lvs oblong, cordate-ovate, 8–12cm long, dark to blue-green; inflorescence compound, flowers clustered in the partial inflorescences, 5-merous, symmetrical; corolla funnel-shaped, white to yellow-green, 4–7mm wide, with a small corona; ovary superior, with the 5 anthers united into a column. Fl.5–8. Poisonous! Oak and pine woods, sunny scrub, warm screes; scattered. Principally C. and S. Europe.

8. Creeping Jenny, *Lysimachia nummularia* L. (Pl.167), plant 10–50cm, prostrate, far-creeping; lvs opposite, roundish or elliptic, obtuse; flowers in the axils of the middle lvs, yellow, 1–2cm broad; calyx teeth cordate. Fl.5–8. Damp woods, river banks, damp meadows, moist hedgebanks; widespread. Almost all of Europe, but absent from the Mediterranean region.

Plate 169   Periwinkle Family, Apocynaceae

1. Lesser Periwinkle, *Vinca minor* L., plant 10–20cm, far-creeping; lvs leathery, evergreen, lanceolate, glabrous, very shortly stalked; flowers solitary in the leaf axils, with stalks 1–2cm long, 5-merous; corolla pale blue, 2–3cm wide; calyx lobes glabrous; ovary superior. Fl.4–5. Deciduous woods, scrub, hedgebanks; cultivated plant; scattered. C. and S. Europe, northwards to Denmark, Britain.

2. Greater Periwinkle, *Vinca major* L., plant 10–50cm, like *V. minor*, but lvs broadest near the base, gradually pointed, ciliate; petioles up to 1.5cm long; corolla 4–5cm across; calyx teeth hairy. Fl.4–5. Deciduous woods, copses, hedgerows; cultivated plant, originally from S. Europe, nowadays often in gardens and naturalised.

Gentian Family, Gentianaceae

3. Willow Gentian, **Gentiana asclepiadea* L., plant 30–80cm; lvs decussate, lanceolate, acute, 4–8cm long, 5-veined; flowers axillary, 1–3 together, often secund when the stems grow obliquely or horizontally on sloping ground (the flowers likewise); calyx tubular, with 5 short, narrow teeth; corolla narrowly campanulate, 3–5cm long, dark blue, red-violet within. Fl.7–9. Protected! Moorland pasture, mountain woods, high-level shrub communities; in the Alps to almost 2000m; rather rare. Mountains of C. and S. Europe. See also *G. cruciata*, Pl.65.

Bedstraw Family, Rubiaceae

4. Crosswort, *Cruciata laevipes* Opiz (*Galium cruciata* (L.)Scop.), plant 10–50cm; stem and inflorescence with spreading hairs; lvs ovate, 3-veined, 1–2cm long, hairy, in whorls of 4, yellowish-green; flowers in axillary whorls; corolla 2–3mm wide, yellow; pedicels hairy. Fl.4–6. Wet woods, hedges, waysides, pastures, weed communities; rather common. Almost all of Europe.

5. Round-leaved Bedstraw, **Galium rotundifolium* L., plant 10–25cm; stem ascending, slender, dainty; lvs broadly ovate, mucronate, 1–2cm long, in whorls of 4; inflorescence lax; corolla white, 3mm wide; fruit with hooked bristles. Fl.6–9. Fir and spruce woods; scattered. Mainly in the mountains of C. and S. Europe.

6. Wood Bedstraw, **Galium sylvaticum* L., plant 30–100cm; stem terete, finely 4-ribbed; lvs blue-green, mucronate, usually 8 in a whorl; inflorescence laxly paniculate, spreading, with slender pedicels; corolla white, 2–3mm wide, with acuminate lobes. Fl.6–8. Deciduous woods, wood margins; rather common. Mainly C. Europe.

7. Woodruff, *Galium odoratum* (L.)Scop. (*Asperula odorata* L.), plant 15–30cm; lvs lanceolate, the lowest 6, the upper 8 to a whorl, mucronate, green on both sides; inflorescence umbellate; corolla funnel-shaped, white, 4-lobed, 4–5mm wide; fruit with hooked bristles. Fl.5–6. Herb-rich beech and mixed deciduous woods; widespread. Almost all of Europe. See also *Asperula*, Pl.66 and *Galium album*, Pl.84.

Plate 170   Jacob's-ladder Family, Polemoniaceae

1. Jacob's-ladder, *Polemonium caeruleum* L., plant 30–80cm, with a horizontal rhizome; stem erect, angled and furrowed, glabrous; lvs imparipinnate, with 15–30 ovate-lanceolate leaflets; flowers 5-merous, in a glandular-hairy panicle 10–30cm long; corolla rotate to campanulate, sky-blue, rarely white, about 2cm across, with rounded lobes; calyx divided to halfway. Fl.6–7. Woods of Grey Alder, high-level shrub communities, thickets and grassland on limestone; rather rare; cultivated plant. Britain, C. and N. Europe, more often naturalised from gardens. Protected!

Borage Family, Boraginaceae

2. Common Gromwell, *Lithospermum officinale* L., plant 30–100cm; stem much branched, densely leafy; lvs lanceolate, 4–8cm long, sessile, the underside with prominent pinnate veins; corolla white or yellowish, 6–8mm wide; corolla tube 4–5mm long, with 5 hairy folds in the throat; nutlets white, shining, very hard. Fl.5–7. Wet woods which are warm in summer, open deciduous woods, hedges and scrub; scattered; medicinal plant. Europe.

3. Purple Gromwell, *Buglossoides purpurocaerulea* (L.)Johnst. (*Lithospermum p.* L.), plant 30–60cm, densely hairy, with creeping sterile shoots; lvs lanceolate, with only the midrib visible beneath (not pinnately veined); corolla 5-merous, red-violet at first, then dark blue, with a tube 14–20mm long; nutlets smooth, shining white. Fl.4–6. Sunny oak scrub, open deciduous woods; rather rare. Britain, C. and S. Europe.

4. Wood Forget-me-not, *Myosotis sylvatica* (Ehrh.)Hoffm., plant 15–45cm, densely hairy; basal lvs shortly stalked, ovate-lanceolate; calyx about 5mm long, lobed to about two-thirds of the way down, lobes linear and with spreading hooked hairs at the tip; corolla flat, lobes spreading, 6–10mm across, blue; fruiting pedicels 5mm long. Fl.5–7. Wood margins, herb-rich deciduous woods, mountain meadows, high-level shrub communities; scattered. Almost all of Europe, northwards to S. Sweden, Britain, southwards to the Apennines and the Balkan peninsula.

5. Sticky Lungwort, **Pulmonaria mollis* Wulff (*P. montana* auct.), plant 15–35cm, bright green; lvs and stem densely glandular and densely soft-hairy; basal lvs oblong-ovate, 4–6cm wide, gradually narrowed into the long stalk; stem lvs ovate-lanceolate, 2–4cm broad, somewhat decurrent; inflorescence sticky above; flowers 15–20mm long, red at first, then blue-violet; calyx densely short-glandular, sticky, 5-veined. Fl.4–6. Wood margins, herb-rich mixed deciduous woods, mountain woods; rather rare. Mainly C. Europe. Similar species: **Velvet Lungwort,** **P. mollissima* Kern., but plant densely velvety-hairy, grey-green; stem lvs not decurrent; mixed oakwoods; rare; Bavaria, C. and S.E. Europe. **Narrow-leaved Lungwort,** **P. angustifolia* L. (*P. azurea* Besser), basal lvs narrowly lanceolate, 2–3cm wide, stem lvs 5–15mm wide; stem, lvs and calyx stiffly hairy, not glandular; calyx 10-veined, not sticky; oak and pine woods, scrub; rare. Mainly E. and C. Europe. Protected!

6. Lungwort, *Pulmonaria officinalis* L., plant 15–30cm; basal lvs cordate-ovate, usually white-spotted; suddenly narrowed into the long stalk; upper stem lvs oval, amplexicaul; inflorescence stiffly hairy; flowers pink, then blue, 1cm across. Fl.3–5. Herb-rich mixed deciduous woods, scrub, wood margins, hedgebanks; widespread. Almost all of Europe.

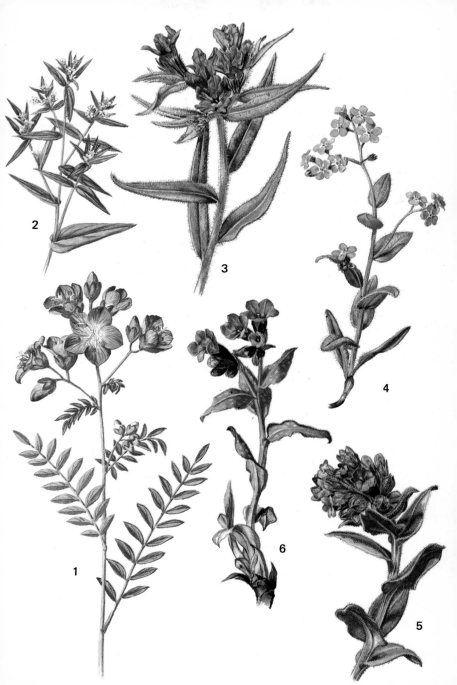

Plate 171 Borage Family, Boraginaceae

1. Tuberous Comfrey, *Symphytum tuberosum* L., plant 25–50cm, with an irregularly thickened, bulbous rhizome, without stolons; stem simple or branched only above; lvs ovate-oblong, 3–12cm long, gradually narrowed into the stalk, scarcely decurrent (in contrast to Common Comfrey); flowers 15–20mm long, pale yellow; scales in the throat of the corolla not projecting. Fl.4–5. Herb-rich deciduous woods, woods by rivers, high-level scrub; scattered or rare. Britain (naturalised), S. Germany, Alps and hinterlands, S. Europe. Similar species: **Bulbous Comfrey,** **S. bulbosum* C. Schimper, but plant with stolons, rhizome slender, with spherical bulbs; lvs suddenly narrowed into the stalk; scales in the throat of the corolla exceeding the corolla; parkland, wood margins, vineyards; naturalised. S. Europe.

See also *Onosma*, Pl.67.

Mint Family, Labiatae or Lamiaceae

2. Bastard Balm, *Melittis melissophyllum* L., plant 20–60cm, densely soft-hairy; lvs ovate, coarsely crenate, 3–9cm long; flowers 2–4cm long, 1–3 together in the axils of the upper lvs; calyx campanulate, 10-veined; corolla 2-lipped, pink or white, with red-violet spots on the lower lip, which is 3-lobed, with a broad, emarginate central lobe; stamens 4. Fl.5–7. Open, warm oak-hornbeam woods, lime woods or mixed beech woods, scrub, hedgebanks; rather rare. Britain, C. and S. Europe.

3. Wood Sage, *Teucrium scorodonia* L., plant 30–60cm, with an unpleasant smell; lvs stalked, ovate, cordate at base, wrinkled, obtuse and irregularly toothed, 3–7cm long; flowers yellowish, 10–14mm long, in lax, slender spikes; calyx 4–6mm long, 2-lobed, upper lip entire, lower lip with 4 lobes. Fl.7–9. Open oak and pine woods on acid soils, heaths, wood margins; widespread. C. Europe, northwards to S. Scandinavia and Britain, southwards to S. France, N. Italy, Croatia (western).

4. Yellow Archangel, *Lamiastrum galeobdolon* (L.)Ehrend. et Pol. (*Lamium galeobdolon* L.), plant 20–50cm, with stolons; lvs ovate-lanceolate, acute, serrate, 3–8cm long; flowers several together in the axils of the upper lvs; corolla 15–25mm long, yellow, lower lip with red spots; in ssp. *galeobdolon*, the base of the stem is hairy only on the angles, the upper lvs are ovate, and the flowers are 2–3 to a whorl; in ssp. *montanum* (Pers.) Ehrend. et Pol., the base of the stem is hairy all the way round, the upper stem lvs are lanceolate and sharply toothed and the flowers are 4–8 to a whorl. Fl.4–7. Deciduous and coniferous woods; widespread. Almost all of Europe.

5. Hedge Woundwort, *Stachys sylvatica* L., plant 30–100cm, with an unpleasant smell; stem with spreading hairs; lvs stalked, coarsely and sharply toothed, hairy; flowers in a spike of whorls of 6, one above the other; corolla 12–15mm long, dirty purple, with the lower lip about twice as long as the upper; calyx 4–7mm long, densely hairy. Fl.6–9. Deciduous woods, woods by rivers, wood margins, hedgebanks, shady waste places; widespread. Europe, in the south only in the mountains.

Plate 172   Mint Family, Labiatae or Lamiaceae

1. Betony, *Stachys officinalis* (L.)Trev. (*Betonica officinalis* L.), plant 20–70cm, woody at base; stem with 1–3 pairs of lvs, appressed-hairy above; lvs long-stalked; leaf blade narrowly ovate, 3–10cm long, slightly crenate, with a network of wrinkles; inflorescence dense, cylindrical; flowers red, with a straight upper lip; calyx with 5 teeth with bristly awns. Fl.6–8. Open deciduous woods, moors and poor pastures in mountains, hedgebanks; common to local; former medicinal plant. Mainly C. and S. Europe (western), northwards to S. Scandinavia and Britain. The **Yellow Betony,** **Stachys alopecuros* (L.)Benth. (*Betonica alopecuros* L.), has pale yellow corollas, lvs broadly cordate, 3–6cm long and stem with rough, spreading hairs; stony pastures, subalpine scree slopes, scrub of Mountain Pine; rare; Alps, Pyrenees, Apennines, Abruzzi, mountains of the Balkan peninsula.

2. Jupiter's Distaff, **Salvia glutinosa* L., plant 50–120cm, glandular-hairy, sticky; lvs long-stalked, blade hastate at base, coarsely serrate, 8–15cm long; inflorescence composed of many whorls of 4–6 flowers each; flowers with pedicels 3–10mm, pale yellow, spotted red-brown, 3–5cm long; stamens 2; calyx narrowly campanulate, sticky with glandular hairs. Fl.7–10. Deciduous woods, mixed mountain woods, clearings, high-level shrub communities; scattered. Mountains of C. and S. Europe.

3. Wild Basil, *Clinopodium vulgare* L. (*Satureja vulgaris* (L.)Fritsch), plant 30–60cm; stem villous-hairy; lvs ovate, 2–4cm, entire, slightly crenate, hairy; flowers 10–20 together in whorls in the axils of the uppermost leaf pairs; calyx teeth with long awns; corolla 1–1.5cm long, pale purple. Fl.7–10. Sunny wood margins, open woods, scrub, hedges, dry calcareous grassland; rather common. Europe.

4. Marjoram, *Origanum vulgare* L., plant 20–60cm, strongly aromatic; lvs shortly stalked, oblong-ovate, the underside dotted with glands, 1–2cm long; flowers in lax panicles and cymes, with roundish, often purplish bracts; corolla pale purple, rarely white, 4mm long. Fl.7–9. Open oak and pine woods, sunny wood margins and hedges, dry grassland, slopes; common. Almost all of Europe.

See also *Glechoma*, Pl.30, *Prunella*, Pls. 68 and 83, and *Ajuga*, Pl.83.

Nightshade Family, Solanaceae

5. Deadly Nightshade, *Atropa belladonna* L., plant 50–150cm; lvs ovate, up to 15cm long, shortly decurrent, opposite in the region of the inflorescence, one of the pair being larger than the other; flowers solitary, axillary; corolla 5-lobed, campanulate, brownish-violet outside, yellow-green inside with violet veins; fruit a shining, black berry. Highly poisonous! Fl.6–8. Woodland clearings, rides, wood margins, hedges, near old buildings; rather rare to fairly common. Britain, C. and S. Europe.

6. Bittersweet, *Solanum dulcamara* L., plant 30–200cm, woody below, often clambering; lvs broadly lanceolate, undivided or with 1–2 lobes at the base; inflorescence of umbellate cymes; corolla 5-lobed, lobes spreading and then recurved, 1cm across, violet; fruit a red berry. Fl.6–8. Damp woods, river banks, woodland clearings, waste ground and on shingle beaches; rather common. Almost all of Europe.

Plate 173 Nightshade Family, Solanaceae

1. Cape Gooseberry, *Physalis alkekengi* L., plant 25–60cm; stem shortly hairy, with simple branches; lvs triangular-ovate, acute, often sinuate-dentate on the margins; flowers solitary; corolla 15–25mm wide, greenish-white; calyx at fruiting time orange-red and inflated, like a Chinese lantern, about 4cm long, enclosing the red, cherry-sized fruit. Fl.5–8. Woods by rivers, vineyards; scattered; garden plant; berries edible. Naturalised in Britain, wild in C. and S. Europe (eastern).

Figwort Family, Scrophulariaceae

2. Heath Speedwell, *Veronica officinalis* L., plant 10–30cm, prostrate, only the inflorescence erect; stem roughly hairy; lvs ovate, short-stalked, dark green, bluntly crenate; flowers in clearly demarcated, axillary racemes with very small bracts; calyx 4-lobed; corolla pale violet, 6–7mm wide; fruiting capsule triangular to cordate, glandular-hairy. Fl.6–8. Woods, wood margins, heaths, dry grassland, on acid soils; common to widespread; former medicinal plant. Europe.

3. Nettle-leaved Speedwell, **Veronica urticifolia* Jacq. (*V. latifolia* L.), plant erect, 20–60cm; stem equally hairy all the way round; lvs broadly lanceolate, acute, sharply serrate, lower lvs shortly stalked, the upper sessile; flowers in many-flowered, long racemes; pedicels 6–8mm long; calyx 4-lobed; corolla 6–8mm wide, lilac or reddish, with darker veins; fruiting capsule roundish. Fl.6–8. Woods in ravines and herb-rich mountain woods, high-level shrub communities; scattered. Mountains of C. and S. Europe. Similar species: **Germander Speedwell,** *V. chamaedrys*, see Pl.83. **Wood Speedwell,** *V. montana* L., plant 15–45cm, creeping; lvs long-stalked, roundish-ovate, wrinkled, incised crenate-serrate; inflorescence a few-flowered raceme; flowers pale lilac, with darker veins; fruiting capsule almost spectacle-shaped; herb-rich deciduous woods with ash or beech; scattered; Britain, S. Sweden, C. and S. Europe.

4. Large Speedwell, **Veronica austriaca* L. ssp. *teucrium* (L.)Webb, plant ascending to erect, 15–60cm; lvs ovate, coarsely toothed, rounded at the base, sessile; flowers in opposite, axillary racemes, which are dense at first and elongate later; calyx 5-lobed; corolla pale blue, 10–13mm wide. Fl.6–7. Sunny wood margins, scrub, open oak and pine woods, poor pastures; scattered. Mainly C. Europe, southwards to C. Spain, Macedonia, eastwards to the Caucasus. Similar species: **Austrian Speedwell,** **V. austriaca* L. ssp. *austriaca*, but lvs lanceolate, narrowed into the short stalk, some of them pinnatifid; corolla dark blue; sunny calcareous grassland; rare. Protected!

5. Foxglove, *Digitalis purpurea* L., plant 40–150cm; stem simple, grey-felted; lvs ovate-lanceolate, crenate, grey-felted below, the lower stalked, the upper sessile; inflorescence secund; corolla tubular-campanulate, with an oblique, 4-lobed margin, 3–5cm long, purple, rarely white, red-spotted inside. Fl.6–8. Poisonous! Mountain woods, woodland rides, wood margins and clearings, heaths and mountain rocks, calcifuge; common to scattered; cultivated and medicinal plant. Mainly W. Europe.

6. Large Yellow Foxglove, **Digitalis grandiflora* Mill. (*D. ambigua* Murr.), plant 60–120cm; lvs oblong-lanceolate, stiffly hairy below; corolla inflated-campanulate, 3–4cm long, pale yellow, with brown markings inside. Fl.6–8. Protected! Deciduous woods, clearings, sunny wood margins, scree slopes; in the Alps to about 1600m; scattered; mainly C. Europe, southwards to the S. Alps, N. Greece, Caucasus. Similar species: **Small Yellow Foxglove,** **D. lutea* L., but stem and lvs usually glabrous; corolla 2–2.5cm long, without brown markings inside; rare. Protected!

Plate 174   Figwort Family, Scrophulariaceae

1. Common Figwort, *Scrophularia nodosa* L., plant 50–150cm; stem 4-angled, not winged; lvs ovate or cordate, doubly serrate; flowers in a terminal panicle; corolla dirty brown, greenish at the base, 7–9mm long. Fl.6–9. Scrub, wood clearings, herb-rich woods, hedgebanks; common to widespread. Europe. Similar species: **Water Figwort,** *S. auriculata* L. (*S. aquatica* L.), but stems and petioles broadly winged; lvs narrowly cordate, blunt, crenate; corolla 8–10mm long, red-brown; reed beds by rivers, ditches, wet woods and meadows; common in Britain but rarer abroad. **Green Figwort,** *S. umbrosa* Dum. (*S. alata* Gilib.), likewise stem broadly winged, but lvs acute, narrowed or rounded at base, sharply serrate at the tip; corolla 6–8mm long, greenish-yellow.

2. Common Cow-wheat, *Melampyrum pratense* L., plant 10–50cm; stem 4-angled; lvs lanceolate, 3–6cm long; inflorescence secund, racemose; calyx teeth lanceolate, less than half as long as the corolla tube; corolla 15–20mm long, with a straight tube, yellowish-white. Fl.5–9. Open woods and margins, moorland, heaths; widespread. Most of Europe. Similar species: **Small Cow-wheat,** *M. sylvaticum* L., but corolla with a short, curved tube, golden-yellow, 6–9mm; calyx teeth triangular-lanceolate, about as long as the corolla tube. Coniferous woods, mountains; rare in Britain.

3. Wood Cow-wheat, *Melampyrum nemorosum* L., plant 10–50cm; stem 4-angled, shortly hairy on 2 opposite faces; lvs lanceolate, long-pointed; bracts broadly cordate to hastate, hairy, with teeth 1–5mm long on each side, blue-violet; inflorescence secund; corolla 16–20mm long, golden-yellow, later orange, with a red-brown, straight corolla tube; calyx villous-hairy. Fl.6–9. Deciduous woods; rare. C. and N. Europe (but not the Arctic or Britain). See also *M. cristatum*, Pl.32.

Broomrape Family, Orobanchaceae

4. Toothwort, *Lathraea squamaria* L., plant 10–25cm, pale violet or reddish, with small, reddish, scale-like lvs, without chlorophyll; flowers 10–15mm long, reddish, in a many-flowered, secund raceme. Fl.3–4. Damp woods, by rivers and in ravines, hedgerows; scattered; parasite on the roots of alder, hazel, beech, elm or poplar. Almost all of Europe.

Honeysuckle Family, Caprifoliaceae

5. Elder, *Sambucus nigra* L., shrub 3–7m; lvs pinnate, bluish-green below, with usually 5 oblong-ovate, finely serrate leaflets, 5–10cm long; pith of stem white; inflorescence an umbellate cyme, 10–20cm wide; corolla rotate, 5-lobed, white or yellowish; berries black, edible. Fl.5–6. Hedges, herb-rich damp woods, waste places, scrub; common. Most of Europe. **Dwarf Elder,** *S. ebulus* L., is herbaceous, 1–2m tall, with unpleasant smell, has 7 or 9 lanceolate, finely serrate leaflets, 5–15cm long, white or pink flowers with red anthers and black, inedible fruits; wood clearings, rides, riverside woods; local to widespread; C. and S. Europe, north to Britain, S. Sweden.

6. Red-berried Elder, *Sambucus racemosa* L., shrub 1–4m; pith of stem yellow-brown; lvs pale green on both sides, with oblong-lanceolate leaflets; inflorescence an ovoid panicle, appearing with the lvs; corolla greenish-yellow; berries red, edible. Fl.4–5. Coniferous and mixed woods, clearings, scree slopes; in the Alps to about 2000m; widespread. C. Europe, south to Pyrenees, N. Italy, Bulgaria, introduced in Britain.

7. Twinflower, *Linnaea borealis*, see p.398.

Plate 175   Honeysuckle Family, Caprifoliaceae

1. Wayfaring-tree, *Viburnum lantana* L., shrub 1–3m; lvs ovate, with a network of wrinkles, finely serrate, dark green above, grey-felted below; flowers, in umbellate cymes 5–10cm wide, are all the same size; corolla white, 6–8mm wide; berries oval, compressed, 7–9mm long, red then black. Fl.5–6. Wood margins, scrub, open oak and pine woods, hedges; common to scattered. Britain, C. and S. Europe.

2. Guelder-rose, *Viburnum opulus* L., shrub 1–4m; lvs roundish, 3–5 lobed, up to 12cm long and wide, green on both sides, glabrous; lobes irregularly toothed; flowers in an umbellate cyme 5–10cm across; flowers 6–8mm wide, the marginal ones sterile, larger, with an unsymmetrical, rotate corolla, 1–2cm wide; berries spherical, red, 8–10mm across. Fl.5–6. Damp woods, edges of woods and streams, scrub and hedges; common. Almost all of Europe.

3. Fly Honeysuckle, *Lonicera xylosteum* L., shrub 1–2m; lvs opposite, broadly elliptic, softly hairy, dark green above, paler beneath, 2–6cm long; flowers in pairs on the same peduncle, which is 1–2 times longer than the flowers; corolla yellowish-white, 10–15mm long; ovaries of the flower-pair united at the base only; berries red, inedible, in pairs, but not united. Fl.5–6. Herb-rich deciduous and mixed woods, hedges; rare in Britain but commoner elsewhere in Europe. Almost all of Europe. Similar species: **Black-berried Honeysuckle,** **L. nigra* L., but peduncle 3–4 times as long as the whitish flowers, berries black, united at the base only; mixed mountain woods; mountains of C. and S. Europe. **Alpine Honeysuckle,** **L. alpigena* L., shrub 50–150cm; flowers dull red; ovaries of the flower-pairs almost completely united; lvs elliptic, 7–10cm long; berries red, shining, united in pairs; beech and mixed mountain woods, high-level shrub communities, in the Alps to 2300m; mountains of C. and S. Europe. The **Blue-berried Honeysuckle,** *L. caerulea* L., has likewise the ovaries and berries joined in pairs, but the berries are black with a blue pruina; flowers yellowish-white; lvs ovate, 2–5cm long; mountain woods on acid soils, dwarf shrub communities, calcifuge; in the Alps to 2600m; N. Europe, Alps, Pyrenees, Carpathians.

4. Honeysuckle, *Lonicera periclymenum* L., twining shrub, up to 5m; lvs ovate-lanceolate, 4–10cm long, shortly stalked, but the topmost leaf-pair under the inflorescence sessile; flowers 4–5cm long, yellowish, in stalked heads, scented; fruit dark red, not united, somewhat poisonous! Fl.6–7. Scrub, woods with oak, birch or hornbeam on acid soils, hedges, shady rocks; common to widespread; cultivated plant. W. and C. Europe, northwards to S. Scandinavia, Britain, southwards to C. Italy, Corsica. Similar species: **Perfoliate Honeysuckle,** *L. caprifolium* L., twining shrub, up to 4m, the topmost leaf-pair on flowering shoots broadly connate and surrounding the stem, lvs blue-green below; flower head sessile; flowers white to yellow, often reddish-tinged; fruits red, not united; cultivated, naturalised; originally from S. Europe.

---

7. Twinflower, *Linnaea borealis* L. (Pl.174), plant 5–15cm high, stem filiform, far-creeping, up to 2m long, glandular-hairy; lvs evergreen, broadly ovate or roundish, 7–12mm long, with 1–2 small teeth on each side, blue-green below; flowers 1–2 together on long stalks, drooping, campanulate, 7–10mm long, with 5 broad lobes, white or pink, red-striped and hairy within; stamens 4. Fl.6–8. Pine, spruce or larch woods, shady rocks, creeping over moss cushions; in the Alps to about 2400m; rare, sometimes more common locally. Britain, N. Europe, Alps, Carpathians, Caucaus.

Plate 176   Moschatel Family, Adoxaceae

1. Moschatel, *Adoxa moschatellina* L., plant dainty, 5–10cm; lvs opposite, basal lvs twice ternate, the 2 stem lvs once ternate; flowers 4–6 together, greenish-yellow, in a terminal, almost cubical head; the top flower with a 4-lobed corolla and 2-lobed calyx, and the lateral flowers with 5-lobed corollas and 3-lobed calyces. Fl.3–5. Scrub, damp deciduous woods, woods by rivers, high-level scrub, hedges, mountain rocks; in the Alps to about 1800m; local. Almost all of Europe.

Valerian Family, Valerianaceae

2. Mountain Valerian, *Valeriana montana* L., plant 10–50cm; stem with 3–8 ovate, green, shining, pairs of lvs; flowers pink or white; inflorescence many-flowered, laxly panicled; bracts lanceolate, green. Fl.5–7. Open, stony mountain woods, calcareous scree slopes; scattered. Mountains of C. and S. Europe.

3. Three-leaved Valerian, *Valeriana tripteris* L., plant 10–50cm, similar to Mountain Valerian, but stem lvs 3-lobed, with dentate lobes; basal lvs cordate, coarsely toothed; flowers whitish to pink; bracts linear, with scarious margins, coarsely toothed. Fl.4–6. Open, stony mountain woods, cracks in rocks, scree slopes; scattered; in the Alps to over 2200m. Mountains of C. and S. Europe. See also *V. officinalis*, Pl.117.

Teasel Family, Dipsacaceae

4. Wood Scabious, *Knautia dipsacifolia* Kreutzer (*K. sylvatica* (L.)Duby), plant 30–100cm; stem often bristly hairy; lvs oblong-ovate, long-pointed, entire or crenate; flowers in flat heads, 3–4cm wide, surrounded by a green involucre; receptacle without scales; corolla of florets 4-lobed, lilac; the outermost flowers rather larger. Fl.6–9. Shady wood margins, woods by rivers, high-level shrub communities; rather common. Mainly the mountains of C. Europe. Similar species: **Hungarian Scabious,** *K. drymeia* Heuffel, but basal leaf rosette with several curved-ascending, flowering, pubescent-hairy stems and broadly ovate lvs; deciduous woods which are warm in summer, rare. C. Germany, S. Alps, S. Europe (eastern). See also *Dipsacus*, Pl.33.

Bellflower Family, Campanulaceae

5. Nettle-leaved Bellflower, *Campanula trachelium* L., plant 30–100cm; stem stiffly hairy, sharply angled; lvs triangular-ovate, toothed like nettle lvs, the lowest long-stalked; flowers funnel-shaped, blue-violet, 3–4cm long, in long, leafy racemes; pedicels with 2 bracts at the base. Fl.7–8. Herb-rich deciduous woods, clearings, scrub; widespread. Almost all of Europe. Similar species: **Giant Bellflower,** *C. latifolia* L., but stem bluntly angled, softly hairy; lvs ovate-oblong, serrate, with a short, winged stalk; herb-rich mixed mountain woods and woods in ravines, hedgebanks; local to rare; Alps and hinterlands, N. and C. Europe, Britain.

6. Peach-leaved Bellflower, *Campanula persicifolia* L., plant 30–80cm; basal lvs rough, dark green, lower lvs ovate-oblong, narrowed into the stalk, upper lvs lanceolate, sessile, finely serrate; flowers broadly campanulate, 2–4cm long and wide, pale blue, in few-flowered racemes; calyx lobes lanceolate. Fl.6–8. Sunny, herb-rich deciduous and coniferous woods, wood margins and waysides; widespread. Introduced in Britain, wild elsewhere in Europe.

See also *C. rapunculoides*, Pl.34, *Campanula*, Pl.72 and *C. patula*, Pl.85.

Plate 177   Bellflower Family, Campanulaceae

1. Spiked Rampion, *Phyteuma spicatum* L., plant 20–60cm; basal lvs long-stalked, ovate-cordate, 1–2 times as long as broad, doubly serrate; stem lvs rather narrower, the lower stalked, the upper sessile; flowers in a cylindrical spike, 4–10cm long; corolla yellowish-white, 1cm long; petals united at the tip. Fl.5–7. Herb-rich woods, thickets, mountain meadows; rare (in Britain) or widespread. Almost all of Europe. Similar species with long-attenuate flower heads: **Black Rampion,** *P. nigrum* F. W. Schmidt, but flower heads dark violet and bracts linear, shorter than the head; basal lvs twice as long as broad, middle stem lvs cuneate-based; mountain meadows, mixed deciduous woods, non-calcareous soils; widespread; C. Europe. **Dark Rampion,** *P. ovatum* Honck. (*P. halleri* All.), but basal lvs as long as wide, middle stem lvs cordate or rounded at base; bracts lanceolate, about as long as the blackish-violet flower heads; mountain meadows, high-level shrub communities; scattered; mountains of C. and S. Europe. **Betony-leaved Rampion,** *P. betonicifolium* Vill., but basal lvs narrow, 2–4 times longer than broad; bracts of the pale blue flower heads very short, bristly and lanceolate; siliceous grassland; scattered; Alps.

See also *Jasione*, Pl.72.

Daisy Family, Compositae or Asteraceae

2. Hemp Agrimony, *Eupatorium cannabinum* L., plant 50–150cm; stem often reddish, densely leafy; lvs opposite, palmately 3–5 lobed, with serrate, elliptic lobes; flower heads small, oblong, 1cm long, in dense umbellate panicles, with tubular florets only, pink. Fl.7–8. Damp woods, ditches, stream banks, clearings, marshes and fens; common. Europe.

3. Alpine Adenostyles, *Adenostyles alpina* (L.)Bluff et Fingerh. (*A. glabra* (Mill.)DC.), plant 30–80cm; lvs roundish-kidney-shaped, regularly serrate, grey-green below, with a close-meshed network of veins, glabrous or with hairs on the veins only; petioles not auricled; flowers pale pink or red-violet, in umbellate panicles; flower heads usually 3-flowered. Fl.6–8. Stony mountain woods, screes, in the Alps at 800–2500m; common. Mountains of C. and S. Europe. Similar species: **Grey Adenostyles,** *A. alliariae* (Gouan) Kern., but lvs cordate to kidney-shaped, irregularly coarsely serrate, the lower side thinly grey-felted, with a fine network of veins; petioles of the upper lvs auriculate at base; flower heads 3–6 flowered; mountain woods, high-level shrub communities; widespread; mountains of C. and S. Europe. The **Felted Adenostyles,** *A. leucophylla* (Willd.)Rchb. (*A. tomentosa* (Vill.) Schinz et Thell.), has lvs felted-hairy below and 12–24 flowered heads with felted-hairy bracts; non-calcareous screes of the W. Alps.

See also *Serratula*, Pl.120.

4. European Michaelmas-daisy, *Aster amellus* L., plant with short stiff hairs, 20–50cm; lvs elliptic, the upper lanceolate, entire, roughly hairy; flower heads several to each stem, with blue-violet ray florets and yellow disc florets; bracts somewhat spreading. Fl.8–10. Open pinewoods, sunny wood edges and scrub, dryish grassland; scattered or rare. C. and S.E. Europe. The **Michaelmas-daisy,** *A. novi-belgii* L., is almost glabrous; lvs lanceolate, numerous; heads numerous. Garden escape in waste places, frequent. Europe, originally from N. America. There are numerous similar species.

Plate 178   Daisy Family, Compositae or Asteraceae

1. Hairy Fleabane, *Inula hirta* L., plant 15–45cm; stem with spreading hairs; lvs ovate, roughly hairy on both sides, prominently net-veined, sessile with a rounded base; flower heads 2–5cm across, 1–3 together; ray florets 15–20mm long, yellow. Fl.6–7. Open oak and pine woods, sunny hedges, dry grassland; rare. Mainly C. and E. Europe. Similar species: **Swiss Fleabane,** *I. helvetica* Web., but stem appressed grey-felted, lvs ovate-lanceolate, narrowed into the short stalk, the underside densely short-hairy, without a network of veins; flower heads 2.5–3cm across, several together in an umbellate panicle; outermost bracts grey-felted; scrub on river banks, margins of wet woods by rivers, likes warm places; rare; Germany, Switzerland, S.W. Europe. **German Fleabane,** *I. germanica* L., but stem lvs cordate at base, sessile and semi-amplexicaul, the lower side long-hairy, glandular; flower heads 1cm wide, in dense umbellate panicles; ray florets scarcely longer than the disc florets; calcareous grassland, sunny places on the edges of thickets; rare; C. and S. Germany, E. and S.E. Europe. **Ploughman's Spikenard,** *I. conyza* DC., flower heads 1cm wide, in umbellate racemes, brownish, without ray florets, or else they are hidden in the bracts, with spreading bracts; lvs ovate-lanceolate, densely short-hairy and net-veined below; open oak and pine woods, sunny wood margins, dry slopes and cliffs; scattered; Britain, C. and S.E. Europe. Compare also with *I. salicina*, Pl.73, and see *Buphthalmum*, Pl.74.

2. Scentless Feverfew, *Tanacetum corymbosum* (L.)Schultz-Bip. (*Chrysanthemum c.* L.), plant 50–100cm; lvs pinnate with 7–15 ovate-oblong segments which are further pinnatisect in linear, toothed lobes; flower heads 1–2cm wide, 6–20 together in umbellate panicles; ray florets linear, narrow, 1–2cm long, white; disc florets yellow. Fl.6–8. Open deciduous woods, wood margins, scrub, likes warm places; scattered. C. and S. Europe. Similar species: **Large-flowered Feverfew,** *Tanacetum macrophyllum* Schultz-Bip. (*Chrysanthemum m.* W. et Kit.), but lobes of the segments ovate-roundish or broadly lanceolate, ray florets broader than long; disc florets brownish-white; flower heads 6–8mm wide; cultivated plant, naturalised in weed communities; S.E. Europe. See also *T. parthenium*, Pl.36.

3. Great Marsh Thistle, *Carduus personata* (L.)Jacq., plant 60–120cm; stem narrowly winged; lvs large, soft, the underside arachnoid-hairy, the upper lvs oval, with bristly teeth, the lower deeply lobed; flower heads several, 1–2cm wide, purple; bracts linear, acute, scarcely prickly, longer than the flowers. Fl.7–8. Woods of Grey Alder, willow thickets, stream banks, high-level scrub; rather rare. Mountains of C. and S. Europe.

4. Heath Cudweed, *Omalotheca sylvatica* (L.)Schultz-Bip. et F. W. Schultz (*Gnaphalium sylvaticum* L.), plant 10–50cm, with numerous non-flowering rosettes at flowering time; stem felted; basal lvs lanceolate, 2–6cm long, 2–5mm wide, shortly stalked, 1-veined like the stem lvs, felted below; flower heads in long spikes, occupying at least one-third the length of the stem; bracts with a pale brown, scarious margin. Fl.7–9. Woodland rides, clearings, heaths, dry acid grassland; common. Almost all of Europe. Similar species: **Highland Cudweed,** *O. norvegica* (Gunn.) Schultz-Bip. et F. W. Schultz (*Gnaphalium norvegicum* Gunn.), but basal lvs 5–12cm long, 3-veined, long-stalked, usually withered by flowering time, all lvs 3-veined, about 5–10mm wide; spikes shorter; bracts with dark brown margins. Fl.7–8. Siliceous grassland, mountain rocks, calcifuge; Britain (rare), Europe, mainly in the mountains.

5. Perennial Cornflower, *Centaurea montana*, see p.406.

Plate 179   Daisy Family, Compositae or Asteraceae

1. Goldenrod, *Solidago virgaurea* L., plant 20–100cm; lvs oblong-elliptic, 3–4 times as long as broad, usually toothed; flower heads 7–8mm long and 10–15mm wide, in erect panicles or racemes, spreading all round the stem; ray florets 6–12, longer than the linear, greenish-yellow, scarious margined bracts. Fl.7–10. Herb-rich deciduous and mixed woods, heaths, hedgebanks, dry pasture, rocks, cliffs and dunes; common to widespread. Europe. Similar species: **Alpine Goldenrod,** *S. virgaurea* ssp. *minuta* (L.)Arc. (*S. alpestris* W. et Kit.), but lvs narrower, 4–6 times as long as wide; flower heads up to 10mm long and over 15mm wide in a spike-like panicle; dwarf shrub communities, alpine scrub, mountain meadows; calcifuge; mountains of N., C. and E. Europe.

2. Purple Colt's-foot, *Homogyne alpina* (L.)Cass., plant 10–30cm; stem almost leafless, 1-headed; basal lvs long-stalked, glabrous below, hairy on the veins only, cordate to kidney-shaped, serrate-crenate; flowers pale violet; bracts woolly, brownish-red at tip; pappus snow-white. Fl.5–8. Mountain spruce woods, dwarf shrub communities, siliceous grassland; widespread. Scotland (very rare), mountains of C. and S. Europe. Similar species: **Woolly Colt's-foot,** *H. discolor* (Jacq.)Cass., but lvs roundish to kidney-shaped, white-felted below; pappus dirty white; places where the snow lies late, at 1400–2600m in the E. Alps.

3. Alpine Ragwort, *Senecio nemorensis* L. ssp. *fuchsii* (Gmel.)Čelak, plant 60–150cm; lvs lanceolate, the upper over 5 times as long as wide, finely serrate, glabrous, short-stalked or sessile with a cuneate base; flower heads 2–3cm across, in an umbellate panicle; involucre usually with 8 bracts, cylindrical; ray florets 5, yellow. Fl.7–8. Wood clearings, herb-rich mixed woods, in the Alps to over 2000m; common. Mainly C. Europe. Similar species: **Wood Ragwort,** *S. nemorensis* L. ssp. *nemorensis*, but the lvs about 3 times as wide, serrate with thickened teeth, strongly ciliate, the upper lvs semi-amplexicaul or sessile with a rounded base; involucre somewhat campanulate, with 10 bracts; ray florets 5–7; mixed mountain woods; C. and E. Europe. In the **Rayless Ragwort,** *S. cacaliaster* Lam., the flower heads are pale yellow and lack ray florets; lvs ovate-lanceolate, sharply serrate; S. Alps, Apennines, Cevennes, S.E. Europe.

4. Stinking Woodsalad, *Aposeris foetida* (L.)Less., plant 5–25cm, stinking; stem simple, 1-headed, leafless; lvs in a basal rosette, pinnatisect, like dandelion lvs; bracts in 2 rows; flower heads 2–3cm broad, with ray florets only, yellow. Fl.6–8. Mixed mountain woods, especially with beech, oak and hornbeam; in the Alps to over 2000m; widespread. Alps and hinterlands, mountains of C. and S. Europe.

---

5. Perennial Cornflower, *Centaurea montana* L. (Pl.178), plant 20–70cm; stem simple, unbranched; lvs ovate, acute, decurrent down stem like wings, thinly felted above, densely felted-hairy below, becoming glabrous later; flower heads solitary; marginal florets greatly enlarged, deep blue, central flowers violet; bracts with comb-like, blackish-brown fringes. Fl.5–7. Woods on mountains and in ravines, grassy scrub and meadows in mountains; scattered; in the Alps to over 2000m. Mountains of C. and S. Europe, a garden escape in Britain.

Plate 180 Daisy Family, Compositae or Asteraceae

1. White Butterbur, *Petasites albus* (L.)Gaertn., plant 15–30cm at flowering time, up to 60cm when fruiting; dioecious; basal lvs appearing after the flowers; stem with pale scale lvs only; lvs roundish-cordate, irregularly toothed, grey-felted below, with a close-meshed network of veins; petioles rounded above (in *Tussilago*, see Pl.34, channelled above); flower heads in dense racemes; flowers whitish. Fl.3–5. Herb-rich mixed deciduous woods, woods in ravines, river banks; scattered. Almost all of Europe, mainly in the mountains, rare garden escape in Britain. Similar species: **Alpine Butterbur,** **P. paradoxus* (Retz.)Baumg. (*P. niveus* Baumg.), but lvs ovate-triangular, pointed, usually longer than wide, almost regularly and sharply serrate, white-felted below; scale lvs on stem red-brown or violet tinged; flowers reddish; scree communities, river shingle, landslips with marly soil; scattered; Alps and hinterlands, Pyrenees, mountains of the Balkan peninsula. **Butterbur,** *P. hybridus* (L.)G.M.Sch. (*P. officinalis* Moench), but lvs roundish, up to 60cm wide, shallowly sinuate and almost regularly dentate, grey-woolly beneath, becoming glabrous later; flower heads scented, reddish; river banks, willow and alder thickets, wet meadows; widespread; almost all of Europe, in the south only in the mountains. See also *Petasites*, Pl.118.

2. Purple Lettuce, **Prenanthes purpurea* L., plant 50–150cm, paniculate-branched above; lvs oblong-ovate, sinuate-dentate, with cordate base, and amplexicaul, glabrous, blue-green; flower heads drooping, 2–5 flowered, purple. Fl.7–8. Herb-rich beech, fir- and oak-beech woods, high-level shrub communities; widespread. C. and S. Europe.

3. Wall Lettuce, *Mycelis muralis* (L.)Dum. (*Lactuca m.* L), plant 40–80cm; stem hollow, paniculately branched above; lvs deeply pinnatisect with angular side lobes and a large terminal lobe, dark green, often red-tinged; heads with 5 flowers, in lax panicles; flowers pale yellow. Fl.7–8. Herb-rich deciduous and coniferous woods, clearings, shady walls and rocks; widespread. Almost all of Europe.

4. Alpine Sow-thistle, *Cicerbita alpina* (L.)Wallr. (*Mulgedium alpinum* (L.)Less.), plant 60–150cm; stem simple, erect, violet-tinged above; lvs irregularly pinnately lobed, with a triangular-hastate terminal lobe, dark green above, blue-green below; flower heads in a narrow, paniculate raceme; involucre and peduncle with spreading, brown, glandular hairs; corolla blue-violet. Fl.7–9. High-level shrub communities, mixed mountain woods, Green Alder scrub, moist rocks on mountains, in the Alps at 1000–2200m; widespread. Scotland (rare), Scandinavia, mountains of C. Europe, the Alps, Pyrenees, Apennines, Carpathians, mountains of the Balkan peninsula. Similar species: **French Sow-thistle,** *C. plumieri* (L.)Kirschl., but inflorescence glabrous; corollas pale blue; heads in umbellate corymbs; leaf terminal lobes ovate-oblong; rare; mountains of France, C. Europe (west), Bulgaria; garden escape in Britain. Protected!

See also *Arctium*, Pl.37.

Plate 181   Daisy Family, Compositae or Asteraceae

1. Few-leaved Hawkweed, *Hieracium murorum* L. (*H. sylvaticum* L.), plant 20–60cm; stem leafless or with 1–2 small lvs; basal lvs ovate, with cordate base, usually long-stalked, coarsely toothed, soft, dark green; flower heads 2–15, in umbellate panicles; peduncles and bracts usually glandular. Fl.5–8. (Aggregate species with a very large number of microspecies.) Deciduous and coniferous woods, shady rocks and walls; common. Europe. Similar species: **Forked Hawkweed,** **H. bifidum* Kit., but inflorescence branched dichotomously, few-flowered; peduncles and bracts stellate-hairy, felted, without glands; lvs rough, sinuate-dentate, the underside often red; open beech woods, rock outcrops, stony calcareous grassland; scattered; Europe, in the south only in the mountains.

2. Common Hawkweed, *Hieracium argillaceum* Jord. (*H. lachenalii* C.C.Gmel.), plant 30–100cm, with a basal leaf rosette at flowering time; stem with 3–8 lvs; lvs green, usually unspotted, ovate to ovate-lanceolate, coarsely toothed to lobed, gradually narrowed into the stalk; stem lvs becoming smaller upwards; peduncles and bracts stellate-hairy and with black glands; bracts not overlapping. Fl.6–8. (Aggregate species with several microspecies.) Open deciduous and coniferous woods on acid soils, heaths, poor grassland, rock outcrops; widespread. Europe. Similar species: **Prenanth Hawkweed,** *H. prenanthoides* Vill., but stem lvs 10–30, amplexicaul with a broadly cordate base, with a network of veins below, panduriform; bracts and inflorescence densely glandular; petals of ray florets ciliate at tip; high-level shrub communities, subalpine pastures, Mountain Pine scrub, streamsides and woods in ravines; local to rather rare; Britain, N. Europe, mountains of C. and S. Europe.

3. Savoy Hawkweed, *Hieracium sabaudum* L., plant 50–150cm, without a basal rosette at flowering time; stem leafy in the same manner throughout, or the lower lvs crowded and sessile; lvs ovate-lanceolate, coarsely serrate, blue-green below, the upper lvs semi-amplexicaul and sessile; flower heads in lax panicles, yellow; bracts overlapping each other in a regular fashion, obtuse, dark. Fl.8–10. Open deciduous woods on acid soils, wood margins, wayside banks, quarries; widespread. Britain, but mainly C. Europe. Similar species: **Umbellate Hawkweed,** *H. umbellatum* L., but flower heads in umbels; bracts stiffly spreading with recurved tips; lvs lanceolate, often inrolled at the margins; oak and pine woods, poor grassland, heaths, fallow ground, waysides, widespread; almost all of Europe, in the south only in the mountains.

4. Sticky Hawkweed, *Hieracium amplexicaule* L., plant 10–50cm, sticky-glandular; stem dichotomously forked, usually with 3–6 stem lvs; basal lvs in a rosette, ovate-oblong, narrowed into the stalk, which is often winged, lvs with large teeth, dark green; stem lvs with cordate, amplexicaul bases; flower heads 2–12; involucre 10–16mm long, glandular-hairy; petals of ray florets ciliate at tip. Fl.6–8. Stony slopes, mountain meadows, rocky outcrops, old walls; widespread. Mountains of C. and S. Europe, Britain (rare introduction).

See also *Crepis paludosa*, Pl.120.

Plate 182   Sedge Family, Cyperaceae

1. Alpine Tufted Sedge, *Carex sempervirens* Vill., plant tufted, 20–50cm, basal sheaths fibrous and dark grey; lvs shining, 2–3mm wide, somewhat shorter than the terete stem; female spikes 2–3, lax-fruited, 1–2cm long; male spike 1, terminal; glumes brown with a white scarious border, shorter than the green fruit. Fl.6–8. Calcareous alpine meadows, 1400–3000m; common. Mountains of C. and S. Europe.

2. Rusty Sedge, *Carex ferruginea* Scop., plant 30–60cm, similar to *C. sempervirens*, but with rhizomes; basal leaf sheaths rusty to purplish-red, not fibrous; lvs 1–2mm wide, flaccid. Fl.7–9. Well-watered, deep soils, mountain meadows, about 1000–2700m; common. Mountains of C. and S. Europe.

3. Cushion Sedge, *Carex firma* Host., plant forming hemispherical cushions, 5–20cm; lvs rough, stiff, densely set, 2–3mm wide, 4–5cm long, spreading almost horizontally; stem leafless; female spikes 6–10mm long, 1–3; styles 3; male spike 1, terminal. Fl.6–8. Stony grassland rich in lime, rock outcrops, about 1500–2500m, occasionally washed down lower; common. Mountains of C. and S. Europe.

4. Curly Sedge, *Carex curvula* All., plant 5–25cm, with remains of the lvs of previous years at the base; basal leaf sheaths yellow-brown; lvs bristle-like, channelled, 1–2mm wide, curved outwards, rough, thinning out above, and there grey-brown to yellow-green; spikes all alike, arranged in a head, female below, male above; styles 3. Fl.7–8. Stony, acid soils, about 2000–3000m; widespread. Mountains of C. and S. Europe.

Rush Family, Juncaceae

5. Three-leaved Rush, *Juncus trifidus* L., plant 10–25cm; stems terete, at the base there are only yellow-brown, almost leafless leaf sheaths, stems usually with 3 filiform lvs in the upper third; inflorescence 1–4 flowered, terminal; perianth segments 6, equally long, chestnut-brown, finely pointed. Fl.7–8. Acid soils, siliceous rocks, stony places, 1000–3000m; scattered. Scotland, N. Europe, mountains of C. and S. Europe.

Grass Family, Gramineae or Poaceae

6. Alpine Meadow-grass, *Poa alpina* L. var. *vivipara* L., plant 5–50cm, tufted; base of stem thickened by the persistent remains of many superimposed old leaf sheaths; lvs flat, 2–5mm wide, green to blue-green; panicle lax; spikelets 5–10 flowered, greenish-yellow and speckled with red-violet, usually growing out into leafy bulbils. Fl.5–9. Deep, clayey soils, rich meadows, alpine meadows, about 1400–2600m (lower in the north); common. Britain (rather rare), N. Europe, mountains of C. and S. Europe.

Orchid Family, Orchidaceae

7. Black Vanilla Orchid, *Nigritella nigra* (L.)Rchb., plant 8–20cm; lvs grass-like, channelled above; flowers blackish-purple, rarely pink, in dense, spherical spikes, 1–2cm long, strongly scented of vanilla; perianth segments lanceolate, 1-veined; lip directed upwards. Fl.6–9. Protected! Poor grassland, about 1600–2800m; scattered. N. Europe and the mountains of C. and S. Europe.

8. Globe Orchid, *Traunsteinera globosa* (L.)Rchb., plant 20–50cm; inflorescence 3–6cm long, at first pyramidal, then spherical; flowers pink; perianth segments at first connivent into helmet, then campanulate-spreading; lip 3-lobed, with dark purple dots. Fl.6–8. Protected! Calcareous grassland, at 1000–2500m; rare. C. and S. Europe.

Plate 183 Willow Family, Salicaceae

1. Net-leaved Willow, *Salix reticulata* L., creeping shrub, with branched, rooting stems; lvs long-stalked, broadly elliptic, 1–4cm long, dark green above, dull, the underside grey to whitish, hairy, net-veined; catkins long-stalked, densely flowered, pinkish-red. Fl.7–8. Calcareous boulder scree, snow valleys, rock ledges, about 1800–2500m in the Alps, lower in N. Europe; scattered. Arctic, Scandinavia, Scotland, Pyrenees, Alps, mountains of the Balkan peninsula.

Dock Family, Polygonaceae

2. Mountain Sorrel, *Oxyria digyna* (L.)Hill., plant 5–25cm; basal lvs long-stalked, kidney-shaped; flowers hermaphrodite, in whorled, racemose inflorescences; perianth segments 4, the 2 inner appressed to the flat side of the fruit, much larger than the 2 outer, which are spreading; fruit lenticular, with reddish-purple wings. Fl.6–8. Non-calcareous scree with late snow cover, rocky places on mountains, beside streams, about 1600–2800m in the Alps, lower in N. Europe; scattered. Arctic, N. Europe, Britain, mountains of C. and S. Europe.

3. Alpine Bistort, *Polygonum viviparum* L., plant 10–25cm; lvs glabrous, dark green above, blue-green beneath, the margin inrolled; flowers in long spikes; the lower part with bulbils, which root after falling; flowers white. Fl.6–8. Mountain and alpine pasture, wet rocks, about 1000–3000m in the Alps, at sea level in Scotland; widespread. N. Europe, mountains of C. and S. Europe.

Pink Family, Caryophyllaceae

4. Alpine Gypsophila, **Gypsophila repens* L., plant 8–25cm, with a bluish pruina; lvs lanceolate, rather fleshy, 1–3cm long; inflorescence racemose; flowers 5-merous; petals 6–10mm long, white or pink. Fl.5–8. Calcareous scree slopes, about 1000–2800m, lower on river shingle of alpine rivers; widespread. Mountains of C. and S. Europe.

5. Alpine Mouse-ear, *Cerastium alpinum* L., plant 5–20cm, loosely tufted, with grey-green shoots with spreading hairs; lvs ovate-lanceolate, hairy, 5–20mm long; flowers white; petals deeply emarginate, twice as long as the villous-hairy sepals. Fl.7–9. Non-calcareous or acid soils, wind-exposed ridges, rock ledges, about 500–2800m; scattered. Britain, N. Europe, mountains of C. and S. Europe.

6. Alpine Pink, **Dianthus alpinus* L., plant 2–20cm, with several basal leaf rosettes; stem with 2–5 pairs of lvs; lvs linear, 1-veined, rough on the margins; flowers solitary; calyx scales half the length of the striped calyx tube; petals irregularly toothed, purple, with scattered white spots at the base. Fl.6–8. Stony alpine meadows, about 1000–2500m; scattered. E. Alps.

7. Moss Campion, *Silene acaulis* (L.)Jacq., cushion-plant, 1–3cm, stems densely leafy with overlapping lvs; lvs linear-subulate, 5–12mm long; flowers solitary; petals dark to pale red, emarginate, 6–14mm long; styles 3. Fl.6–8. Protected! Stony, usually calcareous soils, scree slopes, mountain cliffs and rock ledges, about 1500–3600m in the Alps, but at sea level in Scotland; widespread. N. Europe, mountains of C. and S. Europe.

See also *Sagina saginoides*, Pl.9.

1

2

3

4

5

6

7

Plate 184 Buttercup Family, Ranunculaceae

1. Alpine Buttercup, *Ranunculus alpestris* L., plant 5–15cm; basal lvs long-stalked, 3–5 lobed, lobes coarsely crenate, shining, dark green; stem lvs absent or narrowly linear; flowers 20–25mm across, 1–2 together, white; petals slightly emarginate. Fl.5–9. Fine screes, snow valleys, damp cracks in rocks, about 1500–2700m, occasionally down in the valleys; common. Mountains of C. and S. Europe.

2. Glacier Crowfoot, *Ranunculus glacialis* L., plant 5–15cm, erect or ascending; basal lvs stalked, fleshy, dark green, 3-lobed to the base, lobes again 3 to many divided; stem lvs sessile, palmately 3–5 lobed; flowers 20–30mm across; petals white or pink, usually dark outside, remaining after flowering; sepals densely dark brown hairy outside. Fl.7–8. Scree of siliceous rocks, moraines, about 2000–4200m; scattered. Arctic, Scandinavia, Alps, Sierra Nevada, Pyrenees, Carpathians.

3. Mountain Buttercup, *Ranunculus montanus* Willd., plant 5–30cm, highly variable; stem erect, with 1–3 flowers; basal lvs stalked, shining, 3–5 lobed, with obovate, toothed, dark green lobes; upper stem lvs sessile, with linear lobes; flowers golden-yellow, 12–25mm wide; achenes compressed laterally, with a short, hooked beak. Fl.4–8. Mountain meadows and pastures, also open woods, at about 1000–3000m, also in the hinterlands of the Alps in damp places; common. Alps, Black Forest, Jura, Carpathians. See also *R. aconitifolius* and *Trollius*, Pl.105.

4. Alpine Pasqueflower, *Pulsatilla alpina* (L.)Del., plant 15–45cm; stem lvs whorled, twice ternate, with toothed segments, hairy; flowers solitary, 4–5cm wide, long-stalked; perianth segments 6, ovate, white within, usually tinged violet outside, hairy; style elongated at fruiting time to a length of 5cm, serving as an aid to the dispersal of the seed by wind. Fl.6–8. Protected! Mountain meadows, stony alpine pastures, calcareous soils, at about 1500–2800m; scattered. Mountains of C. and S. Europe.

4a. Yellow Alpine Pasqueflower, *Pulsatilla alpina* (L.)Del. ssp. *apiifolia* (Scop.)Nym., similar to the last, but flowers sulphur-yellow. Fl.6–8. Protected! Non-calcareous or acid soils, alpine meadows, stony poor grassland, dwarf shrub heaths, about 1500–2800m; scattered. Mountains of C. and S. Europe.

5. Spring Pasqueflower, *Pulsatilla vernalis* (L.)Mill., plant 5–15cm, when fruiting up to 35cm; lvs over-wintering, leathery, pinnate, with unequally 2–5 lobed segments; stem felted-hairy, with an involucre of deeply divided, densely hairy bracts; flower solitary, drooping at first and later erect, 3–6cm across; perianth segments usually 6, white or pale lilac within, and violet, pinkish-red or bluish and silkily hairy outside; style of the ripe fruit 3–4cm long and villous. Fl.4–6. Protected! Siliceous grassland, open pine woods, dwarf shrub heaths; from sea level to 3600m; scattered. N. Europe, Denmark, N. Germany, mountains of C. and S. Europe.

Plate 185   Buttercup Family, Ranunculaceae

1. Narcissus-flowered Anemone, *Anemone narcissiflora* L., plant 20–40cm, with spreading hairs; basal lvs palmately 3–5 lobed, the lobes divided into narrow segments; stem lvs similar; flowers 2–3cm across, 3–8 together in an umbel, with 3 deeply and unequally divided sessile bracts beneath; perianth segments 5–6, white, glabrous on both surfaces, often reddish outside. Fl.5–8. Protected! Calcareous mountain and alpine meadows, about 1400–2400m; scattered. Mountains of C. and S. Europe.

See also *Anemone*, Pl.142.

2. Alpine Clematis, *Clematis alpina* (L.)Mill., twining plant, 1–2m; lvs opposite, long-stalked, lvs twice ternate; leaflets coarsely serrate; flowers axillary, long-stalked, violet to pale blue, campanulate; nectaries 10–12, spathulate, white-felted, half as long as the 4 perianth segments; fruit with a feathery style. Fl.5–7. Scrub of Alpenrose, open, shrubby mountain woods, coniferous woods, at about 1000–2400m; scattered. Mountains of C. and S. Europe.

See also *Aconitum*, Pl.142.

Poppy Family, Papaveraceae

3. Rhaetian Poppy, *Papaver rhaeticum* Ler., plant 5–15cm; lvs blue-green, in a basal rosette, 1–2 pinnate; segments ovate, blunt, 2–6mm wide, in 2–4 pairs; flowers golden-yellow, 4–5cm wide; petals 4, stamens numerous. Fl.7–8. Protected! Calcareous scree, rock outcrops, about 1500–3000m; scattered. S.W. and E. Alps, E. Pyrenees. Similar species: **White Alpine Poppy,** *P. sendtneri* Kern., but flowers white; lvs 1–2 pinnate, the segments oval-lanceolate, acute, 1–2mm wide; C. and E. Alps.

Cabbage Family, Cruciferae or Brassicaceae

4. Round-leaved Penny-cress, *Thlaspi rotundifolium* (L.)Grand., plant 5–15cm, malforming, with a deep taproot; stem creeping; lvs bluish-green, rather fleshy, ovate, the lower stalked, the upper broadly auriculate at base and amplexicaul; flowers pale violet, in corymbs; petals 6–8mm long; siliculae elliptic, 4–8mm long. Fl.7–9. Calcareous scree, at about 1300–3300m; widespread. Alps.

5. Chamois Cress, *Hutchinsia alpina* (Torn.)R.Br., plant 5–12cm, with a basal leaf rosette; stems several, simple, leafless; lvs pinnatisect, cut to the midrib; flowers in racemes which are compact at first, elongating later; petals 3–5mm long, white; siliculae ovoid, 4–5mm long. Fl.6–8. Calcareous scree and boulders, rock outcrops, about 1500–3400m; in the shingle of alpine rivers, where it is often swept down to lower altitudes; widespread. Mountains of C. and S. Europe.

6. Yellow Whitlow-grass, *Draba aizoides* L., plant 5–10cm; lvs pale green, narrowly lanceolate, 1–2cm long, ciliate with stiff bristly hairs which resemble the teeth of a comb, in spherical rosettes; stem leafless; flowers in 3–18 flowered corymbs; petals 4, yellow, 4–6mm long; siliculae narrowly elliptic, 6–10mm long, on long, glabrous stalks. Fl.4–8. Calcareous rocks, coarse screes, stony grassland, about 1600–3400m in the Alps; widespread in the mountains of C. and S. Europe, rare in Britain (where it grows near sea level).

See also *Cochlearia pyrenaica*, Pl.198.

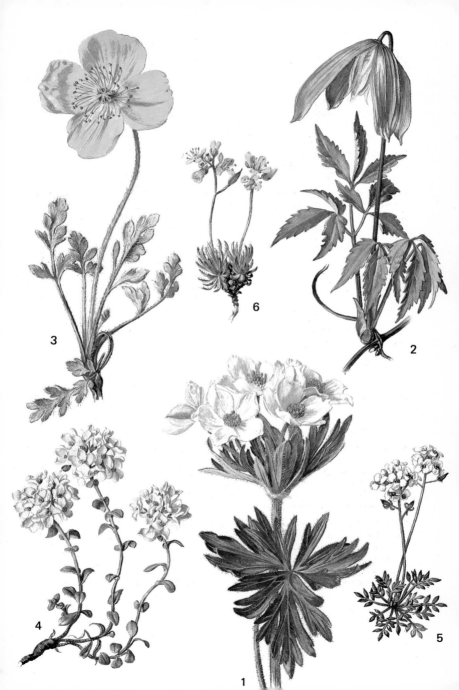

Plate 186   Stonecrop Family, Crassulaceae

1. Mountain House-leek, *Sempervivum montanum L., plant 5–15cm, with rosettes which are spherical at first, lvs spreading like a star later; rosette lvs lanceolate, with a green or reddish tip, surfaces glandular-hairy; flowers 2–8 together, 2–3cm wide; petals usually 12, red-violet, with a darker central vein. Fl.7–9. Siliceous grassland, at 1500–3400m; scattered. Mountains of C. and S. Europe. See Sedum, Pls. 53 and 54.

Saxifrage Family, Saxifragaceae

2. Blue-leaved Saxifrage, *Saxifraga caesia L., plant 2–10cm, forming a cushion of compact, spherical rosettes; lvs blue-green, 3–6mm long, curved and reflexed, inrolled at the margin, with 5–9 chalk-glands above; stem with 2–6 white flowers. Fl.6–9. Calcareous rocks, rough scree, stony grassland, at about 1600–3000m; descending lower on shingle of alpine rivers; widespread. Mountains of C. and S. Europe.

3. Livelong Saxifrage, *Saxifraga paniculata Mill. (S. aizoon Jacq.), plant 10–40cm; lvs spathulate, 3–5cm long, lime-encrusted on the margins; flowers white, often red-spotted, in many-flowered panicles. Fl.5–8. Cracks in calcareous rocks, stony grassland; in the Alps at 1300–3400m; widespread. European mountains.

4. Alpine Saxifrage, Saxifraga nivalis L., plant 5–20cm, with a basal leaf rosette; lvs 15–35mm long, obovate to spathulate, narrowed into the short, winged stalk, bluntly toothed, fleshy, usually red below; stem leafless, glandular above; inflorescence dense, head-like; flowers 5–10mm across; petals white, more rarely pink, rather longer than the sepals. Fl.7–8. Scree of siliceous rocks; wet rocks; rather rare; relict of the ice ages. N. Europe, Britain, the Sudeten, locally in C. European mountains.

5. Mossy Saxifrage, Saxifraga hypnoides L., plant consisting of a loose carpet of rosettes, 10–30cm; rosette lvs long-stalked, rather fleshy, divided into 3–9 linear, mucronate segments; stem curved-ascending, with 3–12 flowers; petals white, 3-veined, twice as long as the acute, glandular-hairy sepals; non-flowering shoots with bulbils in the axils. Fl.5–6. Siliceous rocks, walls, rock ledges, stony grassland; often planted; locally common. N.W. Europe, Britain, Vosges.

6. Purple Saxifrage, Saxifraga oppositifolia L., plant 2–6cm, forming low cushions; lvs elliptic, obtuse, blue-green, rather fleshy, ciliate, opposite, dense, 2–5mm long; flowers solitary, wine-red to violet. Fl.5–7. Calcareous scree, cracks in rocks, stony ground on mountains, about 1500–3500m in C. Europe, lower in the north; widespread. Britain, N. Europe, mountains of C. and S. Europe.

See also Saxifraga and Parnassia, Pl.106.

Rose Family, Rosaceae

7. Creeping Avens, *Geum reptans L., plant 5–15cm, with long stolons; lvs pinnate, the terminal leaflet little larger than the incised lateral lobes (in contrast to G. montanum L., with a large terminal leaflet and no stolons); flowers 3–5cm wide, 6–8-merous, yellow; calyx red-brown, hairy; the style growing out in fruit with feathery hairs, as an aid to wind dispersal. Fl.7–8. Damp siliceous screes, glacial moraines, at about 1500–3400m; widespread. Mountains of C. and S. Europe. See also G. rivale, Pl.106.

8. Golden Cinquefoil, *Potentilla aurea L., plant 5–20cm; basal lvs palmately 5-lobed, shining, lobes sharply serrate at tip; stipules lanceolate; stem lvs ternate; flowers golden-yellow. Fl.6–8. Stony grassland, at 1300–3000m; widespread. C. and S. Europe.

Plate 187 Rose Family, Rosaceae

1. Lax Cinquefoil, *Potentilla caulescens* Turn., plant 10–30cm, hairy; basal lvs palmately 5-lobed, 2–4cm long, each lobe with 2–5 teeth on each side, silkily ciliate on the margins; petiole 5–15cm long; flowers 15–20mm across, in dense or lax cymes; petals 5, white, hardly exceeding sepals; inner and outer sepals the same length. Fl.7–9. Calcareous rocks at about 900–2400m; widespread. Mountains of C. and S. Europe.

2. Pink Cinquefoil, *Potentilla nitida* L., shrub, scrambling over rocks, 2–5cm high, forming silvery-grey carpets, with woody, procumbent branches; lvs ternate, silky-hairy; flowers 2–3cm across; petals 5, obovate, emarginate, pinkish-red, twice as long as the sepals; anthers blackish-purple. Fl.7–8. Sunny rock outcrops, calcareous scree, at about 1500–3200m; scattered. W. Alps, and the calcareous S. Alps.

3. Mountain Avens, *Dryas octopetala* L., plant 2–10cm, creeping over rocks, much-branched; lvs oval, obtusely crenate, stalked, 1–3cm long, the upper side dark green, the underside white-felted; flowers solitary, 2–3cm across, long-stalked; petals 7–9, white; sepals 7–9, brown-felted. Fl.5–8. Calcareous scree, calcareous grassland, open pinewoods, at about 800–2500m; lower in the north or when swept down with river shingle; rare to common. N. Europe, Britain (rare), Alps, Pyrenees, Carpathians, Apennines, mountains of the Balkan peninsula, Caucasus.

4. Alpine Lady's-mantle, *Alchemilla alpina* L., plant 10–20cm; lvs divided palmately to the base in 5–9 narrow lobes; lobes green above, silvery silky-hairy on the margins and below, serrate at tip; flowers yellowish-green, about 3mm wide, in dense inflorescences, which are usually shorter than the lvs. Fl.6–8. Siliceous grassland, rock crevices, screes; calcifuge; scattered. Mountains of C. and N. Europe, Britain. Similar species: **Greater Alpine Lady's-mantle,** *A. conjuncta* Bab., but leaf lobes united from one-fifth to halfway; inflorescence as long or longer than the lvs; calcareous alpine meadows and rocks; widespread. Britain (rare), mountains of C. and S. Europe.

Pea Family, Fabaceae or Papilionaceae

5. Alpine Milk-vetch, *Astragalus alpinus* L., plant 7–25cm, prostrate to ascending; lvs imparipinnate, with 15–25 elliptic leaflets; leaflets obtuse, hairy on both sides at first; flowers 10–15mm long, 5–15 together in almost spherical racemes; standard bluish or violet, wings whitish, keel with a violet tip; pod villous, dark-coloured. Fl.7–8. Stony grassland, alpine meadows, exposed mountain ridges, at 500–2800m; widespread. N. Europe, Scotland (rare), Alps, Pyrenees, Carpathians. See *Astragalus*, Pl.60.

6. Purple Oxytropis, *Oxytropis halleri* Bunge, plant with a stout taproot, 10–25cm; lvs imparipinnate, densely villous-hairy on both sides, with 11–29 ovate-lanceolate leaflets, 10–15mm long; stipules scarious, net-veined, ciliate; flowers 6–16 together in head-like or spike-like racemes; corolla blue to red-violet, 15–20mm long, with a pale, shortly pointed keel. Fl.6–8. Dry, infertile mountain meadows, sea cliffs. Scotland (rare), Alps (absent from Germany), Pyrenees, Carpathians.

7. Yellow Oxytropis, *Oxytropis campestris* (L.)DC., plant 5–15cm; lvs basal, imparipinnate, hairy on both sides, grey-green, with 10–12 pairs of leaflets; flowers 10–18 in head-like, long-stalked racemes; calyx provided with long white hairs and short black ones; corolla yellowish-white, keel long-pointed, often with a violet spot on both sides. Fl.7–8. Stony alpine meadows, mountain ridges, rock ledges; rather rare. Alps, Apennines, Carpathians, Pyrenees, Balkan mountains, Scotland, Scandinavia.

Plate 188 Pea Family, Fabaceae or Papilionaceae

1. Alpine Sainfoin, **Hedysarum hedysaroides* (L.)Sch. et Thell., plant 10–30cm, erect or ascending; lvs imparipinnate, with 11–19 elliptic leaflets, dark green above and pale green below; flowers purple-red, 15–20mm long, 12–35 together in a secund raceme; pod compressed, 2–4cm long, falling apart when ripe into 2–6 1-seeded sections. Fl.7–8. Mountain meadows, dwarf shrub heaths, at 1600–2800m; widespread. Mountains of C. and S. Europe. See also *Trifolium badium*, Pl.59.

Violet Family, Violaceae

2. Yellow Wood Violet, **Viola biflora* L., stem 8–12cm, with 2 pale green, broadly kidney-shaped, crenate lvs and 1–2 yellow flowers with a spur 2–3mm long. Fl.5–8. Damp screes, shady places on loamy soils; widespread. Mountains of C. and S. Europe.

Carrot Family, Umbelliferae or Apiaceae

3. Alpine Lovage, **Ligusticum mutellina* (L.)Crantz, plant aromatically scented, 10–50cm; lvs 2–3 pinnate, with scarious leaf sheaths; umbels 7–10 rayed; bracteoles 3 to many; flowers white, pink to purple. Fl.6–8. Mountain meadows and shrub communities, to 2800m; widespread. Mountains of C. and S. Europe. See also *Meum*, Pl.64.

Heath Family, Ericaceae

4. Trailing Azalea, *Loiseleuria procumbens* (L.)Desv., prostrate, creeping, mat-forming shrub; lvs leathery, evergreen, narrowly elliptic, entire, opposite, 4–7mm long, and with inrolled margins; flowers 2–5 together; corolla 5-lobed, campanulate, pink; sepals 5, red; anthers 5, purple. Fl.6–8. Siliceous scree, mountain ridges, high-level moors, about 1600–3000m (down to 400m in Scotland); widespread. N. Europe, Alps, Pyrenees, Carpathians.

5. Dwarf Alpenrose, **Rhodothamnus chamaecistus* (L.)Rchb., dwarf shrub, 10–30cm, with evergreen, leathery, narrowly elliptic, finely ciliate lvs; flowers usually in pairs, long-stalked, pinkish-red. Fl.5–7. Protected! Calcareous rocks, stony grassland, Mountain Pine scrub; scattered, at 1000–2400m. Mountains of C. and S. Europe.

6. Alpenrose, **Rhododendron ferrugineum* L., dwarf shrub, up to 1m, lvs evergreen, rough, elliptic to oblong, inrolled at the margin, not ciliate, dark green above, the underside covered with yellow-brown glandular scales, which turn rusty-brown later. Fl.6–7. Protected! Dwarf shrub heaths, woods near the tree line, calcifuge, about 1500–2800m; widespread. Mountains of C. and S. Europe.

7. Hairy Alpenrose, **Rhododendron hirsutum* L., dwarf shrub, up to 1m, with evergreen, elliptic, long-ciliate lvs; flowers 3–10 together, calyx teeth lanceolate, acute; corolla funnel-shaped to campanulate, pale red, hairy inside. Fl.7–8. Protected! On calcareous soils, dwarf shrub heaths, woods at the tree line, 1200–2600m. Alps.

Diapensia Family, Diapensaceae

8. Diapensia, *Diapensia lapponica* L., evergreen cushion-plant, 2–6cm; lvs leathery, spathulate, 5–10cm long, in dense rosettes; flowers solitary, on peduncles 5–40mm long; corolla 5-merous, white, 10–15mm wide; style 3-lobed. Fl.5–6. Cracks in rocks, stony mountain pasture. N. Europe, Scotland (very rare).

Plate 189   Primrose Family, Primulaceae

1. Alpine Snowbell, *Soldanella alpina* L., a dainty plant, 5–15cm; lvs basal, roundish and kidney-shaped, 1–3cm wide, with a basal sinus and prominent veins above; inflorescence 2–3 flowered, leafless; corolla 8–15mm long, blue-violet, rarely whitish, funnel-shaped, split to halfway with many linear lobes. Fl.4–6. Calcareous soils, where snow lies late, flushes, 1000–3000m; widespread. Mountains of C. and S. Europe.

2. Dwarf Snowbell, *Soldanella pusilla* Baumg., plant 4–8cm; lvs thin, roundish and kidney-shaped, under 1cm wide; flowers solitary; corolla narrowly campanulate, 10–15mm long, pale violet, split to one-quarter of the way down. Fl.5–8. Non-calcareous soils, where snow lies late, damp screes, poor grassland, about 1500–3000m; scattered. Mountains of C. and S. Europe (eastern). Similar species: **Least Snowbell,** *S. minima* Hoppe, but lvs rather thick, orbicular, without a basal sinus; flowers pale lilac to whitish; flushes, damp screes; rare. E. Alps, Apennines.

3. Alpine Rock-jasmine, *Androsace alpina* (L.)Lam., plant 2–5cm, in loose cushions; lvs thick, lanceolate, rather blunt, 3–10mm long, with stellate hairs; flowers solitary, 5–6mm wide, on stalks 2–12mm long, white to pink, throat yellow; corolla lobes not emarginate. Fl.7–8. Damp siliceous screes, moraines, 2000–4200m; widespread. Alps.

4. Milkwhite Rock-jasmine, *Androsace lactea* L., rosette plant, loosely tufted, 5–15cm; lvs linear, acuminate, 1–2cm long, slightly ciliate; flowers in a 2–6 flowered umbel; corolla 8mm wide, white, with a yellow throat; corolla lobes emarginate. Fl.5–7. Calcareous rock outcrops, stony alpine meadows, about 1600–2200m; scattered. Mountains of C. and S. Europe (eastern).

5. Sticky Primrose, *Primula glutinosa* Wulf., plant 2–8cm; lvs ovate-oblong, usually toothed at tip, gradually narrowed into the short, winged stalk, dark green, very sticky; flowers 2–7, scented, dark blue, later violet; pedicels 1–2mm long; bracts brownish-red. Fl.7–8. Protected! Poor siliceous grassland, stable scree, 1800–3600m; rare. E. Alps.

6. Least Primrose, *Primula minima* L., plant 1–4cm; lvs 10–15mm long, shining, glabrous, cuneate at base, truncate with large, thickened saw-like teeth at the tip; inflorescence 5–15mm long, usually 1-flowered; corolla red, becoming paler later, 1–2cm across, throat white, glandular-hairy; corolla lobes deeply bifid. Fl.6–7. Protected! Poor, siliceous grassland, damp scree, cracks in rocks, about 1200–3000m; scattered. Mountains of C. and S. Europe.

7. Bear's-ear, *Primula auricula* L., plant 5–30cm; lvs thick, fleshy, grey-green, mealy pruinose, 5–12cm long; flowers 8–15mm wide, yellow, 4–12 together in a secund umbel. Fl.4–7. Protected! Calcareous rocks, rocky outcrops, stony grassland, rarely on level moorland, about 1500–2500m; widespread. Mountains of C. and S. Europe.

8. Hairy Primrose, *Primula hirsuta* All., plant 3–10cm, densely set with sticky glands; lvs usually coarsely serrate, rather fleshy, oval, suddenly narrowed into the stalk, 3–6cm long; flowers 1–2cm across, pink; pedicels 3–15mm long; corolla lobes deeply emarginate; corolla tube glandular-hairy. Fl.4–7. Protected! Cracks in siliceous rocks, stable scree, about 1200–3600m. Alps, Pyrenees.

9. Marginate Primrose, *Primula marginata* Curt., plant 5–20cm, with a stout root-stock; lvs fleshy, white-bordered, sharply serrate, ovate, narrowed into the short stalk, 3–8cm long, glabrous; flowers 3–10 together, pink to violet, 15–25mm wide. Fl.4–7. Cracks in calcareous rocks, rock outcrops, about 1000–2600m; scattered. W. Alps.

Plate 190   Gentian Family, Gentianaceae

1. Stemless Gentian, *Gentiana clusii* Perr. et Song., plant 5–10cm; rosette lvs ovate-lanceolate, acute, 2–5cm long, widest at or below the middle, shining; calyx campanulate, 5-lobed; calyx lobes lanceolate, acute, appressed, at least half as long as the corolla tube; sinuses between the calyx lobes acute; corolla blue, campanulate, without green spots inside. Fl.5–8. Protected! Calcareous grassland, moorland pasture, from the valleys up to 2800m; scattered. Mountains of C. and S. Europe.

2. Trumpet Gentian, *Gentiana acaulis* L. (*G. kochiana* Perr. et Song.), plant 5–10cm, similar to *G. clusii*, but rosette lvs obovate to elliptic, obtuse, widest in the upper third; calyx lobes spathulate, somewhat spreading, shorter than half the corolla tube; sinuses between the calyx lobes broad; corolla blue, with olive-green spots inside. Fl.6–8. Protected! Non-calcareous, acid alpine meadows, stony grassland, at about 1200–3000m; scattered. Mountains of C. and S. Europe.

3. Brown Gentian, *Gentiana pannonica* Scop., plant 30–60cm; stem erect, purple-tinged above; lvs opposite, elliptic, 5–7 veined, the lower stalked, the upper sessile; calyx with 5–8 reflexed lobes; corolla dull purple, yellow-green with blackish-red spots at the base, yellowish inside. Fl.7–9. Protected! Non-calcareous, poor grassland, at about 1600–2300m; rare. E. Alps, Bohemia.

4. Great Yellow Gentian, *Gentiana lutea* L., plant 40–140cm, with a very thick rootstock (from which a liqueur is made, and because of which this and other related species have been almost eradicated by collecting in some areas); stem with opposite, elliptic, bluish-green lvs, with prominent, curved veins; flowers 3–10 together, short-stalked, in the axils of bowl-shaped bracts; corolla rotate, 5–6 lobed to the base, golden-yellow. Fl.7–8. Protected! Alpine meadows, rocky slopes; at about 1000–2500m; scattered. Mountains of C. and S. Europe.

5. Spotted Gentian, *Gentiana punctata* L., stem 20–60cm, often with a metallic shine in the upper part; lvs ovate-oblong, pointed; flowers usually several together in the leaf axils, pale yellow, usually spotted with dark violet; calyx with 5–8 unequal teeth. Fl.7–8. Protected! Poor grassland, stony slopes, calcifuge, at about 1400–3000m; scattered. Mountains of C. and S. Europe.

6. Bavarian Gentian, *Gentiana bavarica* L., plant 4–20cm, mat-forming, densely leafy; lvs obovate, obtuse, almost all the same size; calyx tubular, very narrowly winged; corolla deep blue, with a paler tube. Fl.7–9. Protected! Damp alpine meadows, flushes, places where snow lies late, calcifuge, at about 1800–2600m; widespread. Alps, Carpathians. Ssp. *subacaulis* Cust., a high alpine form, has shorter stems with imbricate lvs, which are nearly round and which become smaller below; in the Alps to 3600m, on damp siliceous screes.

7. Alpine Gentian, *Gentiana nivalis* L., plant 2–15cm, dainty, branched; basal lvs in a rosette, small, obtuse; stem lvs ovate, acute; flowers at the ends of the branches, solitary, deep blue, 10–15mm across; calyx keeled or narrowly winged, calyx teeth lanceolate, acute. Fl.6–8. Protected! Calcareous stony grassland, fine scree, rock ledges, at about 1700–3000m; scattered. N. Europe, Scotland (rare), mountains of C. and S. Europe.

See also Pl.65.

Plate 191　Borage Family, Boraginaceae

1. Alpine Forget-me-not, *Myosotis alpestris* Schmidt, plant 5–20cm, roughly hairy; stalks of the rosette lvs clearly distinct from the ovate-oblong leaf blades (in *M. sylvatica*, Pl.170, the leaf and blade grade into one another); inflorescence rather dense; calyx with appressed, curved and hooked hairs; corolla sky-blue, with yellow scales at the throat. Fl.6–7. Mountain meadows, stony slopes; widespread, at 1600–3000m in C. Europe, lower in north. Mountains of N., C. and S. Europe, N. Britain.

2. King of the Alps, *Eritrichum nanum* (Amann.)Schrad., cushion-plant, 2–5cm, silkily shining, densely hairy; lvs ovate, 5–10mm long; flowers 3–6 together; each flower with a bract; corolla 5–8mm wide, sky-blue with yellow scales in the throat. Fl.7–8. Protected! Cracks in siliceous rocks, stable scree, at about 2400–3500m; scattered. Alps. See also *Cerinthe*, Pl.26 and *Onosma*, Pl.67.

Figwort Family, Scrophulariaceae

3. Leafless-stemmed Speedwell, *Veronica aphylla* L., rosette plant, 2–8cm; lvs broadly ovate, 10–15mm long, ciliate on the margin; flowers 2–4 together in a long-stalked, head-like raceme; calyx 4-lobed, glandular-hairy; corolla lilac or deep blue with darker veins, 6–8mm across; fruiting capsule glandular-hairy. Fl.6–8. Stony grassland, rocks, calcareous scree, at 1000–2800m; widespread. Mountains of C. and S. Europe.

4. Rock Speedwell, *Veronica fruticans* Jacq., plant 5–15cm, slightly woody, branched from the ground up; lvs oblong-elliptic, 1–2cm long, slightly crenate, shining, almost glabrous; flowers in 4–6 flowered racemes; calyx 4-lobed, hairy; corolla deep blue, with a purple ring in the throat; fruiting capsule ovoid, scarcely emarginate. Fl.6–8. Rock outcrops, stony alpine meadows, at about 800–2800m; widespread. N. Europe, Scotland (rare), mountains of C. and S. Europe. See also *V. serpyllifolia*, Pl.33.

5. Alpine Toadflax, *Linaria alpina* (L.)Mill., plant 5–15cm; stems prostrate, ascending at the tips; lvs fleshy, bluish-green, glabrous, narrowly lanceolate, 8–15mm long; flowers blue-violet, with a yellow spot on the palate, more rarely uniformly blue-violet, 1–2cm long, with a long spur. Fl.6–8. Calcareous scree, rocks, at 1200–3800m, also lower on shingle of alpine rivers; widespread. Mountains of C. and S. Europe.

6. Beaked Lousewort, *Pedicularis rostrato-capitata* Crantz, plant 5–20cm; lvs lanceolate, 2-pinnatisect, 3–10cm long; basal lvs often violet-tinged; flowers in short racemes; calyx tubular-campanulate, with leaf-like, crenate lobes; corolla purple-red, 15–25mm long; upper lip curved over into a long, straight beak. Fl.6–8. Calcareous, stony alpine meadows, at 1600–2800m; scattered. E. Alps. See also *Bartsia*, Pl.116.

Mint Family, Labiatae or Lamiaceae

7. Alpine Calamint, *Acinos alpinus* (L.)Moench (*Calamintha alpina* (L.)Lam.), plant 10–30cm; stem prostrate to ascending, slightly hairy; lvs ovate, opposite, shortly stalked, serrate towards the tip; flowers in whorls of 3–6, 10–18mm long, red-violet. Fl.6–9. Calcareous grassland, pinewoods, to 2500m; widespread. C. and S. Europe.

8. Pyramidal Bugle, *Ajuga pyramidalis* L., plant 10–20cm; lvs decussate, densely set, becoming smaller above, the lower in a rosette, obovate, slightly crenate, 5–10cm long, the upper often violet-tinged; flowers in whorls of 2–4; corolla pale violet-blue, 10–18mm long. Fl.7–8. Non-calcareous mountain meadows, from almost sea level in the north to 2700m; widespread. N. Europe, mountains of C. and S. Europe.

Plate 192 Globularia Family, Globulariaceae

1. Matted Globularia, *Globularia cordifolia* L., plant 3–10cm, branched, prostrate, woody at the base; lvs in rosettes at the ends of the prostrate shoots, spathulate to obovate, cordate-emarginate at the tip, gradually narrowed into the stalk, leathery; flowering stems with 0–2 scale lvs; flowers in heads 10–15mm wide; corollas of the florets 6–8mm long, blue. Fl.5–6. Sunny, calcareous screes, rock outcrops and crevices, from the valleys up to 2800m; widespread. Mountains of C. and S. Europe.

2. Leafless-stemmed Globularia, *Globularia nudicaulis* L., plant 5–25cm, similar to *G. cordifolia*, but herbaceous, forming solitary rosettes; lvs obovate, rounded at tip, almost as long as the stem; flower heads 15–25mm wide; corollas 10–12mm long, blue. Fl.5–8. Stony, calcareous grassland, calcareous scree, from the valleys to 2600m; widespread. Mountains of C. and S. Europe (western). See also *Jasione*, Pl.72.

Bellflower Family, Campanulaceae

3. Devil's-claw, *Physoplexis comosa* (L.)Schur (*Phyteuma comosa* L.), plant 5–15cm; basal lvs kidney-shaped, deeply and unequally serrate, stalked; stem lvs obovate to lanceolate, sharply serrate; flowers 15–30mm long, 8–20 together in a spherical umbel, corolla swollen at the base, pale lilac, ending in a long blue-violet beak; petals remaining united both at the base and at the tip. Fl.6–8. Rock crevices on limestone and dolomite, 1000–2000m; rare. S. Alps. See also *Phyteuma orbiculare*, Pl.72.

4. Yellow Bellflower, *Campanula thyrsoides* L., plant 10–40cm, erect, densely leafy, roughly hairy; lvs lanceolate, tapered to the base; flowers numerous, in club-shaped spikes; corolla funnel-shaped to campanulate, 15–25mm long, yellowish, hairy. Fl.7–9. Sunny alpine meadows and pastures, about 1500–2600m; scattered. Mountains of C. and S. Europe.

5. Bearded Bellflower, *Campanula barbata* L., plant 10–40cm, stiffly hairy; basal lvs in a rosette, oblong-lanceolate, roughly hairy; flowers shortly stalked, nodding, 2–12 together in a secund raceme; calyx villous-hairy, sinuses between the calyx teeth with reflexed appendages; corolla inflated-campanulate, 15–30mm long, pale blue, the lobes bearded within. Fl.6–8. Acid, poor grassland and dwarf shrub heaths, at about 1200–2800m; widespread. Alps, Sudeten, W. Carpathians, S. Norway.

6. Scheuchzer's Bellflower, *Campanula scheuchzeri* Vill., plant 5–40cm, forming a loose mat; basal lvs long-stalked, roundish and kidney-shaped, crenate, usually withered by flowering time; stem lvs linear-lanceolate, almost sessile, serrate to entire; flowers nodding, broadly campanulate, 15–25mm long, dark blue-violet, calyx teeth linear. Fl.7–8. Non-calcareous, poor grassland, stony alpine pastures, at about 1400–3100m; widespread. Mountains of C. and S. Europe.

7. Fairy's Thimble, *Campanula cochleariifolia* Lam. (*C. pusilla* Haenke), plant 5–20cm, mat-forming; lower lvs long-stalked, roundish-cordate, coarsely toothed, upper lvs narrowly lanceolate, sessile; flowers solitary or in few-flowered, secund racemes, inflated-campanulate, blue, 10–20mm long. Fl.7–9. Rock outcrops, calcareous scree, river shingle, from the valleys to 3000m; common. Mountains of C. and S. Europe. See also *C. rotundifolia*, Pl.72.

Teasel Family, Dipsacaceae. See *Scabiosa lucida*, Pl.71

Plate 193 Daisy Family, Compositae or Asteraceae

1. Yellow Genipi, **Artemisia umbelliformis* Lam. (*A. laxa* (Lam.)Fritsch), plant 10–20cm, aromatically scented; lvs silvery-shining, the lower 2-ternate, the upper digitately divided, segments under 1mm wide; flower heads 5–20 together, arranged in a spike-like raceme, 4–6mm across, 30–40 flowered; flowers yellow. Fl.7–9. Protected! Crevices in siliceous rocks, stony grassland, at about 1600–3700m; rather rare. Mountains of C. and S. Europe. Similar species: **Genipi,** **A. genipi* Web. (*A. spicata* Wulf.), plant 5–10cm, with grey silky-hairy, 2–3 times ternately divided basal lvs; stem lvs sessile, once pinnate; flower heads almost spherical, in secund spikes, which droop at first and are later erect; bracts felted, with a blackish-brown border; siliceous scree, at about 2200–3800m; Alps. Protected!

2. Dark-stemmed Sneezewort, **Achillea atrata* L., plant 10–25cm; lvs once pinnatisect, with 2–3 lobed segments; flower heads 12–18mm wide, 3–12 together; ray florets 7–12, white, disc florets yellow; bracts black-bordered. Fl.7–9. Calcareous scree, at 1700–4200m; widespread. E. Alps. The **Dwarf Milfoil,** **Achillea nana* L., is found on siliceous scree, plant 5–10cm, with a curious smell; lvs woolly-villous, oblong in outline, pinnate; segments 3–5 lobed; flower heads 10mm wide; bracts villous, with a black scarious border; flowers dirty white; 1700–3800m; Alps. See *A. tomentosa*, Pl.73.

3. Saw-leaved Moon-daisy, **Leucanthemum atratum* (Jacq.)DC. (*Chrysanthemum atratum* Jacq.), plant 10–30cm, single-headed; lvs rather fleshy, glabrous, dark green, sharply serrate-dentate; flower head 3–5cm wide; ray florets white, disc florets yellow; bracts black-bordered. Fl.7–9. Calcareous scree, at 1500–2600m; widespread. E. Alps. Similar species: **Alpine Moon-daisy,** **Leucanthemopsis alpina* (L.)Heywood (*Chrysanthemum alpinum* L.), but basal lvs pinnatisect like the teeth of a comb; stem lvs linear, entire or toothed; slightly or non-calcareous, fine screes, at 1800–2800m. Alps, Pyrenees, Carpathians.

4. Alpine Fleabane, **Erigeron alpinus* L., plant 5–20cm; lvs spathulate, appressed-hairy, gradually narrowed into the stalk; stems usually 1-headed; flower heads 15–30mm across, ray florets in several rows; between the pinkish-red ray florets and the yellow, hermaphrodite, central disc florets are slender, female, disc florets; bracts green, hairy. Fl.7–9. Stony grassland, alpine pastures, about 1500–2500m; rather rare. Mountains of C. and S. Europe. Similar species: *E. borealis* (Vierh.)Simm., but basal lvs glabrous or sparsely pubescent; N. Europe, Scotland.

5. Alpine Aster, **Aster alpinus* L., plant 5–15cm, hairy, usually 1-headed; rosette lvs spathulate, narrowed into the short stalk, 3-veined; stem lvs lanceolate, obtuse, sessile; flower heads 3–5cm wide; involucre 8–12mm long, with several rows of bracts; ray florets violet, in 1 row; disc florets yellow. Fl.7–8. Mountain meadows, rock outcrops, calcareous stony grassland, at about 1400–3100m; widespread. N. Europe, Alps, Pyrenees, Harz mountains, Bohemia, Carpathians, Balkan mountains.

6. Edelweiss, **Leontopodium alpinum* Cass., plant 5–10cm, densely woolly-felted; lvs lanceolate, gradually narrowed at the base, felted; inflorescence consisting of 5–10 hemispherical, yellowish heads in an umbel; heads 5–6mm wide, surrounded by 5–15 densely white-felted bracts, which spread like a star and resemble the petals of a flower. Fl.7–9. Protected! Rock outcrops, calcareous, sunny, stony grassland, at about 1700–3400m; scattered to rare. Mountains of C. and S. Europe.

See also *Antennaria carpatica*, Pl.73.

Plate 194   Daisy Family, Compositae or Asteraceae

1. Large-flowered Leopard's-bane, *Doronicum grandiflorum* Lam., plant 15–50cm; stem densely glandular, densely leafy below; lower lvs stalked, broadly ovate, truncate at base or cordate, coarsely sinuate-dentate; upper lvs cordate-amplexicaul; flower heads usually 1, 4–7cm across, yellow; achenes 2mm long, with 10 ribs. Fl.7–8. Calcareous screes, rock outcrops, stony alpine pastures, at about 1500–3100m; widespread. Mountains of C. and S. Europe. Similar species: **Glacier Leopard's-bane,** *D. glaciale* (Wulf.)Nym., but basal lvs oblong, narrowed into the stalk, stiffly ciliate on the margins and with short glands; scree with trickling water. E. Alps. **Heart-leaved Leopard's-bane,** *D. columnae* Ten., but cordate-triangular, long-stalked, regularly toothed lvs; scrubby, stony alpine pastures of the E. and S.E. Alps.

2. Chamois Ragwort, *Senecio doronicum* L., plant 20–50cm; stem arachnoid-woolly; lvs leathery, rough, the lower oblong-ovate, stalked, coarsely toothed, the upper lanceolate, sessile; flower heads 3–6cm wide, yellow, 1–5 together, usually long-peduncled; bracts woolly, surrounded by an outer involucre; achenes 5–6mm long, 10–12 ribbed. Fl.7–8. Calcareous, stony grassland, alpine meadows, dwarf shrub communities, at about 1600–3100m; widespread. Mountains of C. and S. Europe. See also *Arnica*, Pl.74.

3. Fox and Cubs, *Hieracium aurantiacum* L., plant 20–50cm, with stolons; rosette lvs ovate-lanceolate, gradually narrowed at the base; stem lvs 1–4, rapidly diminishing in size; flower heads 2–3cm across, 2–12 together in short corymbs, with yellow-orange to brownish-red ray florets only; bracts narrow, with black glands. Fl.6–8. Poor grassland, pastures, roadsides, about 1200–2400m in C. Europe, where it is wild, to sea level in the north as a garden escape; widespread. Britain, N. Europe, Germany, mountains of C. and S. Europe. See also *Crepis aurea*, Pl.86.

4. Woolly Hawkweed, *Hieracium villosum* Jacq., plant 10–30cm, villous white-hairy; rosette lvs oblong-lanceolate, entire, blue-green; stem lvs 1–4, sessile with rounded bases; flower heads 2–4cm wide, with pale yellow ray florets only; bracts acute, roughly hairy. Fl.7–8. Stony grassland, calcareous meadows, rock outcrops, at about 1200–2800m; widespread. Mountains of C. and S. Europe (eastern).

See also *Carlina*, Pl.75, *Scorzonera*, Pl.76, *Crepis pyrenaica*, Pl.120.

2

3

1

4

Plate 195   Grass Family, Gramineae or Poaceae

1. Townsend's Cord-grass, *Spartina townsendii* H.et J.Grev., plant 30–130cm, with far-creeping rhizomes; inflorescences 10–25cm long, with 3–6 slender, erect spikes; spikelets 12–20mm long, arranged in 2 rows, closely appressed to the spike axis, 1-flowered; lvs stiff, flat, 5–8mm wide. Fl.7–8. (Hybrid derived from *S. maritima* (Curt.) Fern. and *S. alterniflora* Lois.) Tidal, marine mudflats, often planted. S. England and North Sea coasts.

2. Lyme-grass, *Leymus arenarius* (L.)Hochst. (*Elymus arenarius* L.), plant 60–120cm, a striking blue-green, with long rhizomes; lvs stiff, with sharp points, inrolled, rough above, 8–20mm wide; ligules very short; spikes dense, 10–30cm long; spikelets 2–3cm long, usually with 2 short-stalked lateral spikelets, these 'triplets' of spikelets sessile and alternate. Fl.6–8. Sand dunes; common on European coasts, rare or planted inland. Similar species: **Sand Couch,** *Elymus farctus* (Viv.)Runemark (*Agropyron junceiforme* A.et D.Löve), but spikelets without lateral spikelets, 15–30mm long, 5–8 flowered; spike 2-rowed, stiff, 5–20cm long; spike axis very brittle; plant 30–60cm, blue-green; lvs 8mm wide, rough above, flat at first, later much inrolled; coasts of Europe; widespread. Similar to the last is **Sea Couch,** *Elymus pungens* (Pers.) Melderis (*Agropyron p.* (Pers.) R.et S.), but spikelets 10–18mm long, spike axis not brittle; lvs glabrous above. Sand dunes and saltmarshes. Britain, W. and S. Europe.

3. Common Saltmarsh-grass, *Puccinellia maritima* (Huds.)Parl., plant 20–60cm, with long stoloniferous shoots in spring; lvs inrolled, fleshy, rush-like, smooth; spike branches smooth, somewhat spreading, later all erect, the lowest usually in pairs; spikelets 5–9 flowered, oblong, 5–10mm long, often violet. Fl.6–9. Saltmarshes and muddy estuaries; widespread. North Sea, Baltic and Atlantic coasts. Similar species: **Reflexed Saltmarsh-grass,** *P. distans* (Jacq.)Parl., but stolons absent, spike branches rough, branches horizontally spreading, later reflexed, the lower usually 4–5 together; spikelets 4–5 flowered, 4–5mm long; saltmarshes, coastal spray zone, waste places inland (rarely); scattered to rare. European coasts.

4. Marram, *Ammophila arenaria* (L.)Link, plant 60–100cm; lvs grey-green, stiff, acute, usually inrolled, ribbed above; ligule 1–2cm long; spike dense, cylindrical, whitish, 10–20cm long; spikelets 10–15mm long, with fine hairs within. Fl.6–8. Sand dunes; common on the coasts of Europe; occasionally planted inland.

5. Sea Barley, *Hordeum marinum* Huds. (*H. maritimum* With.), plant 10–40cm, tufted and branched at the base, grey-green; stem geniculate, ascending, leafy almost to the tip; lower leaf sheaths glabrous or softly hairy, the upper inflated; spikes 4–6cm long; axis of spikes brittle; spikelets usually in 3s (triplets); glumes rough, narrowly lanceolate or bristle-like; lemma with a long awn. Fl.5–7. Grassy places near the sea; widespread on the North Sea coast and the coasts of W. and S. Europe, local in Britain; rarely inland in waste places.

1

2

3

4

5

Plate 196 Tasselweed Family, Ruppiaceae

1. Beaked Tasselweed, *Ruppia maritima* L., plant submerged, 15–40cm, with filiform stems which root at the nodes, and lvs 1mm wide, in 2 rows, sheathing at the base; flowers inconspicuous, without bracts, in a 2-flowered spike; stamens 2. Fl.6–10. Coastal and brackish water, salty inland waters; scattered to rare. Europe.

Sea-grass Family, Zosteraceae

2. Eelgrass, *Zostera marina* L., plant submerged in the sea (except sometimes at low tide), 30–100cm, with grass-like, 3-veined lvs, 4–10mm wide; inflorescence 9–12cm long, flowers small, bractless, 1 male and 1 female flower enclosed in a sheath. Fl.6–9. Coastal seawater, down to 10m, often over large areas; widespread. Europe.

Arrow-grass Family, Juncaginaceae

3. Sea Arrow-grass, *Triglochin maritimum* L., plant 15–70cm; lvs linear, semi-circular in cross-section, basal; flowers in a raceme, dense above; perianth segments 6, green, 2–3mm long; styles 6; fruit ovoid, falling apart into 6 sections. Fl.6–8. Grassy places by the sea, saltmarsh turf; widespread on the coasts of Europe, rare inland.

Rush Family, Juncaceae

4. Saltmarsh Rush, *Juncus gerardi* Loisel., plant 15–50cm, with a terete stem, and linear lvs 1mm wide; inflorescence terminal, with short bracts; each flower with 2 scarious bracteoles; perianth segments 6, dark brown, with a green midrib and a white margin, 2–3mm long. Fl.6–7. Saltmarshes, salty places; scattered and locally common on coasts, rare inland. Europe. Protected!

Sedge Family, Cyperaceae

5. Sea Club-rush, *Scirpus maritimus* L. (*Bolboschoenus maritimus* (L.) Pall.), plant 30–100cm; stem sharply triquetrous, rough; lvs flat, keeled, 3–8mm wide; bracts leaf-like, much longer than the inflorescence; spikes 1–2cm long, red-brown, arranged in heads. Fl.6–8. River banks, ditches, ponds near the sea, in nutrient-rich, often salty water; locally common to rare. Europe.

6. Sand Sedge, *Carex arenaria* L., plant 15–40cm, with far-creeping rhizomes; lvs stiff, channelled, 3–4mm wide; inflorescence 4–6cm long, consisting of 6–16 similar-looking spikes, the lower female, the upper male, those in between male at base and female at tip; fruits 4–5mm long, yellowish, winged; styles 2. Fl.5–6. Sand dunes, dry sandy grassland, dry pinewoods; widespread on coasts of Europe, rare inland.

Goosefoot Family, Chenopodiaceae

7. Sea-purslane, *Halimione portulacoides* (L.)Aellen (*Obione p.*(L.)Moq.), plant 30–80cm, woody at base, prostrate to ascending; lvs opposite, clustered below, oblong-ovate, thick, fleshy, pruinose; flowers greenish, in axillary and terminal compound spikes, unisexual; bracteoles sessile in fruit, 3-lobed, 3–4mm long. Fl.7–9. Salt-marshes that are flooded at high tide; coasts of Europe; widespread. Similar species: **Stalked Orache,** *H. pedunculata* (L.)Aellen, but plant herbaceous; lvs alternate, lower lvs opposite; bracteoles long-stalked in fruit, 3-lobed, with a very small central lobe; salty soils, also inland; scattered to rare. Britain (extinct?), W. Europe. Protected!

Plate 197   Goosefoot Family, Chenopodiaceae

1. Grass-leaved Orache, *Atriplex litoralis* L., plant 30–80cm, at first grey and mealy; branches rather brush-like, erect; lvs linear to linear-lanceolate, cuneately narrowed at base, entire or sinuate-dentate; the 2 bracteoles of each flower united only at the base during fruiting, ovate to triangular-rhomboid, often toothed. Fl.7–9. Salty, sandy and muddy soils near sea; widespread on European coasts; rarely inland. **Sea Beet,** *Beta vulgaris* L., is usually perennial and decumbent; lvs thick, leathery, rhombid, dark green; flowers green in small clusters. Sea shores; widespread. Coasts of Europe.

2. Prickly Saltwort, *Salsola kali* L., plant 15–60cm, grey-green, usually much-branched; branches stiff and spreading; lvs fleshy, terete, prickly pointed, mucronate; flowers solitary, axillary, green, with 2 acute bracts; perianth segments 5, transversely thickened about middle. Fl.7–9. Sandy shores, grassy places near sea; widespread.

3. Annual Sea-blite, *Suaeda maritima* (L.)Dum., plant 10–40cm, glabrous, blue-green, often red-tinged; lvs linear, fleshy, not mucronate; flowers small, 2–3 together in the axils; perianth segments 5, green. Fl.7–9. Saltmarshes and seashores, on muddy soils, widespread on the European coasts; rare inland on salty soils.

4. Glasswort, *Salicornia europaea* L., plant 5–30cm, very variable, fleshy, green, red in the autumn; stem usually much-branched, segmented; lvs opposite, fleshy, closely appressed to stem, which thus appears leafless; flowers minute, buried in stems in terminal inflorescences. (Many similar species which are hard to tell apart.) Fl.8–10. Saltmarshes, mudflats on European coasts, common; rare inland on salty soils.

Pink Family, Caryophyllaceae

5. Sea Sandwort, *Honkenya peploides* (L.)Ehrh., plant 5–25cm, with creeping stems which root at the nodes; lvs fleshy, sessile, oval, densely set in 4 rows, yellowish-green, 5–20mm long; flowers unisexual, white, 6–10mm across. Fl.6–7. Sandy shores and damp sand dunes on the coasts of N. and W. Europe; common to scattered.

6. Greater Sea-spurrey, *Spergularia media* (L.)C.Presl., plant 5–30cm; lvs fleshy, linear, 10–15mm long; flowers 8–12mm wide; petals pale pink, fading gradually to white at the base; stamens 10; seeds with a broadly winged margin. Fl.7–9. Salty mud and sand, saltmarshes of the coasts of Europe; scattered, rarely inland. Protected!

7. Lesser Sea-spurrey, *Spergularia marina* (L.)Griseb. (*S. salina* J. et C. Presl.), similar to *S. media*, but flowers 6–8mm across; petals deep pink, suddenly white at the base; stamens 2–5. Fl.5–9. Coasts of Europe, widespread, rarely inland. Protected!

Poppy Family, Papaveraceae

8. Yellow Horned-poppy, *Glaucium flavum* Crantz, plant bluish-green, pruinose, 20–80cm; lvs pinnatisect, 6–10cm long; petals 4, yellow, 2–3cm long; fruit a curved, rough capsule, 15–30cm long. Fl.6–7. Dry shingle banks near the sea; widespread in Britain, S. and W. Europe, otherwise rare, northwards to S. Sweden.

Cabbage Family, Cruciferae or Brassicaceae

9. Sea Rocket, *Cakile maritima* Scop., plant 15–60cm; lvs fleshy, blue-green, deeply pinnatisect; flowers in bractless racemes, scented, 1–2cm across; petals 6–14mm long, twice as long as the calyx, violet, pink or whitish; siliqua 10–25mm long, in 2 sections, the upper section much the longest. Fl.7–10. Seashores; widespread. Europe.

Plate 198   Cabbage Family, Cruciferae or Brassicaceae

1. Common Scurvy-grass, *Cochlearia officinalis* L., plant 5–30cm; lvs fleshy; lower lvs broadly cordate or kidney-shaped; flowers 8–10mm across; petals white, 4–9mm long; siliculae ovoid to spherical, 4–7mm long. Protected! Fl.5–8. Formerly eaten as a source of vitamin C, especially for sailors. Saltmarshes, grassy places by the sea, cliffs; scattered, also sometimes inland. N. and W. Europe. Similar species: **Alpine Scurvygrass,** *C. pyrenaica* DC. (*C. alpina* Bab.), but lvs not fleshy, fruit ovoid, tapered at both ends; flushes, rock ledges in mountains; scattered; Britain, C. Europe.

2. Danish Scurvy-grass, *Cochlearia danica* L., plant 5–10cm; lower lvs cordate; stem lvs stalked, 3–7 lobed, like ivy lvs; flowers 3–5mm wide, white or pink; silicula 3–6mm long, ellipsoid, finely net-veined. Fl.5–6. Saltmarshes and coastal sand; widespread. Coasts of W. and N. Europe.

Pea Family, Papilionaceae or Fabaceae

3. Strawberry Clover, *Trifolium fragiferum* L., plant 5–10cm, with stolons and rooting at the nodes; lvs trifoliate, finely serrate; flower heads long-stalked, 1–2cm across; corolla 5–7mm long, pink; the upper lip of the calyx is strongly inflated after flowering, pale brown or reddish, hairy, so that the head resembles a strawberry. Fl.5–9. Saltmarshes, meadows on salty soils, river banks; widespread on the coasts of Europe, otherwise rarer. Protected!

Spurge Family, Euphorbiaceae

4. Sea Spurge, *Euphorbia paralias* L., plant 30–60cm, stiff, blue-green; stem with scale lvs at the base; lvs thick, fleshy, ovate, acute, imbricate, 5–20mm long; umbels with 3–6 rays; bracts broadly ovate, not united; involucral glands with short horns; fruit finely warty. Fl.5–9. Sand dunes; scattered. On the coasts of W. and S. Europe.

Carrot Family, Umbelliferae or Apiaceae

5. Rock Samphire, *Crithmum maritimum* L., plant 15–50cm, glabrous, blue-green, much-branched; lvs pruinose, 2–3 pinnate, with linear, acute segments; umbel 8–20 rayed; bracts and bracteoles numerous; flowers yellowish or greenish-white; fruit ovoid, with prominent ribs. Fl.7–9. Lvs are pickled and eaten. Cliffs and rocks by the sea; local. Britain, W. and S. Europe.

6. Sea-holly, *Eryngium maritimum* L., plant 20–60cm, with white pruina, often bluish; lvs sinuate-toothed, rigid, roughly prickly, the upper amplexicaul; flower heads 2–3cm across, surrounded by ovate, almost 3-lobed, prickly bracts; bracteoles narrow, usually bluish, longer than the flowers; flowers about 8mm, blue. Fl.6–8. Protected! Sand dunes on the coasts of Europe; widespread.

Primrose Family, Primulaceae

7. Sea-milkwort, *Glaux maritima* L., plant 5–15cm; lvs ovate, fleshy, overlapping and arranged in 4 rows, sessile, 4–12mm long; flowers sessile, axillary, 4–6mm across; petals absent; calyx 5-lobed, campanulate, pink. Fl.5–8. Protected! On salty soils, grassy saltmarshes; widespread on the coasts of Europe, rare inland.

8. Oysterplant, *Mertensia maritima*, see p.446.

Plate 199   Sea-lavender Family, Plumbaginaceae

1. Thrift, *Armeria maritima* (Mill.)Willd., cushion-plant, 5–30cm, densely tufted, with perennial rosettes; lvs fleshy, linear, 1–2mm wide, 1-veined; flowers in dense, spherical heads, 1–3cm across; peduncle leafless; inflorescence surrounded by scarious bracts, 4–7mm long; flowers 5-merous; corolla 6–8mm long, pink. Fl.5–11. Saltmarshes, old sand dunes, grassland by the sea, sea cliffs and rocks, on acid and heavy-metal soils, also on mountains inland; common to rare. N. and W. Europe.

2. Common Sea-lavender, *Limonium vulgare* Mill. (*Statice limonium* L.), plant 20–50cm; stem branched; lvs obovate, entire, glabrous, gradually narrowed into the stalk; inflorescence secund, cymose, with scarious bracts; corolla blue-violet, 8mm long. Fl.7–9. Saltmarshes; locally common; coasts of the Atlantic, North Sea, Baltic. (Related species in the Mediterranean.)

Plantain Family, Plantaginaceae

3. Sea Plantain, *Plantago maritima* L., plant 15–30cm, blue-green, woody at the base; lvs linear, fleshy, channelled at first, weakly 3–5-veined, 2–6mm wide; spikes 4–10cm long; corolla bluish; bracts and calyx teeth glabrous or very shortly ciliate. Fl.6–9. Saline soils, saltmarshes, grassland by the sea; common in coastal districts; also less commonly by streams on mountains. Europe.

4. Buck's-horn Plantain, *Plantago coronopus* L., plant 5–30cm; stem leafless, unbranched; lvs in basal rosettes, pinnatisect or coarsely toothed; flowers yellowish, in spikes 0.5 to 4cm long and 3–4mm wide; corolla tube hairy. Fl.5–10. Seaside, dry sandy places, waysides, rock crevices, less commonly inland; from widespread to rare. Almost all of Europe.

Daisy Family, Compositae or Asteraceae

5. Sea Aster, *Aster tripolium* L., plant glabrous, 15–60cm; stem branched, many-headed; lvs fleshy, oblong-lanceolate, usually entire, 7–12cm long; flower heads 1–3cm wide; ray florets lilac or whitish; disc florets orange-yellow; bracts obtuse, appressed, 1–3mm long. Fl.7–9. Wet meadows on saline soil, saltmarshes, sea cliffs and rocks; widespread on the coasts of Europe, rare inland.

6. Sea Wormwood, *Artemisia maritima* L., plant 30–60cm; lvs 2–3 pinnatisect, with linear, obtuse segments, 1–2mm wide, white or grey-felted on both sides; flower heads ovoid, 2–3mm long; bracts felted; flowers hermaphrodite. Fl.9–10. On saline soils, widespread on the coast of Europe, rare inland.

See also *Matricaria*, Pl.36.

Borage Family, Boraginaceae

8. Oysterplant, *Mertensia maritima* (L.)S. F. Gray (Pl.198), plant prostrate, blue-green, fleshy, 10–60cm; lvs lanceolate to spathulate, 1–6cm long, rough above; the lower lvs long-stalked, the upper sessile; flowers pink, then blue and pink; corolla tube cylindrical; calyx glabrous; nutlets smooth. Fl.6–8. On shingle by the sea. N. Europe, Britain.

Index of English names

There are no page numbers in this index. All the figures given refer to PLATE numbers.

449

Index of scientific names

There are no page numbers in this index. All the figures given refer to PLATE numbers.

459